Food Safety
A Practical and Case Study Approach

ISEKI-FOOD SERIES

Series Editor: Kristberg Kristbergsson, *University of Iceland*
Reykjavík, Iceland

Volume 1 FOOD SAFETY: A Practical and Case Study Approach
Edited by Anna McElhatton and Richard J. Marshall

Volume 2 ODORS IN THE FOOD INDUSTRY
Edited by Xavier Nicolay

Volume 3 UTILIZATION OF BY-PRODUCTS AND TREATMENT OF WASTE IN THE FOOD INDUSTRY
Edited by Vasso Oreopoulou and Winfried Russ

Volume 4 PREDICTIVE MODELING AND RISK ASSESSMENT
Edited by Rui Costa and Kristberg Kristbergsson

Volume 5 EXPERIMENTS IN UNIT OPERATIONS AND PROCESSING OF FOODS
Edited by Maria Margarida Cortez Vieira and Peter Ho

Volume 6 CASE STUDIES IN FOOD SAFETY AND ENVIRONMENTAL HEALTH
Edited by Maria Margarida Cortez Vieira and Peter Ho

Food Safety

A Practical and Case Study Approach

Edited by

Anna McElhatton
University of Malta
Msida, Malta

Richard J. Marshall
London Metropolitan University
London, United Kingdom

Springer

Anna McElhatton
Institute of Health Care
(Environmental Health Division)
University of Malta
Tal-Qroqq
Msida MSD 06
Malta
anna.mcelhatton@um.edu.mt

Richard J. Marshall
Department of Health and
Human Sciences
London Metropolitan University
Holloway Road
London N7 8DB
UK
r.marshall@londonmet.ac.uk

Series Editor
Kristberg Kristbergsson
University of Iceland
Department Food Science
Faculty of Science
Hjarðarhaga 2-6
107 Reykjavík, Iceland

Library of Congress Control Number: 2006926456

ISBN-10: 0-387-33509-9 e-ISBN 0-387-33957-4
ISBN-13: 978-0387-33509-4

Printed on acid-free paper.

9 8 7 6 5 4 3 2 1

springer.com

SERIES ACKNOWLEDGEMENTS

ISEKI-Food is a thematic network on Food Studies, funded by the European Union as project N° 55792-CP-3-00-1-FR-ERASMUS-ETN. It is a part of the EU programme in the field of higher education called ERASMUS which is the higher education action of SOCRATES II programme of the EU.

SERIES PREFACE

The single most important task of food scientists and the food industry as a whole is to ensure the safety of foods supplied to consumers. Recent trends in global food production, distribution and preparation call for increased emphasis on hygienic practices at all levels and for increased research in food safety in order to ensure a safer global food supply. The ISEKI-Food book series is a collection of books where various aspects of food safety and environmental issues are introduced and reviewed by scientists specializing in the field. In all of the books a special emphasis was placed on including case studies applicable to each specific topic. The books are intended for graduate students and senior level undergraduate students as well as professionals and researchers interested in food safety and environmental issues applicable to food safety.

The idea and planning of the books originates from two working groups in the European thematic network "ISEKI-Food" an acronym for "Integrating Safety and Environmental Knowledge In to Food Studies". Participants in the ISEKI-Food network come from 29 countries in Europe and most of the institutes and universities involved with Food Science education at the university level are represented. Some international companies and non teaching institutions have also participated in the program. The ISEKI-Food network is coordinated by Professor Cristina Silva at The Catholic University of Portugal, College of Biotechnology (Escola) in Porto. The program has a web site at: http://www.esb.ucp.pt/iseki/. The main objectives of ISEKI-Food have been to improve the harmonization of studies in food science and engineering in Europe and to develop and adapt food science curricula emphasizing the inclusion of safety and environmental topics. The ISEKI-Food network started on October 1st in 2002, and has recently been approved for funding by the EU for renewal as ISEKI-Food 2 for another three years. ISEKI has its roots in an EU funded network formed in 1998 called Food Net where the emphasis was on casting a light on the different Food Science programs available at the various universities and technical institutions throughout Europe. The work of the ISEKI-Food network was organized into five different working groups with specific task all aiming to fulfill the main objectives of the network.

The first four volumes in the ISEKI-Food book series come from WG2 coordinated by Gerhard Schleining at Boku University in Austria and the undersigned. The main task of the WG2 was to develop and collect materials and methods for teaching of safety and environmental topics in the food science and engineering curricula. The first volume is devoted to Food Safety in general with a practical and a case study approach. The book is composed of fourteen chapters which were organized into three sections on preservation and protection; benefits and risk of microorganisms and process safety. All of these issues have received high public interest in recent years and will continue to be in the focus of consumers and regulatory personnel for years to come. The second volume in the series is devoted to the control of air pollution and treatment of odors in the food industry. The book is divided into eight chapters devoted to defining the problem, recent advances in analysis and methods for prevention and treatment of odors. The topic should be of special interest to industry personnel and researchers du to recent and upcoming regulations by the European Union on air pollution from food processes. Other countries will likely follow suit with more strict regulations on the level of odors permitted to enter the environment from food processing operations. The third volume in the series is devoted to utilization and treatment of waste in the food industry. Emphasis is placed on sustainability of food sources and how waste can be turned into by products rather than pollution or land fills. The Book is composed of 15 chapters starting off with an introduction of problems related to the treatment of waste, and an introduction to the ISO 14001 standard used for improving and maintaining environmental management systems. The book then continues to describe the treatment and utilization of both liquid and solid waste with case studies from many different food processes. The last book from WG2 is on predictive modeling and risk assessment in food products and processes. Mathematical modeling of heat and mass transfer as well as reaction kinetics is introduced. This is followed by a discussion of the stoichiometry of migration in food packaging, as well as the fate of antibiotics and environmental pollutants in the food chain using mathematical modeling and case study samples for clarification.

Volumes five and six come from work in WG5 coordinated by Margarida Vieira at the University of Algarve in Portugal and Roland Verhé at Gent University in Belgium. The main objective of the group was to collect and develop materials for teaching food safety related topics at the laboratory and pilot plant level using practical experimentation. Volume five is a practical guide to experiments in unit operations and processing of foods. It is composed of twenty concise chapters each describing different food processing experiments outlining theory, equipment, procedures, applicable calculations and questions for the students or trainee followed by references. The book is intended to be a practical guide for the teaching of food processing and engineering principles. The final volume in the ISEKI-Food book series is a collection of case studies in food safety and environmental health. It is intended to be a reference for introducing case studies into traditional lecture based safety courses as well as being a basis for problem based learning. The book consists of thirteen chapters containing case studies that may be used, individually or in a series, to discuss a range of food safety issues. For convenience the book was divided into three main sections on microbial food safety; chemical residues and contaminants and a final section on risk assessment and food legislation.

The ISEKI-Food books series draws on expertise form close to a hundred universities and research institutions all over Europe. It is the hope of the authors, editors, coordinators and participants in the ISEKI network that the books will be useful to students and colleagues to further there understanding of food safety and environmental issues.

March, 2006 Kristberg Kristbergsson

ACKNOWLEDGEMENTS

The editors wish to thank the contributors for all the hard work they have put into the various chapters of the book. Thanks are also due to the ISEKI Food Coordinator Prof. C.L. Silva and Work package leaders Professors K. Kristbergsson and G. Schleining for entrusting us with the task of editing this book. Finally we would like to express our gratitude to the European Union for funding this project. ISEKI-Food is a thematic network on Food Studies, funded by the European Union as project N° 55792-CP-3-00-1-FR-ERASMUS-ETN. It is a part of the EU programme in the field of higher education called ERASMUS, which is the higher education action of the EU SOCRATES II programme.

January 2006

Anna McElhatton
Richard J. Marshall

PREFACE

Food quality and safety has become a major concern to governments, industry and consumers. We are fundamentally all consumers and surely all give priority to the need for high quality and safe foods. The whole of the journey that our food takes, from farm to fork, is fraught with the risks of contamination and / or spoilage. These risks have been investigated at length, and continue to be investigated, by both Industry and Academia and a body of knowledge, experience and expertise has been built up. The major objective of this book is to demonstrate, using contributions from people currently working in the field, how food quality and safety are interrelated and how they impinge significantly on the quality of our daily lives. Practical examples in the form of case studies are used to give the reader a tangible view of the theory discussed.

This book attempts to bring together salient and topical aspects of food quality and safety and the editors and authors hope that the book may help the reader obtain a clear overview of the intricacies of the science involved and its application to the production of nutritious, healthy and tasty food.

THE EDITORS
JANUARY 2006

CONTENTS

Series Acknowledgements v

Series Preface vii

Acknowledgements xi

Preface xiii

Contributors xvii

PART I PRESERVATION AND PROTECTION

1. METHODS OF FOOD PRESERVATION 3
Tsvetko Prokopov and Stoyan Tanchev

2. THE CHALLENGE OF MYCOTOXINS 26
Armando Venâncio and Russell Paterson

3. PREVENTIVE MEASURES FOR FOOD SAFETY 50
Gerhard Schleining

4. PACKAGING 68
Mona Popa and Nastasia Belc

PART II BENEFITS AND RISKS OF MICROORGANISMS

5. HACCP IN THE CHEESE MANUFACTURING PROCESS, A CASE STUDY 91
Eleftherios H. Drosinos and Panagiota S. Siana

6. GENETICALLY MODIFIED ORGANISMS AND FOOD SAFETY 112
Dezider Toth

7. NUTRITIONAL STRATEGIES TARGETING THE BENEFICIAL MODULATION OF THE INTESTINAL MICROFLORA WITH RELEVANCE TO FOOD SAFETY: THE ROLE OF PROBIOTICS AND PREBIOTICS 133
Konstantinos C. Mountzouris

8. EXPLOITATION OF MICROORGANISMS BY THE FOOD AND BEVERAGE INDUSTRY 153
George Kalantzopoulos, Manuela Pintado, and Ana Gomes

9. PATHOGENIC, COMMENSAL AND BENEFICIAL MICROORGANISMS IN FOODS 177
Ana M.P. Gomes, Manuela E. Pintado, and F. Xavier Malcata

10. FOODBORNE VIRUSES: AN EMERGING RISK TO HEALTH 202
Leen Baert, Mieke Uyttendaele, and Johan Debevere

PART III PROCESS SAFETY

11. SAFETY MODELS: HACCP AND RISK ASSESSMENT 225
Gabriela Rotaru and Daniela Borda

12. APPLICATION OF HACCP IN SMALL FOOD BUSINESSES 239
Vassilis Georgakopoulos

13. CLEANING AND DISINFECTION PROCEDURES IN THE FOOD INDUSTRY GENERAL ASPECTS AND PRACTICAL APPLICATIONS 253
Giuliano Sansebastiano, Roberta Zoni, and Laura Bigliardi

14. ENSURING BIOSAFETY THROUGH MONITORING OF GMO IN FOOD WITH MODERN ANALYTICAL TECHNIQUES, A CASE STUDY 281
Aurora Rizzi, Claudia Sorlini, Saverio Mannino, and Daniele Daffonchio

SUBJECT INDEX 295

AUTHOR INDEX 302

CONTRIBUTORS

Leen Baert
Ghent University, Ghent, Belgium B-9000. e-mail Leen.Baert@ugent.be

Nastasia Belc
Institute of Food Bioresources Bucharest – Romania.

Laura Bigliardi
Department of Public Health, Hygiene Section, University of Parma. Parma
Italy 43100.

Daniela Borda
Gabriela Rotaru, University Dunărea de Jos, Faculty of Food Science and
Engineering, 111 Domneasca Str, 800201, Galati, Romania.
e-mail daniela.borda@ugal.ro

Daniele Daffonchio
Department of Food Science and Microbiology, Via Celoria 2, 20133, Milan, Italy.

Johan Debevere
Ghent University, Ghent, Belgium B-9000.

Eleftherios H. Drosinos
Laboratory of Food Quality Control and Hygiene, Department of Food Science and
Technology, Agricultural University of Athens, 75, Iera Odos, Str., Votanikos,
Hellas (Greece), EL-118 55. e-mail ehd@aua.gr

Vassilis Georgakopoulos
10 Miniati street, Athens, 11636, Greece. e-mail vgeorgakopoulos@hotmail.com

Ana Gomes
Esola Superior de Biotecnologia. Rua Dr. António Bernardino de Almeida.
4200-072. Porto. Portugal. e-mail anagomes@esb.ucp.pt

George Kalantzopoulos
Agricultural University of Athenas 118.55 Botanilcos Athens Greece.
e-mail kalatz@aua.gr

F. Xavier Malcata
Escola Superior de Biotecnologia, Universidade Católica Portuguesa P-4200-072
Porto, Portugal.

Saverio Mannino
Department of Food Science and Microbiology, Via Celoria 2, 20133, Milan, Italy.
e-mail saverio.mannino@unimi.it

Mieke Uyttendaele
Ghent University, Ghent, Belgium B-9000.

Konstantinos C. Mountzouris
Department of Animal Nutrition, Agricultural University of Athens, Iera Odos 75, 118 55, Athens, Greece. e-mail kmountzouris@aua.gr

Russell Paterson
Departamento de Engenharia Biológica, Universidade do Minho, Campus de Gualtar, 4710-057 Braga, Portugal. e-mail russell.paterson@deb.uminho.pt

Manuela Pintado
Esola Superior de Biotecnologia. Rua Dr. António Bernardino de Almeida. 4200-072 Porto, Portugal. e-mail mpintado@esb.ucp.pt

Mona Popa
AGRAL Programme, University of Agronomical Sciences, and Veterinary Medicine Bucharest,59 Marasti Blvd., 011464, Bucharest 1, Romania.
e-mail monapopa@agral.usamv.ro

Tsvetko Prokopov
University of Food Technologies, 26 Maritsa Blvd., 4002 Plovdiv, Bulgaria.
e-mail uft_plovdiv@yahoo.com

Aurora Rizzi
Department of Food Science and Microbiology, Via Celoria 2, 20133, Milan, Italy.
e-mail aurora.rizzi@unimi.it

Gabriela Rotaru
Gabriela Rotaru, University Dunărea de Jos, Faculty of Food Science and Engineering, 111 Domneasca Str, 800201, Galati, Romania.
e-mail gabriela.rotaru@ugal.ro

Giuliano Sansebastiano
Department of Public Health, Hygiene Section, University of Parma. Parma Italy 43100. e-mail giulianoezio.sansebastiano@unipr.it

Gerhard Schleining
DLWT -Department of Food Science and Technology, BOKU -University of Natural Resources and Applied Life Sciences Vienna, Muthgasse 18, A-1190 Vienna, Austria. e-mail gerhard.schleining@boku.ac.at

Panagiota S. Siana
Hellenic Ministry of Rural Development and Food, Veterinary Laboratory of Tripolis, Pelagos, Arcadia, Hellas (Greece), EL-221 00.

Claudia Sorlini
Department of Food Science and Microbiology, Via Celoria 2, 20133, Milan, Italy.

Stoyan Tanchev
University of Food Technologies, 26 Maritsa Blvd., 4002 Plovdiv, Bulgaria.
e-mail stoyan_tanchev@hotmail.com

Dezider Toth
Slovak Agricultural University of Nitra, Tr.A.Hlinku 2, SK-94976 NITRA, Slovak Republic. e-mail dezider.toth@uniag.sk

Roberta Zoni
Department of Public Health, Hygiene Section, University of Parma, Parma, Italy 43100.

Armando Venâncio
Departamento de Engenharia Biológica, Universidade do Minho, Campus de Gualtar, 4710-057 Braga, Portugal. e-mail avenan@deb.uminho.pt

Part I
Preservation and Protection

1
Methods of Food Preservation

TSVETKO PROKOPOV[1] AND STOYAN TANCHEV[2]

1. Introduction

Virtually all foods are derived from living cells from animals and plant origin and in some cases from some microorganisms by biotechnology methods. Thus, foods are for the most part composed of "edible biochemicals".

One of the most important goals of the food scientist is to make foods as safe as possible whether they are used fresh or processed. The judicious application of food processing, storage and preservation methods helps prevent outbreaks of foodborne illness, that is the occurrence of disease or illness resulting from the consumption of contaminated food. The processed food industry has an outstanding record preventing such cases when it is considered that billions of cans, jars, packets and pouches of processed and fresh food products are consumed annually. Occasionally, however, this excellent record has been broken by limited outbreaks in which persons do succumb to the effects of toxic foods.

Food preservation is an action or method of designed to maintain foods at a desired level of quality. A number of new preservation techniques are being developed to satisfy current demands of economic preservation and consumer satisfaction in safety, nutritional and sensory aspects (Potter and Hotchkiss, 1995).

2. Why Do We Need to Preserve?

The preservation, processing and storage of the food are vital for the continuous supply of foods during seasons and off-seasons. One very important consideration that differentiates the agricultural from all other industrial processes is their

[1]Tsvetko Prokopov, University of Food Technologies, 26 Maritsa Blvd., 4002 Plovdiv, Bulgaria, e-mail: uft_plovdiv@yahoo.com
[2]Stoyan Tanchev, University of Food Technologies, 26 Maritsa Blvd., 4002 Plovdiv, Bulgaria, e-mail stoyan_tanchev@hotmail.com

seasonal nature. The main reasons for food processing and preservation are: to overcome seasonal production in agriculture; to produce value-added products; and to provide variety in diets. People like to eat wide varieties of foods, having different tastes, flavours, nutritional, dietetic and other characteristics. Unfortunately it has been estimated that as many as 2 billion people do not have enough to eat and that perhaps as many as 40 000 die every day from diseases related to inadequate diets, including the lack of sufficient food, protein or specific nutrients. Inadequate nutrition in extreme cases can produce in children an advanced state of protein deficiency known as kwashiorkor or the more widespread protein.

Major processes of food deterioration are caused by environmental factors such as temperature, humidity, oxygen and light which can be reason for several reaction mechanisms that may lead to food deterioration to such an extent that they are either rejected by or harmful to the consumer. Microbial effects are the leading cause of food deterioration and spoilage (Desai, 2000).

2.1. The necessity to preserve

Foods are perishable or deteriorative by nature. Based on the mode of action, major food preservation techniques can be categorised as: slowing down or inhibiting chemical deterioration and microbial growth; directly inactivating bacteria, yeast, moulds and enzymes and avoiding recontamination before and after processing. A number of techniques or methods from these categories are presented in Figure 1.

3. Conventional Food Preservation Methods

3.1. Food preservation by heat treatment

Heat is by far the most commonly used method of food preservation. There are various degrees of preservation by heating that ultimately dictate the type of final product manufactured, the terms used are pasteurisation and sterilisation. However, to be effective, these processes must be carried out under a combination of strict temperature and time control to ensure the killing of pathogenic and non-pathogenic microorganisms. These same factors also cause thermal inactivation of food enzymes and some destruction of food constituents (Heldman and Lund, 1992).

3.1.1. Heat resistance of microorganisms

Heat resistance of microorganisms is a basic topic of thermobacteriology, which is a very important part of microbiology including food microbiology. The most heat resistant pathogen found in foods, especially those that are canned and held under anaerobic conditions is *Clostridium botulinum*. It is spore forming, proteolytic anaerobe, which is able to produce the most harmful known toxin since

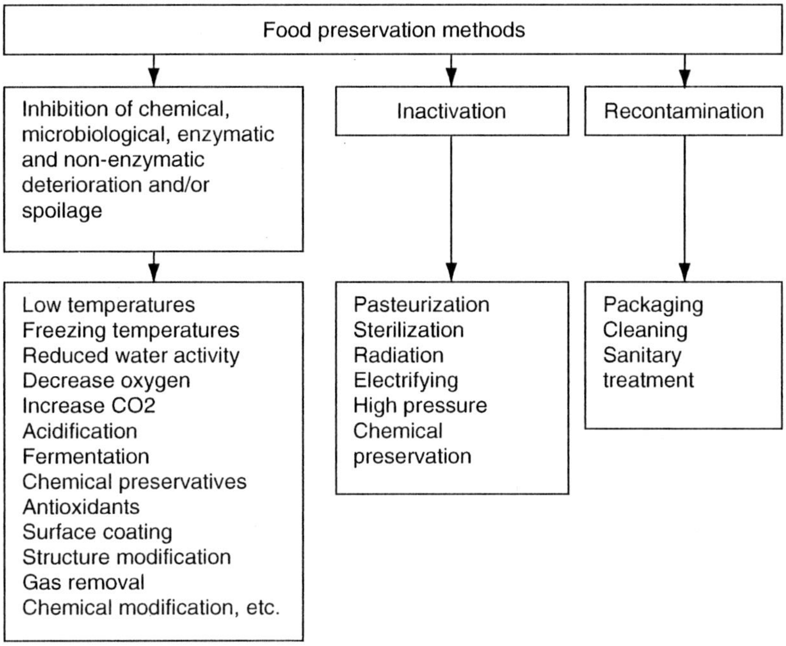

FIGURE 1. Major food preservation methods.

amount of about $10^{-6} - 10^{-8}$g is able to kill one person. However, there are non-pathogenic, spore forming food spoilage bacteria, such as the putrefactive anaerobe *Clostridium sporogenes* 3679 (PA3679) and *Bacillus stearothermophilus* (FS1518) which are more heat resistant than spores of *Cl. botulinum*. This means that if a heat treatment inactivates spores of these spoilage microorganisms, the spores of *Cl. botulinum* and all others pathogens will be also killed (Bell and Kyrakides, 2000).

3.1.2. Kinetics of heat destruction of microorganisms

- **Thermal death time**

The thermal death time is the time of heating required to kill all vegetative cells of microorganisms. Theoretically this is not possible but this expression is used in thermobacteriology for practical purposes.

Microorganisms are killed by heat at a rate that is very nearly proportional to the number of cells of a specified organism (expressed on a logarithmic basis) present in the system (food, laboratory nutritive medium, water, etc.) being heated. This is referred as a logarithmic order of death. A typical thermal death rate curve is shown in Figure 2. It provides data on the rate of destruction of specific microorganisms in specific media or food at specific constant temperature,

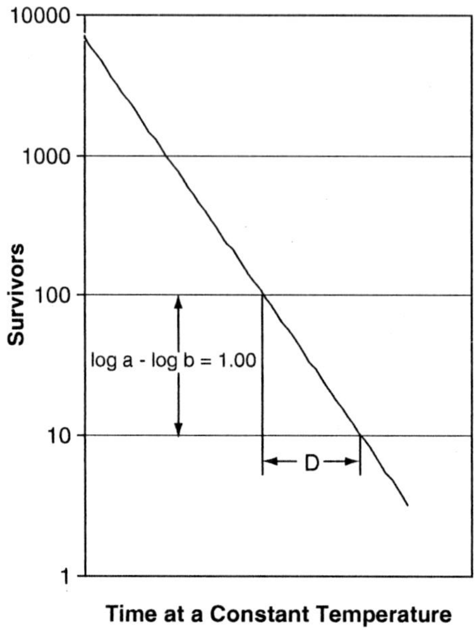

FIGURE 2. Bacterial destruction rate curve showing logarithmic order of death. D, decimal reduction time.

which is able to kill the corresponding microorganism – pathogenic, toxicogenic, or spoilage organism of the specific food.

Figure 2 shows the logarithmic dependence between the time (τ) and number of the killed cells (C) at constant temperature, or $\lg C = f(\tau)$ at t°C = constant. The logarithmic order of thermal killing is valid for all spores and vegetative cells but latter are killed faster. It is valid also for yeast and moulds. The D-value is real kinetic constant determined at t°C = constant. That is why it is normally written as D_t. For example, if t = 100°C then it should be D_{100} which means that the value of D is determined at 100°C and it is valid only for this temperature and product in which the cells have been suspended (Ray, 2000).

This "D-value", or decimal reduction time, is defined as the time, in minutes, at specified temperature required to destroy 90% of the cells at the respective microbial population. During each time interval (1 min or 3 min or 6 min) numbers of the cells is reduced 10 times, let say from 1000 to 100 or from 100 to 10, etc. This means that 90% of the cells are killed during each interval. In each case 1000/100 = 100/10 = 10 which is the reason it is called the decimal reduction time. In other words, the D-value represents the time for the number of cells to be reduced by one logarithmic cycle, for example from 10^6 to 10^5 cells per 1 g.

TABLE 1. Effective time-temperature relationships for destruction of *Clostridium botulinum* spores

Temperature (°C)	Time (min)
100	330
104	150
110	36
116	10
118	5.27
121	2.78
124	1.45
127	0.78

- **Dependence between the thermal death time and temperature**

For example, if the time-temperature combinations required for destruction of *Clostridium botulinum* spores in low-acid media (i.e pH >4.5) are taken from this type of relationship. The time-temperature relationships that will be equally effective are shown in table 1.

From such data, the dependence of thermal death time on the temperature can be presented graphically in semi-logarithmic co-ordinates called an Arrhenius plot where the heating time is plotted in a logarithmic scale. This shows (Figure 3) that for any one initial concentration of the cells, time-temperature relationship is linear and can be described by the equation:

$$\lg \tau = f\,(t^{\circ}C), \text{ at } C = \text{constant}$$

This figure illustrates two terms or kinetic constants, the "Z-value" and the "F-value". The "Z-value" is the number of degrees required to pass through one log cycle, which means that the thermal death time is changed by factor of 10, let say from 100 to 10 min or from 10 to 1 min. The "F-value" is defined as the number of minutes at a specific temperature, required to destroy the desired number of cells of any microorganism. The "F-value" is a measure of the capacity of any heat treatment applied to a specific food product in order to sterilise it. When it is written as F_0 it means that the tested microorganism is spores of *Cl. botulinum* when they are treated at 121.1°C (Ray, 2000).

The dependence of thermal death times on the temperature has been determined for many important pathogens and food spoilage microorganisms. Such curves for putrefactive anaerobe *Clostridium sporogenes* (PA3679) and *Bacillus stearothermophilus* (FS1518) are shown in Figure 4. This figure shows the length of time it takes to kill these microorganisms at a chosen temperature. For example it would take about 60 min at 105°C to kill the specified number of spores of PA 3679. At 121.1°C, the same numbers of spores are killed in a little over 1 min (Shapton and Shapton, 1991).

It has been shown that the criterion for commercial sterilisation is that the population of spores of *Cl. botulinum*, should be reduced by 12 log cycles or 12D. This means that if one can contains 10^6 spores before heating, which is unusually high, then after a 12D heat treatment, out of 1 million cans, 999 999 cans will be

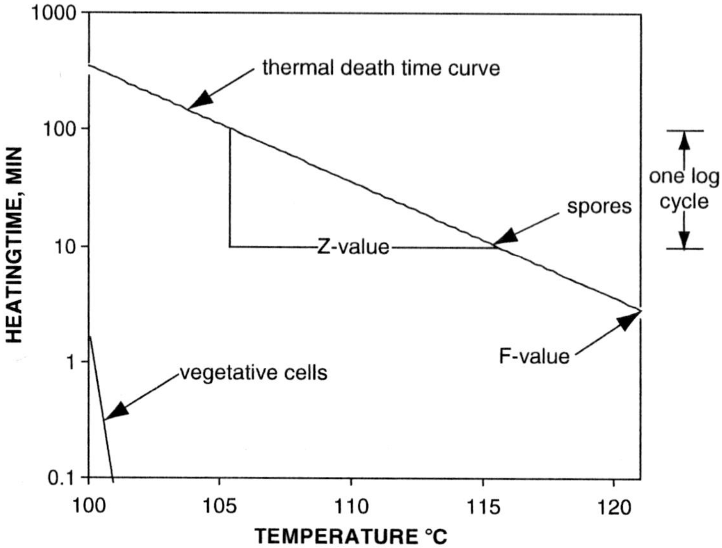

FIGURE 3. Typical thermal death time curves for bacterial spores and vegetative cells.

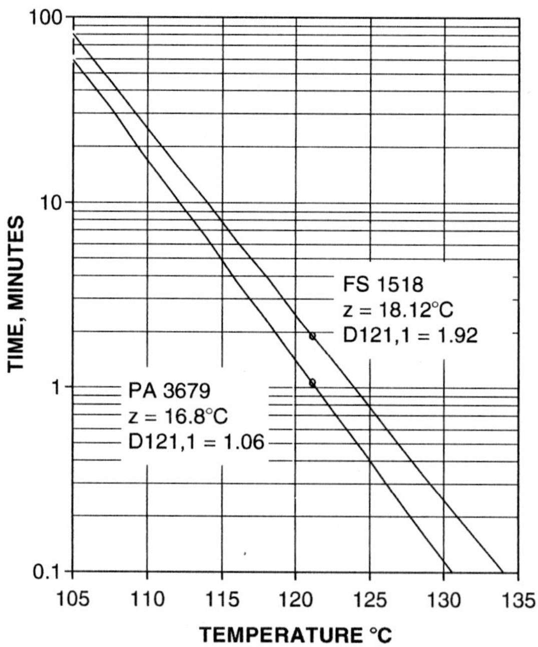

FIGURE 4. Thermal death time curves for test microorganisms PA 3679 and FS 1518.

sterile. For spores of *Cl. sporogenes* PA 3679 and *Bacillus stearothermophilus* FS 1518, in low acid foods, a 5D heat treatment is equivalent to 12D values against *Cl. botulinum*.

For food with pH < 4.6 (higher acid foods) requirement 12D is not valid since *Cl. botulinum* does not grow in these foods (although spores may survive).

In container sterilisation time required to sterilise food is influenced by:

- heat resistance of microorganisms and/or the enzymes in the food, when pH < 4.6; heating method – steam, water, flame, etc.;
- pH of the food;
- the size of the container;
- chemical composition of the food;
- physical state of the product;
- mechanism of heat exchange (convection or conduction);
- initial product temperature;
- temperature of sterilisation;
- state of the containers during sterilisation – static, shaking, rotating, etc.

The mechanisms of heat inactivation and injury of microorganisms is not very well identified since heat will bring about so many changes in biological material, such as microbial cells, that is why identification of the event that causes death or injury of the cells is difficult to predict (Larousse and Brown, 1997).

Thermal death time depends on the microbial cell concentration (Figure 5).

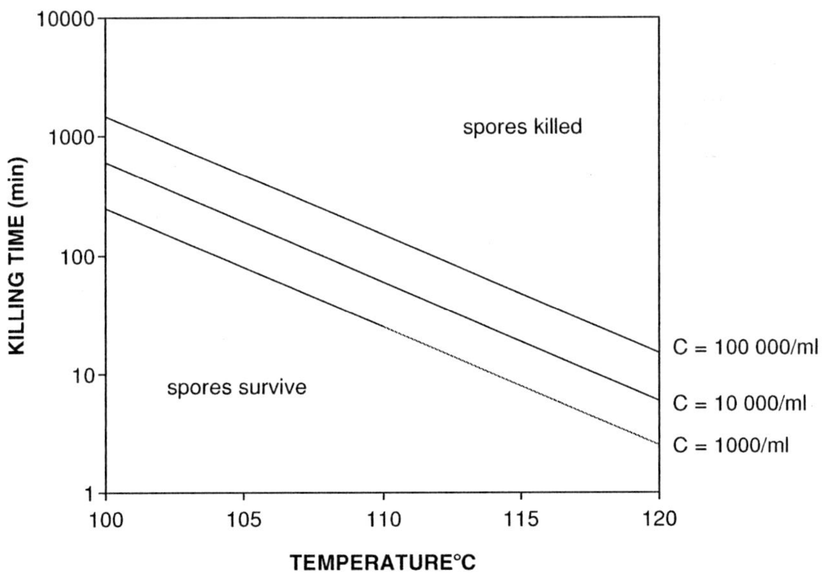

FIGURE 5. Thermal death curves for bacterial spore suspensions of different initial concentrations.

3.2. *Preservation by low water activity (a_w)*

Water activity can be reduced by partial removal of the water (drying, reverse osmosis, concentration) or by adding substances which increase the osmotic pressure of the food or media such as sugars, ethanol, glycerol, salts, etc. (Booth, 1998). The majority of microorganisms are sensitive to the water status in their immediate environment and they can remain metabolically active only in a narrow range of high water activities. There is a lot of information about low water activity limits for the growth of microorganisms. However it is typical that those organisms that are tolerant to low a_w be also tolerant of very high osmotic pressures. The lowest a_w limits for growth recorded up to now illustrate the enormous range of tolerances that exist. The most a_w tolerant species are able to grow when osmotic pressure is as high as about 800 MPa. They can grow slowly below a_w 0.62. The nature of the solute exerts additional affect on potential for growth. Ionic solutes such as NaCl and KCl are more inhibitors than non-ionic solutes such as sugars. Solutes such as glycerol, unlike the salts and sugars, rapidly permeate most bacteria but not yeast, e.g. *Saccharomyces ronxii* and *Debaryomyces hanseni*. However, for the more low a_w-tolerant species this simple relationship is no longer valid. *Staphylococcus aureus*, for example, is extremely salt-tolerant and more sensitive at a higher a_w in glycerol than in sodium chloride (Shapton and Shapton, 1991).

Lowering the a_w by various means may also influence the rate of enzymatic and chemical changes in foods. Whilst all microbiological growth is completely stopped below about $a_w = 0.6$, some enzymatic reactions that cause food spoilage continue and some reactions, such as lipid oxidation, may even be accelerated at very low a_w values (Shapton and Shapton, 1991).

3.3. *Preservation by low pH and organic acid*

3.3.1. Preservation by low pH

Foods are classified according to their acidity as follows: non-acid – 7.0-5.3; low or medium acid – 5.3-4.6; acid I – 4.6-3.7 and acid II – 3.7 and lower.

Microorganisms have a characteristic range of pH values within which they can grow. Most bacteria have an optimum pH near 6.8 and may grow at pH values ranging from 4.0 to 8.0. A small number bacterial species can multiply when pH < 4.0 or pH > 8.0. Yeast and moulds can sometimes grow at pH less than 2.0. Usually the growth rate decreases as the pH drops below the optimum value. Approaching the lower limiting pH for growth, cells are first inhibited and eventually killed. The degree of inhibition increases as pH decreases and this relationship is linear. The differences in inhibition and/or lethal effects of organic acids used for pH reduction having different pK values are well known. The pK values of some acids are as follows: citric – 3.08; malic – 3.4; tartaric – 2.98; acetic – 4.75 (it is usually used as an effective preservative). The theory of food canning accepts a pH of 4.5 or 4.6 (for USA) as the bor-

mortis in which the muscle tissue becomes firm and inextensible. Cooling during anaerobic respiration is necessary to produce the required texture and colour of meat and to reduce bacterial contamination. However, cooling must not be too rapid otherwise cold shortening can occur, which results in tough meat (Potter and Hotchkiss, 1995).

A reduction in temperature below the minimum necessary for microbial growth extends the generation time of microorganisms and prevents or retards reproduction. There are four broad categories of microorganisms based on the temperature range for growth, as follows:

- Thermophilic (minimum 30-40°C, optimum 55-65°C);
- Mezophilic (minimum 5-10°C, optimum 30-40°C);
- Psychrotrophic (minimum <0-5°C, optimum 20-30°C);
- Psychrophilic (minimum <0-5°C, optimum 12-18°C).

Chilling prevents the growth thermophilic and many mesophilic microorganisms. The main microbiological concerns with chilled foods are a number of pathogens that can growth during extended refrigerated storage below 5°C or as a result of any increase in temperature (temperature abuse) and this may cause food poisoning. But it is now known that some pathogenic species can either growth to large numbers at these temperatures or are sufficiently virulent to cause poisoning after ingestion of only a few cells. For example *Aeromonas hydrophilica*, *Listeria* spp., *Yersinia enterocolitica*, enteropathogenic *E. coli*, etc. *E. coli* O157:H7) may cause haemorrhagic colitis after ingestion of as few as 10 cells.

Freezing is method of food preservation in which the temperature of food is reduced below it freezing point and a proportion of the water undergoes a change in state to form ice crystals, resulting concentration of dissolved solutes in unfrozen water respectively reducing the water activity (a_w) and pH values. Preservation is achieved by a combination of low temperatures, reduced water activity in some foods, pre-treatment by blanching and also retarding growth of microorganisms (Ray, 2000).

The proportion of water that remains unfrozen at the temperature used in commercial freezing depends on the type and composition of the food and temperature of storage. For example, at storage temperature −20°C the percentage of water frozen is 88% in lamb, 91% in fish and 93% in egg albumin.

4. Potential Food Preservation Methods

Consumer trends and food markets are changing and will change more in future towards foods with high quality and more fresh like attributes, consequently fewer external treatments and/or fewer additives will be used. This means less heat and chill damage, more freshness, less acid, salt, sugar and fat, etc. To satisfy these demands some changes in the traditionally used preservation techniques must be achieved. From microbiological point of view, these changes have important and significant implications since food safety is an aspect of increasing importance. The safety and quality of food should be well balanced in every one of the new methods of food preservation. From scientific and technological point of view, the following methods have quite interesting prospects:

4.1. *Pulsed electric fields*

The inactivation of microorganisms is attributed to a shock wave generated by an electric arc that prompts the formation of highly reactive free radicals from various chemical species in the food. In general, an increase in the electric field intensity and number of pulses has been found to lead to an increase in the inactivation of microorganisms. Formation of pores in cell membranes may be a reason for killing of microorganisms. Mechanisms of pore formation are not entirely understood. Applying the dielectric rupture theory, it is concluded that membrane rupture is caused by an induced transmembrane potential approximately 1V larger than the natural potential of the cell membrane. The reversible or irreversible rupture, termed electroporation, of the cell membrane depends on factors such as intensity of electric field, number of pulses and duration of the pulses (electroporation is used to open cell membranes temporarily in order to allow the entry of various molecular species including genetic material). For example, when the pulsed electric field is 9.2 kV/cm the number of survivors of *Staphylococcus* cells is 100%, at 14.25 it is 35%, at 19.50 it is 0.9%, at 27.50 it is 0.6%. Other factors that influence microbial inactivation by pulsed electric fields are the treatment temperature, pH, ionic strength and conductivity of the medium containing the microorganisms.

Raw and reconstituted apple and peach juices, skim milk, beaten eggs and pea soup have been exposed to pulsed electric fields at 25-45 kV/cm. *E. coli* inoculated in skim milk and exposed to 60 pulses of 2 μs at 45 kV/cm and 35°C was reduced by 2D. A reduction of 6D was observed in liquid egg inoculated with *E. coli* and treated with pulsed electric fields of 25.8 kV/cm and 100 pulses of 4 μs at 35°C. *E. coli* and *B. subtilis* inoculated in pea soup and exposes to electric field of 25-33 kV/cm (10-30 pulses of 2 μs) provided a limited inactivation (< 1.5D) when the process temperature of pea soup was below 53°C while microbial inactivation was 4.4D with process temperature was between 53°C and 55°C (Barbosa-Canovas et al., 1998).

4.2. *High pressure*

One "new" or emerging technology receiving a great deal of attention is high hydrostatic pressure. The basis of this process is Le Chatelier's principle, according to which, any reaction, like conformational change or phase transition that is accompanied by decrease in volume will be favoured at high pressure, while reactions involving an increase in volume will be inhibited (Knor, 1996).

Hydrostatic pressure technology is novel non-thermal food processing technology, whereby foods are subjected to high hydrostatic pressure, generally in the range of 100-600 MPa at or around room temperature. In general, the vegetative microbial forms are inactivated by pressure between 400 and 600 MPa, while spores of some spaces may resist pressures higher than 1000 MPa at ambient temperatures. Gram-positive bacteria are more pressures resistant than gram-negative ones. Among the gram-positive bacteria, *Staphylococcus* is one of the most resistant and can survive treatment at 500 MPa for more than 60 min. High pressure can inactivate also enzymes. The extend of microbial inactivation achieved at a particular pressure depends on a number of interacting factors including type and number of microorganisms, magnitude and duration of high pressure treatment, temperature and composition of the suspension media, etc.

The first historical data on effects of high pressure on microorganisms were most likely compiled by Certes (1883) who found viable bacteria in water samples obtained from 5100 m depth. (Approx. 50 MPa). Hite (1899) applied hydrostatic pressure-treatment of food for the first time and he reported the reduction of spoilage microorganisms in milk and meat products and later on (1914) in fruit and vegetables.

The higher resistance of spores to high pressure than vegetative cells was shown in 1965 and that pressure can induce germination of bacterial spores was demonstrated in 1970. High hydrostatic pressure induces a number of changes to the morphology, biochemical reactions, genetic mechanisms, cell membrane and wall of microorganisms. At 0.6 MPa intracellular gas vacuoles can collapse (Cheftel and Culioli, 1997).

Most bacteria are capable to growth at pressure around 20-30 MPa. B. aerophiles can grow at pressure higher than 40-50 MPa and baroduric or barotolerant can survive for prolonged periods at pressures > 200 MPa. In pork slurry, vegetative forms of *E. coli* are not affected at pressure lower than 203 MPa but more than 6 log cycles are reduced at pressures higher than 405 MPa for 10 min. For *S. cerevisiae* less than 2 log cycles reduction is observed at pressures lower than 304 MPa and more than 6 log cycles at pressures higher than 405 MPa. *B. cereus* spores are not reduced considerably (less than 1 log cycle) even with treatment at 608 MPa for 10 min.

Generally an increase in pressure increases microbial inactivation. However, increasing the duration of the treatment does not necessarily increase the lethal effect. Above 200-300 MPa, inactivation of vegetative cells increases with pressure or process time. The temperature during pressurisation can have significant effect on the inactivation of microbial cells (Barbosa-Canovas et al., 1998).

Pressure inactivation of yeast and moulds has been reported in citrus juices. Juices pressure treated at 400 MPa for 10 min at 40°C did not spoil during 2-3 months of storage. A population of 10^6 CFU/ml of *Listeria monocytogenes* was inactivated by exposure to 340 MPa at 23°C in UHT milk and the same population of *Vibrio parahaemolyticus* was killed by exposure to 170 MPa within 10 min at 23°C in clam juice. *Salmonella senftenberg* 775 W, a very heat resistant vegetative microorganism, could be reduced in chicken puree by high-pressure treatment but was shown to recover in chicken based medium.

It is reported that a linear relationship was found between the water activity depression factor ($1-a_w$) and the log of *Z. bailii* survival fraction (N/N_0) after a high-pressure treatment at 345 MPa for 5 min. It was found that $\log(N/N_0) = -4.599 + 45.538(1-a_w)$. This relationship indicates that as the a_w of the model system decreases, the number of surviving *Z. bailii* increase. For *Saccharomyces cerevisiae* in mandarin juice at 250 MPa for 5 min, the decimal reduction (D) is about 2D, after 10 min it is about 4D and after 30 min it is about 6D. After 5 min at 300 MPa it is 5D and after 10 min is 6D.

The patterns of high hydrostatic pressure inactivation kinetics observed with different microorganisms are quite variable. Some investigators indicate first order kinetics in the case of several bacteria and yeast. Other authors observed a change in the slope and a two phases inactivation phenomenon, since the first fraction of the population being quickly inactivated, whereas the second fraction to be much more resistant. Pressure, temperature and composition of the medium influence the patterns of inactivation kinetics. Some authors have stated that there is clear evidence that pressure mediating death is not first order and inactivation curves often present pronounced survivor tails. Thus the D and Z concepts commonly used in thermal processing cannot be usefully applied to describe pressure processes.

Current industrial applications of high pressure treated food include:

- Jams, fruit dressing, fruit sauce (topping), yoghurt, fruit jelly by the following processing conditions: 400 MPa, 2-20 min, 20°C;
- Grapefruit juice, 120-400 MPa, 2-20 min, 20°C;
- Mandarin juice, 300-400 MPa, 2-3 min, 20°C;
- Beef (tenderization only), 100-150 MPa, 30-40 min, 20°C;
- Rice cake, 400 MPa, 45-70°C, 10 min.

4.3. Ultraviolet radiation

Ultraviolet fraction of sunlight has long been known to be the major factor in the bactericidal action of light. It is mainly used in sterilising air and thin liquid films due to low penetration dept of the waves. When it is used in high dosage there is a marked tendency toward flavour and odour deterioration before satisfactory sterilisation of the foods is achieved. But low-level radiation at carefully applied doses can often usually extend the shelf life of foods, without damaging quality. Using UV radiation to kill off bacteria in water is well known. This radiation is safe, environmentally friendly and more cost effective to install and operate that conventional chlorinating. It does not affect the taste of the water, as does chlorine. High intensity UV-C lamps have become available which can increase the potential of destroying surface bacteria on food.

Ultraviolet irradiation is being applied commercially in the use of bactericidal ultraviolet lamps in various food applications: tenderizing of meat, curing and wrapping of cheeses, prevention of surface mould growth on bakery products, air purification in bottling and food processing establishment and over pickle vats.

The lethal effects of ultraviolet light on microorganisms have been well documented. The practical application of this has been controversial because of the type and intensity of radiation, methods of estimating lethality of microorganisms and other factors. It is found that 3-83% of the yeast and 33-72% of the moulds are killed in apple cider through layers varying from 2 to 25 mm in thickness. Colouring agents of the foods absorbed a greater part of the light. Incidents energy levels of light at λ = 253.7 nm, inhibit 90% of *Bacillus megatherium* at 1100 mWs/cm^2 and 90% of *Sarcina lutea* at 19800 mWs/cm^2. There was a 90% or 1D reduction in microbial count of apple juice. Coupled with refrigeration, this could be of commercial significance.

It is generally agreed that the wavelength for maximum germicidal effect is 260 nm. The lethal action varies with the time of exposure and intensity of light and other factors like temperature, pH and the number of cell unit area exposed. The relative humidity affects the death rate of bacteria suspended in air but this is more valid for humidity greater than 50% (Barbosa-Canovas et al., 1998).

Spores of bacteria are generally more resistant to UV light than vegetative bacteria. Spores of *B. subtilis* are reported to be 5-10 times more resistant than *E. coli*. Moulds are more resistant than vegetative bacteria, while yeast differ less from bacteria in this respect. Dark pigmented spores are more resistant to UV irradiation than non-pigmented.

This type of radiation affects bacteria and fungi such as *Penicillium* and *Aspergillus* in grapes and oranges. Onion that had been UV irradiated showed the greatest percentage of marketable product and reduction in post-harvest rot compared with gamma rays and electron beams. Sprouting is also controlled. In addition UV radiation dose of 1.50×10^{-3} J/mm^2 controls soft and dry rot diseases of potato tubers in short term storage of 3 months (Gould, 1995).

UV radiation at a wavelength of 253.7 nm was effective in destroying surface bacteria on fresh meat by 2 log cycles (99% reduction) on a smooth surface beef after a radiation dose of 150 mWs/cm^2 but dose level to 500 mWs/cm^2 reduced bacteria 3 log cycles. A significant increase in shelf life may be obtained by the exposure of beef muscle and fat surface to UV light for 2 min at λ = 366 nm

UV radiation (0.2-0.24 μW/cm^2) at 0°C and relative humidity 99.3%, resulted in extension of log phase of *Pseudomonas* and of the moulds *Thamidium* and *Penicillium* but not of the yeast *Candida scotii*.

The use of UV radiation at λ = 254 nm and dose 300 mWs/cm^2 reduced surface microbial count of mackerel by 2 or 3 log cycles.

UV radiation does not penetrate most opaque materials and it is less effective on rough surfaces.

4.4. Sound in food preservation

Ultrasound is sound energy with frequency range from 20 kHz (upper limit of human hearing) to hundreds of MHz.

Bacteria and especially spores are very resistant to ultrasound requiring hours of ultrasonification to have any effect. At low frequency an increase in total amount of cells was reported during treatment of milk.

It has been found that the reduction of *Salmonellae* during treatment follows a logarithmic relationship. Ultrasound is used to free bacteria adhering to surfaces to facilitate the removal of flora contaminating foods and for machinery washing. It has been shown that bacteria are firmly attached to poultry skin and, although bactericides in processing water are lethal to Salmonella, they do not seem to access bacteria that are firmly attached to or entrapped in poultry skin. Most bacteria from milk films on metal surfaces can be affected by ultrasonic waves of 80 kHz. Salmonella cells attached to broiler skin are reduced 1-1.5 log cycles by sonication in peptone at 20 kHz for 30 min and by less than 1 log cycle by chlorine alone but by 2.5-4 log cycles by sonicating skin in chlorine solution (Gould, 1995).

The mechanical destruction of cells by very intense currents and cavitation generated by ultrasound is the main lethal effect on microorganisms rather than any chemical effect (Barbosa-Canovas et al., 1998).

Strains of *Staphylococcus aureus* are much more susceptible to damage by the combined affect of ultrasonic (20 kHz) and heat treatments than by either treatment separately. There is a marked decrease in heat resistance (without killing) of *Bacillus stearothermophillus* spores by ultrasound at 20 kHz regardless of the heating temperature and the storage time between both treatments. After sonication its heat resistance is reduced from one half to one third of its original value. The heat resistance at 105°C of *Bacillus cereus* and *Bacillus licheniformis* decrease after a previous treatment with ultrasound at 20 kHz.

4.5. Application of bacteriocins in food preservation

A bacteriocin is defined as a protein that has a bactericidal action against a limited range of microorganisms, which are mostly closely related. They were discovered many decades ago. Bacteriocins produced by pathogenic bacteria (for example certain *E. coli* produce colicin, some *Staphylococcus aureus* strains produce epidermin) are not suitable for food application. More interesting sources of bacteriocins are the lactic acid bacteria. For centuries these microorganisms have been used in food fermentation to produce stable food products ranging from dairy (cheese), meat (sausages) and vegetables (sauerkraut) mainly through the preservative action of the lactic acid they produce (Davidson and Branen, 1998).

Ever since the identification of the inhibitory activity of a strain of *Lactococcus lactis* subsp. *lactis* in 1928, lactic acid bacteria have been increasingly scrutinised for bacteriocin production. The inhibitory agent was later termed nisin, the first known and most extensively studied bacteriocin of these bacteria. Nisin has been used for:

- Prevention of blowing faults of Swiss type cheese, caused by *Clostridia*;
- Extension of the shelf life of milk;

- Lower heat processing of tomato juice;
- Control of flat sour spoilage of some canned foods caused by thermophilic spoilage bacteria;
- Optimising starter function during sauerkraut production by improving competitiveness;
- Inhibition of beer and wine spoilage by lactic acid bacteria.

Pediocin is the name given to bacteriocins produced by bacterial genus Pediococcus. The first report on pediocin production dates back to 1975, when it was found that *Pediococcus pentosaceus* inhibited the growth and acid production of *Lactobacillus plantarum*, which is undesirable competitor in mixed brine cucumber fermentation. The active agent was designated as pediocin A and it inhibits a broad range of lactic acid bacteria, as well as several *Clostridia*, *Staphylococcus aureus* and *Bacillus cereus*. Pediocin PA-1 produced by a strain of *Pediococcus acidilactici* inhibits growth of *Listeria monocytogenes*, when inoculated into cottage cheese, half and half cream and cheese sauce for 1 week at 4°C but rapid growth to high cell densities was observed in the control samples.

Sakacins, a group of bacteriocins produced by *Lactobacillus sake*, owe their discovery probably to the intensive search for natural antimicrobial compounds capable of increasing shelf life of raw meat by inhibiting growth of meat spoilage microorganisms and controlling *Listeria monocytogenes* (Gould, 1995).

Today about 30 different bacteriocins, produced by some 17 species of lactic acid bacteria have been identified and their bactericidal activity has been studied. For food preservation, the advantageous features of several bacteriocins are their relatively high heat resistance and inhibition of gram-positive foodborne pathogens and spoilage microorganisms. Much attention has been given to the inhibition of *Listeria monocytogenes* which may give rise to a high mortality rate and may occur in many different foods causing problems, specifically in dairy (soft cheese) and meat products (paté, sausages). The bacterial impact of several bacteriocins on spore forming bacteria has been subject of research for many decades and indicates the great potential these bacteriocins could have in food preservation.

Bacteriocins can be used in three basic ways:

- A pure culture of the viable bacteriocin producing lactic acid bacteria;
- A (semi-) purified preparation of the bacteriocin;
- A crude bacteriocin preparation.

The production of lactic acid and bacteriocins by lactic acid bacteria may well provide for a safe and natural means of (bio) preservation.

5. Combined Methods for Food Preservation

The microbial stability and safety of most traditional and novel food is based on the combination of several preservative factors (called hurdles) which microorganisms present in the food are unable to overcome. This is illustrated by so-called hurdle effect, introduced by Leistner (1978). The hurdle effect is of fundamental importance for the preservation of foods, since hurdles in a stable product control microbial spoilage, food poisoning, as well as desired fermentation processes. In fact, the hurdle concept illustrates only the well-known fact that complex interaction of temperature, water activity, pH, redox potential, etc. are significant for the microbial stability of foods. The relation between technology and homeostasis of microorganisms is well established (Leistner and Gorris, 1994). Foods preserved by this method are safe, stable, nutritious, tasty and economical. From an understanding of the hurdle effect, hurdle technology was derived, which allows improvements in the safety and quality of foods using deliberate and intelligent combination of hurdles. In industrialised countries, hurdle technology is currently of practical interest for minimally processed foods, whereas in developing countries foods storable without refrigeration, due to stabilisation by hurdle technology, are at present of paramount importance. The application of deliberate and intelligent hurdle technology is increasing rapidly world-wide. This concept is also referred to as food preservation by combined methods, combined processes, combination of preservations or combination of techniques. At present, the term hurdle technology is most often used.

Many preservation methods are used for making foods stable and safety, e. g. heating, chilling, freezing, drying, curing, salting, sugar addition, acidification, fermentation, smoking, oxygen removal, etc. However these processes are based on relatively few parameters or hurdles, i. e. high temperature (F value), low temperature (t°C value), water activity (a_w), acidification (pH), redox potential (Eh), preservatives, competitive flora, etc. In some of the preservation methods mentioned, these parameters are of major importance; in other they are only secondly hurdles (O'Connor-Shaw, 1995; Tapia de Daza et al., 1996)).

The critical values of these parameters for the death, survival or growth of microorganisms in foods have been determined in recent decades and are now the basis of food preservation. However it must be kept in mind that critical value of a particular parameter changes if other preservative factors are presented in food. For example, the heat resistance of bacteria increases at low a_w and decreases in the presence of some preservatives, whereas a low Eh increases the inhibition of microorganisms caused by a reduced a_w. Furthermore, as mentioned earlier, the microbial stability and safety of many foods is based on the combined effects of hurdles. For instance heated canned foods called "Half-preserved" or "three-quarter preserved", need refrigeration during storage; fermented sausages are only stable and safe if both the a_w and pH are in an appropriate range (Ray, 2000).

Therefore, in food preservation the combined effect of preservative factors must be taken into account, which is illustrated by the hurdle effect. More than 40 hurdles of potential use for foods of animal or plant origin which improve the stability and/or qualities of these products have hitherto been identified (Hudson, 1990).

6. References

Barbosa-Canovas, G.V., Pothakamury, U.R., Palou, E., and Swanson, B.G., 1998, Non-*thermal Preservation of Foods*, Marcel Dekker Inc., New York.

Bell, C. and Kyrakides, A., 2000, *Clostridium Botulinum. A Practical Approach to the Organism and its Control in Foods*, Blackwell Science, London.

Booth, I.R., 1998, Bacterial Responses to Osmotic Stress: Divers mechanisms to Achieve a Common Goal in "The Properties of Water in Food". ISOPOW-6.

Cheftel, J.C. and Culioli, J., 1997, Effect of High Pressure on Meat: *Review of Meat Science*, **46**(3):211-235.

Davidson, P.M. and Branen, A.L, 1998, Antimicrobials in Foods, Second Edition, Marcel Dekker Inc., New York.

Desai, B.B., 2000, *Handbook of Nutrition and Diet*,, Marcel Dekker Inc., New York.

Gould, G.W., 1995, *New Methods in Food Preservation*, Blackie Academic and Professional, London.

Heldman, D.R. and Lund, D.B., 1992, *Handbook of Food Engineering*, Marcel Dekker, New York.

Hudson, B.J.F., 1990, *Food Antioxidants*, Elsevier Applied Science, London.

Knor, D., 1996, Advantages, Opportunities and Challenges of High Hydrostatic Pressure Application to Food Systems.in: *High Pressure Bioscience and Biotechnology* Hayashi, R. and Balny, C. eds., Elsevier Science, London.

Larousse, J. and Brown, B.E., 1997, *Food Canning Technology*, WILEY-VCH, NewYork.

Leistner, L. and Gorris, L.G.M., 1994, Food Preservation by Combined Processing. Final report, FLAIR Concerted. Action No. 7, Subgroup B, EUR 15776 EN.

O'Connor-Shaw, B., 1995, Shelf Life and Safety of Minimally Processed Fruit and Vegetables, in:: *Sauce Vide and Cook-chill Processing for the Food Industry*, S. Ghazala ed., Chapman and Hall, New. York.

Potter, N.N. and Hotchkiss, J.H., 1995, *Food Science*. Chapman and Hall. New York.

Ray, B., 2000, *Fundamental Food Microbiology*. Second Edition. CRC Press, London.

Rooney, M.L., 1995, *Active Food Packaging*, Blackie Academic and Professional, London.

Rozum, J., 1995, Smoke Flavouring in Processed Meat,in *Flavour of Meat, Meat Products and Seafood*, Shahidi, F. ed., Blackie Academic and Professional, London.

Russel, N.J. and Gould, G.W., 1991, *Food Preservatives* Blackie Academic and Professional, London.

Shapton, D.A. and Shapton, N.F., 1991, *Principles and Practices for the Safety Processing of Foods*, Butterworth, London.

Tapia de Daza, M.S., Alzamora, S.M., and Chanes, J.W., 1996, Combination of Preservation Factors Applied to Minimal Processing of Foods. *Critical Reviews in Food Science and Nutrition*, **36**(6):629-659.

2
The Challenge of Mycotoxins

ARMANDO VENÂNCIO AND RUSSELL PATERSON*

1. Introduction

"Mycotoxins are fungal metabolites which, when ingested, inhaled or absorbed through the skin, cause lowered performance, sickness or death in man or animals, including birds" (Pitt, 1996). They are poisonous substances, which occur naturally and are perhaps the most prevalent source of food related health risk in field crops. However, they are almost unrecognized publicly. Consumption of food with high levels can be fatal. Long-term exposure can, amongst other things, increase cancer risk and suppress the immune system. An invaluable web site http://www.inchem.org is available as "a means of rapid access to internationally peer reviewed information on chemicals commonly used throughout the world, which may also occur as contaminants in the environment and food". Mycotoxins are a significant component of the site and should be consulted for further information.

Filamentous fungi are ubiquitous in nature and responsible for producing mycotoxins in agricultural crops. The fungi and mycotoxins are considered generally as being natural and unavoidable. It is accepted that St Anthony's fire (Holy fire), a disease that affected many parts of Europe in the tenth century, was caused by the consumption of rye contaminated with ergot alkaloids, a known mycotoxin of *Claviceps purpurea* (CAST, 2003). Mycotoxins have plagued mankind before and since the beginning of organized crop production, i.e. always. They attract worldwide scientific, political and economic attention because of the significant economic losses associated with impacts on human health, animal productivity and international trade. Unfortunately, they are also considered as potential weapons and so could enter the food chain by this route. A recent report (CAST, 2003) has a wealth of information regarding mycotoxins, and has been referred to liberally in this overview. It is also available on the World Wide Web.

*Armando Venâncio (avenan@deb.uminho.pt), and Russell Paterson (russell.paterson@deb.uminho.pt), Departamento de Engenharia Biológica, Universidade do Minho, Campus de Gualtar, 4710-057 Braga, Portugal.

TABLE 1. Historic overview on mycotoxicoses

Year	Toxicosis	Toxin	Species
994	Ergotism	Ergot alkaloids	*Claviceps purpurea*
1890	Cardiac beri-beri	Citreoviridin	*Penicillium citreonigrum*
1913 & 1944	Alimentary toxic aleukia	Trichotecenes (mainly T-2 toxin)	*Fusarium sporotrichioides*
1952	Balkan endemic nephropathy	Ochratoxins	*Penicillium verrucosum*
1960	Turkey X disease	Aflatoxins	*Aspergillus flavus*
1989	Hole in the head syndrome	Fumonisins	*Fusarium verticillioides*

Food contamination with mycotoxins has only been fully appreciated since the 1960's, when an unknown disease of animals occurred in England (CAST, 2003). Over 100 000 turkeys and a large number of ducks and pheasants died. After a huge research effort, a toxic compound produced by the microscopic fungus *Aspergillus flavus* was discovered in the feeds and was named aflatoxin after the fungus. It is currently thought that many diseases have been caused by mycotoxins (Table 1). Indeed, the consumption of foods with high levels of certain mycotoxins can cause severe illness characterised by vomiting, abdominal pains, pulmonary oedema, convulsions, coma, and (rarely) death. Notable outbreaks were the death of 3 people in Taiwan in 1967 and 100 people in India in 1974, from aflatoxin in rice and corn respectively. Of equal or even greater importance, are long term effects particularly cancer and immune deficiency. Aflatoxins are of particular concern in populations with high levels of hepatitis B as they are shown to be 60 times higher when people have been exposed to the metabolite. The impact of mycotoxins on human and animal health is now recognized and it is estimated that they cause high economic losses. For example, the annual cost in the US alone was estimated as US$1.4 billion (CAST, 2003).

The field of mycotoxins is complex and interdisciplinary. It involves 5 interrelated systems (Paterson et al. 2004):

- commodity system consists of all aspects of production, marketing, and utilisation;
- spoilage system has many interrelated factors, e.g. biological, chemical, physical, micro- and macro-environmental;
- mycotoxin system involves 5 subsystems -productivity, metabolism, toxicology, health and wealth;
- control system comprises prevention, mycotoxin analysis and segregation of contaminated material, and detoxification. A hazard analysis and critical control points (HACCP) approach is considered the optimal system for control; and
- socio-economic system described as cultural, political and economic.

2. Mycotoxins and Relevant Fungi

Filamentous fungi, taken as a whole, produce thousands of toxic compounds. However, the more important mycotoxins belong to species of *Aspergillus*, *Fusarium*, and *Penicillium*. Only a small number of mycotoxins is considered of prime importance, namely, aflatoxins, ochratoxin A (OTA), patulin, deoxynivalenol (DON), fumonisins, nivalenol (NIV), T2-toxin, and zearalenone (ZEA) (CAST, 2003). The chemical structures of selected compounds are demonstrated in Figure 1. Identification of the fungi is carried out by microscopic examination of the isolated fungus with the conidiophore (the sporing structure) being of particular importance in this respect (Figures 2 and 3). This is perhaps not adequately appreciated as an important factor in mycotoxicology. However, it is relevant to the detection of areas within the commodity system where contamination occurs. Biochemical methods, including nucleic acid analysis, are being increasingly used to confirm existing identifications. Figure 4 demonstrates how fungal identifications are relevant to the field of mycotoxicology. A review of these issues is available, which should be consulted (Paterson et al., 2004). It considers predominately the penicillia. However, the points raised apply equally to other mycotoxigenic fungi.

Mycotoxins are classified by the International Agency for Research on Cancer (IARC) according to their toxicity, aflatoxins being the only ones proven to be carcinogenic to humans (IARC, 1993). Others have provided evidence that enables their classification in various other lesser categories (Table 2). Only the more important mycotoxins will be considered in this chapter.

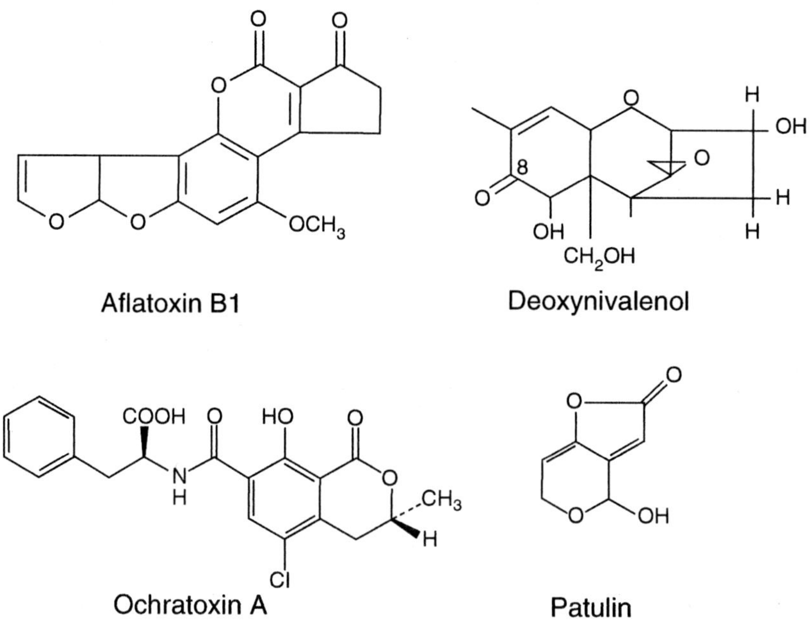

FIGURE 1. Chemical structure of some mycotoxins.

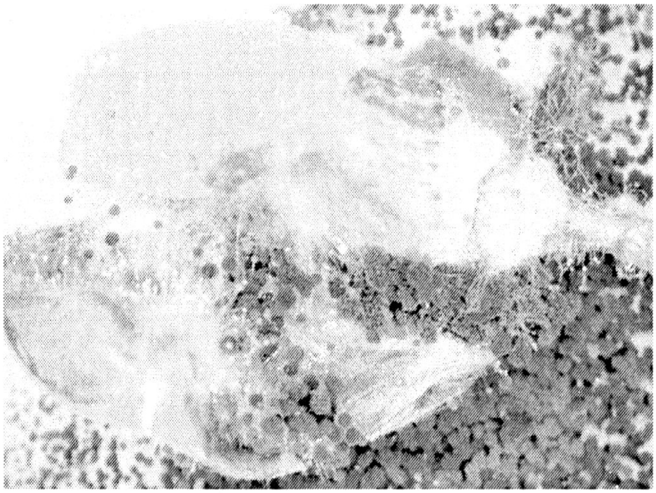

FIGURE 2. Black *Aspergillus* on the surface of grapes.

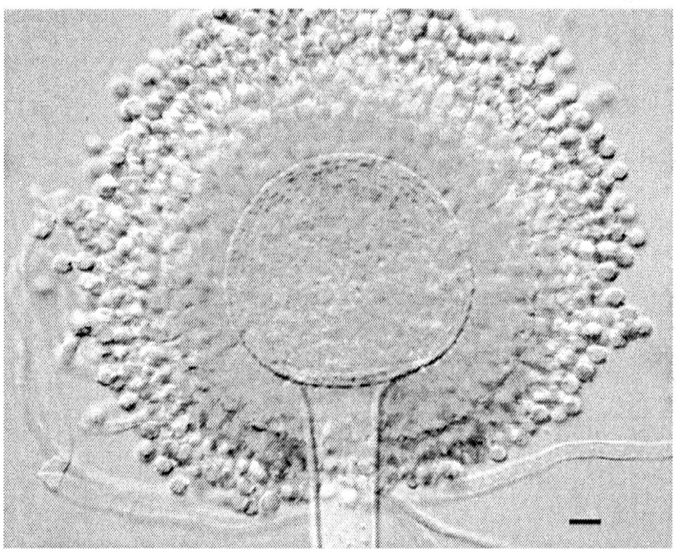

FIGURE 3. SEM image of a black *Aspergillus* conidiophore (sporing structure).

2.1. Aflatoxins

Aspergillus flavus and *A. parasiticus* are the main producers of aflatoxins. Aflatoxins B_1 and B_2 are produced by both species but only *A. parasiticus* is able to produce detectable amounts of G_1 and G_2. B and G refer to the blue and green colours observed after thin layer chromatography (TLC). The numerical

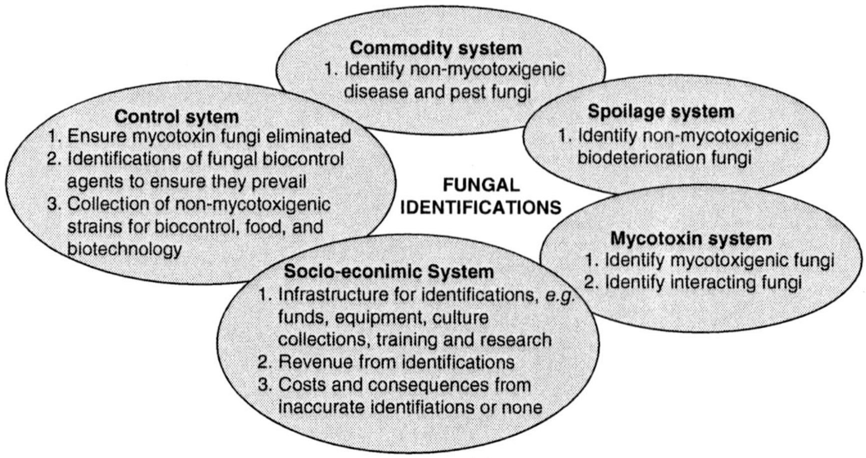

FIGURE 4. Relevance of mycotoxigenic fungi identification to the control of mycotoxins. (adapted from Paterson et al., 2004).

designation is for the different mobility (Rf value) of the compounds. Aflatoxin M_1 (M for milk) is a very important toxin in milk and products and is a minor metabolic product of these *Aspergillus* species. However, it apparently occurs mainly in dairy products as a transformation product by cows of the aflatoxin B_1 contained in animal feeds. Aflatoxins were the first mycotoxins to have their content regulated in foodstuff. In fact, in all countries where regulation exists for mycotoxins, there are tolerances concentrations for aflatoxins in foods or animal feedstuffs (CAST, 2003).

2.2. *Ochratoxin A*

Ochratoxin A (OTA) is produced by particular *Aspergillus* and *Penicillium* species (Table 3). The penicillia are probably the more important producers on grains (CAST, 2003). In other commodities, the most common are *A. carbonarius*,

TABLE 2. Classification of the principal mycotoxins according to their toxicity

	Carcinogenic to[a]		IARC
Mycotoxin	Human	Animal	Group[b]
Aflatoxins	C	C	1
Fumonisins	I	C	2B
OTA	I	C	2B
ZEA	-	L	-
Patulin	-	I	3

[a] (C) clear evidence; (L) limited evidence; (I) insufficient evidence
[b] Group 1 -carcinogenic agent to humans; Group 2A -probably a carcinogenic agent to humans; Group 2B -possibly a carcinogenic agent to humans; Group 3 -not classifiable

TABLE 3. The main mycotoxigenic fungi and most relevant associated mycotoxins

Filamentous fungi	Mycotoxins
Aspergillus	
A. carbonarius	Ochratoxin A
A. flavus	Aflatoxin B_1, B_2
A. ochraceus	Ochratoxin A
A. parasiticus	Aflatoxin B_1, B_2, G_1, G_2
Byssochlamys	
B. fulva	Patulin
B. nivea	Patulin
Fusarium	
F. cerealis	Nivalenol
F. culmorum	Deoxynivalenol, Nivalenol
F. equiseti	Zearalenone
F. graminearum	Deoxynivalenol, Nivalenol
F. poae	Nivalenol
F. sporotrichioides	T-2 toxin
F. verticillioides (=F. moliniforme)	Fumonisin B_1
Penicillium	
P. expansum	Patulin
P. verrucosum	Ochratoxin A

A. ochraceus and *P. verrucosum*. OTA has been associated with Balkan endemic nephropathy, which is potentially fatal. However, its significance has been enhanced by it being classified as possibly carcinogenic to humans by the IARC (group 2B – Table 2).

2.3. Patulin

Patulin is produced by different species of *Aspergillus* and *Penicillium* (Paterson, 2004a). *Byssochlamys* and *Paecilomyces* can also be involved. *P. expansum* is the most well known for causing a blue rot of apples and is associated with patulin production in apple products. *Byssochlamys* spp. are noteworthy as they may produce heat resistant spores which survive food processing with the potential to produce patulin in processed foods. The significance of this mycotoxin has been increased by the recent statutory European Union (EU) regulations in fruit products and its presence in baby foods is of particular concern (Commission Regulation EC n° 1425/2003).

2.4. Fumonisins, trichothecenes, and zearalenone

Fumonisins, trichothecenes, and zearalenone are the most important mycotoxins produced by *Fusarium*, a producer of a wide range of toxic compounds. Tricothecenes are related compounds and the more important are grouped into type A or type B. Type A are more toxic than type B. However, the latter are more common and therefore more relevant to food safety. Type B contains DON and NIV as the main examples, whilst a type A example is T-2 toxin. T-2 toxin was probably the cause of "alimentary toxic aleukia" which affected thousands of people in Siberia during the Second World War. It is responsible for haemorrhagic disease in animals and is associated with the formation of oral lesions and neurotoxic effects in poultry. However the most significant effect of this and other trichothecenes is the immunosuppressive activity, which is probably linked to the inhibitory effect on the biosynthesis of macromolecules (e.g., DNA, proteins). Fumonisins are a group of very closely related compounds (fumonisin B_1, B_2 and B_3), produced predominately in maize by *F. verticillioides* (*F. moniliforme*). Another important *Fusarium* toxin is zearalenone (ZEA), which has oestrogenic properties.

3. Occurrence of Mycotoxins and their Geographical Distribution

As mentioned previously, filamentous fungi are ubiquitous and some produce mycotoxins. They are regarded as natural contaminants of agricultural commodities (CAST, 2003). Specific mycotoxins appear to be restricted to certain environmental loci and crops; others are produced in a wide variety of commodities (Table 4). Mycotoxins are found in processed food from contaminated commodities. They are consumed by animals, resulting in residues in meat, eggs, and milk. These can be used as biomarkers for contaminated feedstuff as an alternative to analysis of feed, which is problematic due to sample variation. In the USA, the FDA analyse raw agricultural products for selected mycotoxins via a compliance programme and exploratory surveillance activities.

Mycotoxin contamination, with the exception of aflatoxins, is poorly studied. Consequently, it is important to establish realistic goals for management. Some general statements are as follows:

- potentially toxic fungi are found frequently;
- toxins occur in a wide diversity of agricultural commodities;
- knowledge about frequency of food/feed contamination is rather limited, but mycotoxins occur in human foods;

TABLE 4. Food commodities associated with the main mycotoxins

Mycotoxins	Foods / Feeds
Aflatoxin B_1, B_2, G_1, G_2	Peanuts, most cereals and cereal products (*e.g.*, edible oil), dried figs, pistachio
Aflatoxin M_1	Milk and Dairy products
Fumonisin	Maize and maize based foods and feeds
Ochratoxin A	Cereals and cereal products (e.g., beer, breakfast cereals), Coffee, dried grapes, wine
Patulin	Fruits (mainly apple) and their products (e.g., juices, nectars, jams, baby foods)
Trichotecenes	Most cereals (e.g., barley, maize, oat, wheat) and cereal products (e.g., beer, breakfast cereals)

- humans are less exposed in developed countries than developing; and
- the major problem in developed countries is animal health as feeds are more likely to contain toxins.

3.1. *Aflatoxins*

Aflatoxin (Figure 1) producing fungi have a worldwide distribution in sub/tropical and tropical climates (CAST, 2003). The toxins are especially common on edible nuts, oilseeds, and cereals. Many other foods and feeds are susceptible. They can be produced before and after harvest. There is a high incidence in Southeast Asian countries (Thailand, Indonesia, and the Philippines). However, they are not necessarily the most severely affected countries. They are detected also in the USA, South America, and Africa.

3.2. *Ochratoxin A*

A. ochraceus and *P. verrucosum* are associated with cereal products, and are the primary producers of OTA. The chemical structure of OTA is provided in Figure 1. *P. verrucosum* is more prevalent in regions with temperate, and *A. ochraceus* with tropical climates. A third species, *A. carbonarius* is associated with grape products and is the main species responsible for the presence of OTA in grapes and products. OTA is found in corn, wheat, sorghum, oats, rice, wine, beer, and green coffee. OTA in coffee (Paterson et al., 2001), corn, grapes, dried fruits and wheat is generally below 500 $\mu g\ kg^{-1}$. In contrast, barley and oats in Denmark and Scandinavian have high levels of OTA. Animal feeds in Canada and Europe can be highly contaminated with more than 5000 $\mu g\ kg^{-1}$ having been recorded. Highest levels have been detected in barley, oats, wheat and corn in (a) Northern Europe, i.e., UK, Denmark, Sweden and in the Balkans (former Yugoslavia) and (b) India (CAST, 2003). In these examples concentrations approach the mg kg^{-1} range. Levels are much lower in the USA. OTA appears to occur mainly in wheat and barley growing areas in the temperate zones of the northern hemisphere. High levels have been recorded in Canadian wheat, and UK barley and wheat. It is also in maize, rice, peas, beans and cowpeas. Developing country origins include Brazil, Chile, Egypt, India, Indonesia, Senegal and Tunisia. Pork is a significant dietary source and OTA has been found in the blood and milk of people in Bulgaria, Denmark, France, Germany, Italy, Poland, Sweden, and the former Yugoslavia (CAST, 2003).

3.3. *Fumonisins, trichothecenes and zearalenone*

After aflatoxin, the *Fusarium* toxins are the most frequently reported in raw agricultural commodities in US; DON (Figure 1) may be the most widely occurring *Fusarium* toxin. Grain containing DON is detected frequently. In Japan, there are high levels of DON in corn, wheat and barley. North America, Japan, and Europe are susceptible to low concentrations of DON in grains, whereas higher levels may occur intermittently in developing countries. Other vulnerable countries include China and India. There are high levels of the related compound NIV in barley, corn, and wheat in Japan. T-2 toxin is produced on cereals in many parts of the world, and is associated with prolonged wet weather at harvest (CAST, 2003).

ZEA mainly occurs in maize in low concentration in Europe, Japan and North America. High concentrations have been detected in developing countries, especially in the more temperate Highland regions (CAST, 2003). ZEA is frequently found in corn, wheat, barley and grain sorghum from the USA.

Low levels of fumonisins are extremely common throughout the world (CAST, 2003). However, in some cases more than 5 mg kg^{-1} is observed. They are produced by the fungus *F. verticillioides* which occurs worldwide and is found frequently on maize. It is prevalent in the USA, Canada, Brazil, South Africa, Austria, Italy, and France. Fumonisins especially occur when maize is grown under warm dry conditions. These compounds are associated with oesophageal cancer in southern Africa, and China.

3.4. Patulin

Occurrence of patulin (Figure 1) is likely to be more frequent than is generally realised as numerous fungi produce this compound and in a wider range of fruits and vegetables (Paterson, 2004a). The toxin is associated with *P. expansum* blue rot of apples and grapes. Anywhere that fruit products such as juice are manufactured will have a potential problem. Patulin producing fungi are isolated frequently from vegetables and so products are likely to be contaminated in some cases, and similarly for silage and malt. The presence of patulin has been reported from many EU countries in relation to the recent regulations imposed by the EU. It occurs frequently in the USA where the FDA operates mandatory action levels in apple juice. It is a problem in parts of South America (see case study below), in South Africa and in Portugal (Smith and Solomons, 1994).

3.5. Co-occurrence of mycotoxins

Fusarium toxins often co-occur with aflatoxins but not with OTA. Production of aflatoxins in stored grain is enhanced by the presence of trichothecenes. Synergistic effects enhance the toxicity of naturally occurring trichothecenes: the effect of DON is enhanced by T-2 toxin; potent synergism of DON with non-toxic *Fusarium* metabolites has been reported. In Japan, ZEA sometimes occurs as a co-contaminant when there are high levels of either DON or NIV in corn, wheat, and barley (CAST, 2003). More information is required generally about this important area.

3.6. Biocontrol agents (BCAs)

Natural pathogens of certain pests and diseases are being considered as alternatives to synthetic chemical pesticides with potential benefits in terms of environmental contamination and resistance of target organisms. However, a possible source of mycotoxins is from fungal BCAs. These may produce toxins and are sprayed onto crops pre- and (potentially) post-harvest. *Metarrhizium anisopliae* is one such and produces the highly toxic cytochalasins amongst other things (Paterson, 2004b). These compounds have been detected in agricultural commodities (CAST 2003) indicating the need to carefully regulate the use and production of these products. In addition, enniatins and beauvaricins are metabolites of insect and plant pathogens and have been detected in wheat and maize infected with *Fusarium* with negative implications for using the insect pathogens as biological control agents. The fact that BCAs are often for use in developing countries compounds the disproportionate mycotoxin problem that already exists.

3.7. Warfare

There are reports of mycotoxins having been used as biochemical weapons. Recent events have indicated that they are still actively being considered. CAST (2003) recommends that assessments are required of mycotoxins as biological weapons. Obviously, levels of mycotoxins in the environment used for this purpose will be at higher concentrations than occur naturally. Drinking water is an obvious target (Paterson and Lima, 2005).

4. Control

The control and detection of mycotoxins is a continuous process in commodity production. They can become established and remain within the commodity anywhere throughout the production, storage, transportation and processing chain (CAST, 2003). Control of mycotoxins is constantly evolving. No absolute controls are available to eliminate mycotoxins, and total control is probably not economically feasible. The most effective control is analysis and subsequent removal of the contaminated lot. It is difficult to make general statements, because of the diversity of fungi producing mycotoxins pre- and post-harvest and in a wide variety of crops. Additional control of mycotoxin contamination involves either the prevention of its synthesis prior to harvest, or the prevention and decontamination (deactivation or removal of the mycotoxin) after harvest. An HACCP approach is recognised generally as the optimal procedure for controlling mycotoxins as is the case for other food safety issues (Coker, 1999). A gene probe for patulin is available with implications for HACCP procedures (Paterson, 2006). Effective decontamination of mycotoxins if used as biochemical weapons requires consideration in the authors' opinion.

Mycotoxins in commodities are unavoidable and to minimize their occurrence in the food chain requires management strategies to prevent contamination entering food/feed processing facilities (CAST, 2003). Careful testing is required to accomplish this, i.e. sampling, sample preparation and analysis. Sampling and sample preparation are of paramount importance and it is crucial to obtain a representative sample. The distribution of toxins in a commodity is heterogeneous, with pockets of high concentration coupled with low or no amounts in other parts. It is extremely difficult to determine with confidence what the true concentration is. An analogous sampling problem exists for isolating the producing fungi (Paterson, et al., 2004).

Analysis is normally by chromatography. Initially, thin layer chromatographic (TLC) methods were employed (see aflatoxins) (CAST, 2003). However, this has been superseded, in many cases, by high performance liquid chromatography (HPLC). Gas chromatography is often used for *Fusarium* mycotoxins. So-called hyphenated techniques are employed such as diode array and mass spectroscopic detection. The use of immunoaffinity columns for analysis of most of the main mycotoxins has greatly improved analysis in terms of quantification, purification and convenience. Immunological test kits such as ELISA are available but they are not as accurate as the chromatography-based methods. TLC is optimal for the detection of previously unknown "new" compounds because of the more universal detection methods employed (CAST, 2003).

4.1. Good agricultural practice (GAP)

The prevention of mycotoxin formation is achieved by influencing environmental conditions through management of agricultural practices prior to harvest (CAST, 2003). In the case of peanuts, it is known that the formation of aflatoxins in the plant takes place when ambient temperatures of 25 to 32 °C occur simultaneously with low humidity in the soil. Therefore, it is possible to control the biosynthesis of aflatoxins in peanuts by controlled irrigation at critical periods of the day. The use of recommended crop management practices and harvesting crops when they are mature decreases, for instance, aflatoxins. Something as prosaic as adjustment of combine harvesters to prevent excess damage to kernels can be effective (CAST, 2003). Careful adjustments to the machine may actually eliminate contaminated kernels, although sound ones may still contain high levels of mycotoxins and so the removal of the damaged ones will not eliminate mycotoxins completely. Care is required to clean storage bins, auger pits, maintain clean lorries, trailers and combine harvesters. Overall decreases of total fumonisins of 60% by screening and gravity settling corn can be obtained.

4.2. Good management practices (GMP)

After harvest, two overriding factors for storage are water activity (a_w) and temperature. Others are cleaning, insect control, use of antifungal agents and maintaining integrity of the seed coat. These parameters, if not kept at the desired levels, may lead to the growth of fungi and the accumulation of mycotoxins in stored crops (CAST, 2003). The source of this inoculum could be the crop itself, mainly when it is of a poor quality, or colonised plant debris, when storage facilities are not properly cleaned.

Harvested grain, coffee beans, fruits, and oilseeds crops should be dried immediately. The final safe moisture depends on crop and climatic conditions where stored. Corn which is contaminated with *A. flavus* and subjected to increased moisture by 18 % for 4-6 h can increase aflatoxin levels (Sinha and Bhatnagar, 1998). Early harvesting of corn followed by drying may avoid increased aflatoxin contamination (but drying may not be economical). Grain needs to be dried to 15 % or less before storing. Drying is critical to coffee (Paterson et al., 2001) and peanuts to avoid OTA and aflatoxins respectively. Contamination of coffee with OTA fungi from soil apparently occurs in the drying process if beans are kept for more than 4 days at an a_w above 0.8 (about 16 % humidity in grain) (Sinha and Bhatnagar, 1998).

If the product is dry and kept dry, no further biodeterioration will occur. However, insects, rodent activity, moisture migration, condensation and/or water leaks may re-hydrate the grain, for example, and lead to fungal growth and possible toxin production. Insects and rodents need to be controlled as they create conditions suitable for fungal growth, which once started, the water of metabolism can be sufficient for more fungal growth and mycotoxin production. Low toxicity antifungal agents are used to supplement good management practices but are not a substitute. Air treatment in storage facilities may prevent growth and contamination, e.g., propionic acid, mixtures of propionic and acetic acid, or ozone (CAST, 2003).

4.3. *Removal of mycotoxins after harvesting*

Another way of controlling mycotoxins is by removal after harvest. This may be achieved from granulated products, such as cereals, bean or others, since contaminated kernels have distinctive characteristics. Damaged kernels may exhibit differences in colour, density, shape or size; they may also exhibit fissures, mould growth, or may be broken. Peanut kernels containing aflatoxin are usually shrivelled and discoloured and can be separated from sound ones by sieving, electronic sorting and hand picking. These contaminated kernels have a lower density, as does contaminated corn, and thus removal is possible by flotation and density sorting. These procedures are reported to remove more than 90% of aflatoxins (Sinha and Bhatnagar, 1998).

A separation of kernels based on different size is also useful for those situations where broken kernels occur, as in the case of fumonisins in corn. These mycotoxins are usually present in amounts ten times higher in broken kernels than in whole ones. Other products, such as almonds, may not be broken but have fissures. From these fissures oil may leak, which is detected by its fluorescence under UV light. For cereals, cleaning and polishing (scouring) may be useful for the removal of OTA (see case study below) or DON (Skudamore and Banks, 2004).

4.4. Removal during processing

The amount of mycotoxins still present in raw material, after the trimming process, may be further reduced during processing. Physical, chemical or biological processes often employed in food processing may be useful in the removal or deactivation of mycotoxins: these include (a) the transformation of mycotoxins during heat treatments (e.g. roasting coffee beans), (b) exposure to sun light, (c) adsorption with clarifying agents (e.g. during refining of edible oils), (d) extraction of immiscible solutions, (e) cross reaction with food additives (e.g. sulphur compounds used in fruit juices production) or (f) reduction during alcohol production by *Saccharomyces cerevisiae.*

Aflatoxins and OTA are rather stable when exposed to dry heat and less stable when exposed to humid heat (Sinha and Bhatnagar, 1998). However, at high temperatures they can be transformed into other products, some of them less toxic, although there may be others formed that are of unknown toxicity. Boiling rice at normal pressure reduces aflatoxins up to 50 %. Increasing the temperature increases the level of aflatoxin transformation: a pressure-cooking can degrade aflatoxins by 70 %. Frying can reduce aflatoxins by 60 % when oil temperature is at 150 °C for 20 minutes. OTA is partially removed from coffee beans during roasting. Degradation of OTA to a maximum of 90 % is possible, depending to the conditions of operation while roasting (e.g. temperature usually between 200 and 250 °C and resident time from 5 to 20 minutes).

Aflatoxins present in edible oil after its extraction from seeds are mainly dispersed in fine particles (Sinha and Bhatnagar, 1998). The removal of these particles is usually aided by the addition of adsorbents, which consequently remove mycotoxins.

OTA appears to be associated with the solid phase in the vinification process of wine (Fernandes et al., 2003). The resulting juice for fermentation contains a considerably lower concentration than the original material. However, OTA, when present in grapes, may be carried over to the juice and persist after fermentation. Wine is usually a cloudy suspension that is clarified with the aid of chemical adjuvant (fining agents). OTA binds to these agents and so is partially removed.

Solvent extraction is employed in various food processes, such as the refining of edible oils or the removal of caffeine to produce decaffeinated coffee. During these processes, mycotoxins are co-extracted with the solvent. As examples, a partial removal of aflatoxins is obtained from edible oil while refining and OTA is removed from beans in the decaffeination of coffee (Pittet et al., 1996). Water is a good solvent for OTA in certain food matrixes, such as coffee: so brewing coffee is the perfect way of extracting OTA from the coffee grains to the cup!

"Phase changes" can often be employed in food processing e.g. cheese production where two immiscible phases (cheese and cheese whey) are obtained from milk. Aflatoxin M_1 is distributed between these phases; since the cheese fraction is much smaller than the whey fraction (10 litres milk for 1 kg cheese) aflatoxin M_1 usually has a higher concentration in cheese than in milk (van Egmond, 1989). However, aflatoxins are generally insoluble in water and so may also dissolve more readily in the lipid rich cheese fraction.

Enzymes and microorganisms are used frequently in food processing and may transform mycotoxins into other compounds with less or unknown toxicity (CAST, 2003). This is the case with proteases. They can hydrolyze the peptide bond of OTA yielding phenylalanine and the much less toxic ochratoxin alpha. *S. cerevisiae* transforms patulin into other compounds during alcohol production in cider production. Patulin is also degraded by (a) sulphur dioxide, and (b) sulphur containing proteins. This approach should be considered as a method for reducing patulin in other commodities. Indeed many of the above processes may find uses in other commodity systems but have simply not been tried.

4.5. Degradation of mycotoxins

The technology for the chemical degradation of aflatoxin in animal feedstuffs (Coker, 1999) is most developed for remediating contaminated material. Numerous oxidising agents, aldehydes, acids, and bases (inorganic and organic) have been investigated as chemical detoxification agents. Patented procedures exist that use ammonia, calcium hydroxide, methylamine and a mixture of calcium hydroxide and methylamine. Most interest is in the use of ammonia in the anhydrous form and in aqueous solution. There have been many studies on the efficiency of the ammonia detoxification and on the nutritional and toxicological properties of the resulting feeds. Ammoniation is an accepted method for animal feeds in some states of USA, Brazil, Mexico, Senegal, South Africa and Sudan (CAST, 2003).

4.6. Others

Considerable effort has been applied to breeding resistance to fungi into crops. For example, groundnuts and maize have been made resistant to *A. flavus*. However, trials were unsuccessful in the field (CAST, 2003). Biocontrol of fungi by the use of nontoxigenic fungal competitors is attractive in theory but in practice there is little evidence that it works, or that it is intrinsically safe.

Food additives that absorb some mycotoxins in the animal gut and hence decrease bioavailability to host animals are available (CAST, 2003). Commercial absorption materials (e.g. clays) have been included in animal feed to absorb aflatoxins and make it less accessible to the animal's digestive system. An example is the absorption of aflatoxin by adapted chicken feed (CAST, 2003). It is unclear whether this works effectively. It obviously adds cost to the feed and the absorbent would presumably always have to be added as it would not be necessarily be known whether the feed actually contained aflatoxins. In the authors opinion there is nothing to suggest the mycotoxin problem has been solved by this procedure.

5. Legislation and Regulation

Legislation and regulation are constantly evolving issues. Regulatory bodies continually assess levels of permissible exposure to humans by risk assessment processes to establish tolerable daily intakes of selected mycotoxins. In some cases biomarkers are available to determine exposure, which are needed to establish problem areas throughout the world (CAST, 2003). Recent developments in metabolomics (Paterson, 2004c) will facilitate greatly the detection of mycotoxin biomarkers.

Countries and even continents have various required levels for mycotoxins and differing capabilities of implementing them. Some of the critical issues have more to do with politics and economics than science. In general, developed countries have more stringent regulations and have the infrastructure to implement them. The different limits and applications vary greatly from country to country. Some examples drawn from two recent reviews (CAST, 2003; EMAN, 2004) on the subject, illustrate the complexity of the situation.

Mycotoxins affect approximately one quarter of global food and feed crop output. Many countries have applied regulations to limit exposure of mycotoxins to consumers. Controlling mycotoxins often involves product and not process standards. These are intended to be based on sound risk assessments. The levels of economic development and the susceptibility of a nation's crops to contamination affect perceptions of how tolerable are the health risks and are leading to widely different national or multilateral agencies standards. For example, the number of countries with mycotoxin regulations has increased from 56 in 1987 to 77 in 1995 (+38%) and 98 in 2002 (+27%) (CAST, 2003; van Egmond and Jonker, 2003). The total population in these 98 countries represents approximately 90% of the world's inhabitants. The 1995 study has revealed that at least 77 countries have specific regulations for mycotoxins (CAST, 2003). No data are available for approximately 50 counties and 13 are known to have no specific regulations, many in Africa. Reaching consensus is difficult as levels vary so widely worldwide and perceptions differ broadly with respect to international trade. Countries with low levels do not want high-level material being imported. In contrast, countries where high levels are common want standards where these levels are permitted so they can trade on the world market.

The Codex Alimentarius Commission of the United Nations (Codex) sets advisory standards on mycotoxins and is multinational. The Commission maintains that these are science-based, appropriate and non-discriminatory. The ultimate goal is to facilitate trade without compromising consumer protection. However, this is controversial and problematic. The standards impose unfair economic and safety burdens on lower-income food-exporting countries, as compliance is too costly and/or unachievable from technical, infrastructure and management perspectives. Also, stricter control implies that the material remaining in exporting countries has higher overall concentrations. The economic losses associated with mycotoxin contamination are difficult to assess in a consistent manner and no comprehensive analyses exist (CAST, 2003). This is partly due to difficulty in establishing cause and effect relationships.

Several studies have established that setting tolerance levels involves transparent tradeoffs between human health and economic opportunity (CAST, 2003). It has been estimated that deaths from a standard of 10 or 20 μg kg^{-1} aflatoxin in the human diet would result in 39 cancer or 41 cancer deaths respectively per year per billion people for a European diet. In Asia a change from 20 to 10 μg kg^{-1} would result in 300 fewer deaths.

In terms of variations in legislation, the USA has a limit of 20 μg kg^{-1} for total aflatoxins in all products, except milk, for human consumption, while those in the EU there are for various food commodities and range from 2 to 10 μg kg^{-1} (see Table 5). The limits set by the US FDA and the EU for milk are of 0.5 and 0.05 μg L^{-1} for aflatoxin M1, respectively.

The US FDA advisory limit for DON in wheat and wheat products for human consumption is 1000 μg kg^{-1}; in the EU this limit varies from 200 μg kg^{-1}, in cereal based food for infants, young children and babies, to 1750 μg kg^{-1} in unprocessed cereal products (see Table 6). Russia has a limit of 1000 μg kg^{-1} for hard wheat and Canada of 2000 μg kg^{-1} for uncleaned soft wheat (EMAN, 2004).

Even within Europe there are differences between the EU and other countries. The EU set limits for aflatoxins, *Fusarium* toxins, OTA, and patulin in several food commodities (Tables 5 to 8). Other countries in Europe have independent limits and in the EU each country may have other limits for other food commodities

TABLE 5. Aflatoxins regulations (Commission Regulation (EC) No 466/2001, and 683/2004)

Commodities	Maximum Admissible Levels of Aflatoxins (μg kg^{-1})		
	B_1	B_1+B_2+G_1+G_2	M_1
Groundnuts, nuts, dried fruit and processed products thereof for direct human consumption or as a food ingredient	2	4	
Groundnuts to be subjected to sorting, or other physical treatment, before human consumption or as a food ingredient	8	15	
Nuts and dried fruit to be subjected to sorting, or other physical treatment, before human consumption or as a food ingredient	5	10	
Cereals and processed products thereof for direct human consumption or as a food ingredient	2	4	
Chillies, chilli powder, cayenne pepper, paprika, white and black pepper, nutmeg, ginger and turmeric	5	10	
Milk (raw milk, milk for the manufacture of milk based products and heat treated milk			0.05
Baby foods and processed cereal based foods for infants and young children	0.1		
Infant formulae and follow-on formulae, including infant milk and follow-on milk			0.025
Dietary foods for special medical purposes intended specifically for infants	0.1		0.025

TABLE 6. *Fusarium* toxins regulations (Commission Regulation (EC) No 856/2005)

Commodities	Maximum Admissible Levels ($\mu g\ kg^{-1}$)
Deoxynivalenol	
unprocessed cereals, other than durum wheat, oats and maize	1250
durum wheat and oats	1750
maize	-*
cereal flour	750
bread, biscuits, cereal snacks and breakfast cereals	500
pasta (dried)	750
cereal based food for infants, young children, and baby food	200
Zearalenone	
unprocessed cereals, other than maize	100
cereal flour except maize flour	75
bread, pastries, biscuits, (other than maize) cereal snacks and breakfast cereals	
Maize, maize flour, and other maize based foods	-*
Fumonisins	-**
T-2 and HT-2 Toxin	-*

*limits to be fixed before 1 July 2007
**limits to be fixed before 1 October 2007

TABLE 7. Ochratoxin A regulations (Commission Regulation (EC) No 472/2002, 683/2004, and 123/2005)

Commodities	Maximum Admissible Levels of Ochratoxin A ($\mu g\ kg^{-1}$)
Cereals (including rice and buckwheat) and derived cereal products. Raw cereal grains (including rice and buckwheat)	5
All products derived from cereal (including processed cereal products and cereal grains intended for direct human consumption)	3
Dried vine fruits (currents, raisins and sultanas)	10
Baby foods and processed cereal based foods for infants and young children	0.5
Dietary foods for special medical purposes intended specifically for infants	0.5
Roasted coffee beans and ground roasted coffee with the exception of soluble coffee	5
Soluble Coffee (instant coffee)	10
Grape juice, grape must and wine	2

and/or for other mycotoxins. Presumably this will become harmonised when more countries become members. However, there are almost certainly considerable differences between EU member countries in their ability to implement the regulations, which relates to the relative economic status of each country. There is a clear requirement for the EU to further assist these countries.

TABLE 8. Patulin regulations (Commission Regulation (EC) No 1425/2002)

Commodities	Maximum Admissible Levels of Patulin ($\mu g\ kg^{-1}$)
Fruit juices and nectar (in particular apple juice, and fruit juice ingredients in other beverages), and concentrated fruit juice after reconstitution	50
Spirit drinks, cider and other fermented drinks derived from apples or apple juices	50
Solid apples products (apple compote, apple puree or direct consumption)	25
Apple products for infants and children, and other baby foods	10

5.1. Process standards vs. product standards

It is particularly apparent that differences in tolerance levels are potential sources of conflict between countries. The use of good agricultural practices (GAPs) at preharvest and good manufacturing practices (GMPs) at processing and distribution can be used to minimise risk, and provide procedures for reducing mycotoxins levels acceptable to more countries (CAST, 2003). They compliment product standards and potentially reduce overall economic losses. The Codex recommended that GAPs and GMPs be used to establish HACCP safety systems throughout the food production chain (CAST, 2003). HACCP principles are likely to be among the most effective means of lowering risks and economic losses since prevention is more practical than decontamination (at least theoretically).

Mycotoxins are economically important although not publicly prominent. Balancing economic cost and health benefits has become a source of friction especially for export-reliant developing countries. Trade disputes are likely to persist with respect to regulatory standards. Many factors that influence contamination are difficult or impossible to control. Tolerable health risks appear to depend on level of economic development and susceptibility of a nation's crops. Mycotoxins standards can be set without internationally accepted risk assessments because of the "precautionary principle". To minimise risk, stakeholders should consider implementing GAPs, GMPs and HACCP principles, although developing countries will require assistance with implementation (CAST, 2003).

6. Case Studies

6.1. Experimental reduction of OTA in wheat by scouring

This procedure is for the reduction of OTA that has already formed in wheat and so is not an HACCP approach. Maximum permissible levels of OTA in cereals and derived products are provided in Table 7. Equivalent limits are expected for other food commodities. The EU funded a project aimed at prevention and management of OTA in cereals (see www.mycotoxin-prevention.com). One aspect examined the fate of OTA in wheat through the food chain including milling, baking and extrusion processing. Results demonstrate how processes such as surface scouring of the whole-wheat grain can reduce concentration. Wheat was inoculated with *P. verrusosum* and grown to yield 5 μg kg^{-1} as an experimental sample. Samples were taken at each relevant stage and analysed by HPLC. One batch was untreated, another was scoured to reduce weight by approximately 1%, and another was reduced by 2% by removing the outer coat. The result was a significant reduction in the level of OTA. A reduction of 65% in bread from white flour was achieved. However, this needs to balance by the economic loss involved and changes in the characteristics of the flour so processed. In wholemeal flour more OTA is present in the bran and offal fractions, which are not removed, and so the bread usually contains higher levels of OTA. This was not an industrial scale experiment and any wider implications of the process require careful economic assessments and as to whether it is practical in large scale (Skudamore and Banks, 2004).

However, HACCP is generally regarded as optimal for controlling mycotoxins as for other food safety problems. The following procedures are similar to those that would be implemented for other commodity and mycotoxin systems.

6.2. *Copra cake and meal—southeast Asia*

Dried coconut flesh (copra) is extracted to produce coconut oil. The residue is called copra cake or copra meal, which is used as protein sources in animal feeds. The European Union imposed aflatoxin B_1 regulations on dairy feed of 5 μg kg^{-1} and copra by-product to 20 μg kg^{-1}. Many oil mills would have become non-viable because of these, which would have caused great hardship to millions of coconut farmers. To save the European market, an HACCP approach was used to raise confidence in the product.

An HACCP team was established consisting of an HACCP specialist, oilseeds specialist, socio-economist, mycotoxicologist, mycologist, and drying engineer. Representatives of the coconut oil industry from the public and private sectors were included.

Aflatoxin was found to be produced within 10 days of splitting coconuts (a_w>0.82) when aflatoxin-producing fungi grew. Premature splitting, during harvest or de-husking, resulted in contamination prior to drying.

Uniform drying within 48 hours of splitting the nut was found to be key to control the aflatoxins. Also, smoke drying was correlated with low-aflatoxin copra. Sun-dried copra had very high concentrations, and was discouraged. Premiums to farmers were increased to encourage them to produce dry copra. The HACCP steps are provided in Table 9 (see also Coker, 1999).

6.3. *Apple juice—South America*

Apple juice in South America was at risk of exceeding a 50 μg kg^{-1} target level. An HACCP team was formed to address this issue involving equivalent specialists to the copra example above. A product description and the intended use were

TABLE 9. HACCP Strategy for reducing aflatoxin in coconuts

HACCP steps
Step 1: Harvesting and dehusking – Critical Control Point (CCP) 1. Eliminate split nuts to isolate any aflatoxin already present by the use of trained harvesters or de-huskers. Validate by determining the aflatoxin concentration of batches of accepted nuts
Step 2: Splitting nuts – GAP. It is Good Agricultural Practice (GAP) to ensure that the coconut meat is protected from contact with soil, which is a rich source of inoculum
Step 3: Drying – CCP2. Dry to a safe moisture content within 48 hours to prevent growth of fungi and production of aflatoxin. The CCP can be validated by measuring the moisture content of the product
Step 4: Primary trader, procurement and drying -GMP/GSP. It is GMP for primary traders to purchase Grade 1 copra with <1% yellow-green fungus (characteristic of *A. flavus* or *A. parasiticus*) and meeting a 12% moisture limit and separating from lower grade. Good storage practice ensured that the copra remained dry
Step 5: Secondary traders, procurement and storage -GMP/GSP. It is also GMP to procure Grade 1 copra, separate from other grades and marketed as low-aflatoxin copra. GSP such as palleted storage in a store with good ventilation and a sound roof, will prevent re-wetting and subsequent contamination with mould and aflatoxin
Step 6: Oil mills, procurement – GMP. Procurement of Grade 1 copra is essential to produce copra by-products containing acceptable levels of aflatoxin, and this is considered to be GMP. It is important to have adequate aeration because 'hot spots' can develop and even result in spontaneous combustion
Step 7: Oil mill, expelling/extracting/pelleting -CCP3. Control measures are unnecessary during expelling of oil and solvent extraction. The high temperatures present will sterilise the copra meal. The pelleting process was classified as a CCP with a critical limit of 12% moisture in the cooled pellets. This CCP can be validated by determining the moisture content of cooled pellets. An example of GSP of pelleted copra by-product is by placing in bags or in bulk until shipment is possible
Step 8: Shipment -GMP/GSP. If copra by-product is loaded at a moisture content of 12% no increase in aflatoxin contamination is likely during shipping. Sea-water damage must be avoided. Validation procedures provided by the fully quantitative aflatoxin results on the pre-loading samples, taken immediately prior to export. The HACCP Plan will be fully documented, including appropriate records at the farmer and primary trader steps

verified. As a result of these, a Commodity Flow Diagram (CFD) was prepared and verified by visits to orchards and processing plants.

For the identification of control measures the steps in the CFD where patulin contamination was most likely to occur were identified. Each step was considered in turn. Patulin contamination was likely to be produced in the orchard during growing and bulk storage. There was little risk of further contamination during transportation but damage to apples at this stage can increase the risk of subsequent contamination. In the factory, patulin contamination is most likely to increase during storage. There was likely to be patulin contamination present in the apples, or the resultant apple juice, at every step in the commodity chain. Hence it was important to minimise contamination, and reduce levels of contamination to the acceptable level.

TABLE 10. HACCP procedures for reducing patulin in apple juice

HACCP steps
Step 1: Growing in the orchard – GAP. Growth of the mould *P. expansum*, and subsequent patulin contamination, can occur pre-harvest, where it is associated with damaged and over-ripe fruit. GAP will minimise insect and bird damage
Step 2: At harvest -CCP1. The control measure at this step is to reject rotten and damaged apples during harvesting. Application of this control measure at Step 2 is considered a CCP because it will reduce mould contamination to an acceptable level. The critical limit for this CCP will relate to the percentage of visibly mouldy apples remaining after sorting
Step 3: Farm, bulk storage – GAP. Application of GAP and GSP is necessary to minimise rotting of fruit and subsequent patulin production during bulk storage. Storage of sound apples is important and the length of storage should be minimised, unless refrigerated storage facilities are used
Step 4: Transportation – GAP. There is little risk of patulin contamination during short duration journeys, but any physical damage sustained during transportation, including loading and unloading, will predispose the fruit to subsequent mould attack and possible patulin contamination. The correct handling of fruit is therefore required. For long journeys low temperatures are required
Step 5: Factory procurement – GMP. Procurement of batches of low-grade apples, with a high percentage of damaged and rotten fruit, are to be avoided
Step 6: Factory sorting -CCP2. The control measure is sorting to remove visibly mouldy apples. The critical limit for this CCP will be the acceptable percentage of mouldy apples remaining after the sorting procedure, and monitoring will be by use of a trained supervisor
Step 7: Factory, washing -CCP3. The control measure is washing the apples using high-pressure water spraying to remove rotten apple flesh, and patulin, from the fruit. Patulin levels will be reduced at this step, but spores will be suspended in the water. This inoculum will increase the risk of mould growth during bulk storage
Step 8: Bulk storage of whole apples -CCP4. The control measure is to prevent mould growth and patulin production by storing at reduced temperature. If refrigerated storage is not available, then storage time must be minimised. These critical limits for temperature are monitored
Step 9: Pressing/extraction process – GMP. Good Manufacturing Practice will ensure that the presses are cleaned regularly to prevent a build-up of mouldy apple waste which could be a source of patulin contamination
Step 10: Filtration -CCP5. The control measure is the removal of fine, patulin-rich particles held in suspension in the crude juice. Critical limits are set for the size and quantity of particles remaining in the apple juice after filtration. These critical limits are monitored by microscopic examination of samples of apple juice
Step 11: Pasteurisation – CCP. This step is a CCP for the control of bacterial hazards. However, it can also be considered as a CCP for control of the patulin hazard since pasteurisation will destroy spores of *P. expansum* (although other thermotolerant species are needed to be considered)
Step 12: Asceptic packaging process – GMP. Following pasteurisation, it is important to prevent the re-introduction of microorganisms, including mould spores, during packaging. These procedures are covered by GMP. Packaging is selected which will protect the juice from contamination by micro-organisms, e.g. tetra packs, or glass bottles with air-tight seals for the lid
Step 13: Storage and dispatch – GMP. No subsequent contamination with patulin is likely

N.B. It is noted that analysis for patulin is not recommended which is to the detriment of the HACCP protocol. The current authors recommend that validation analyses are required immediately after step 2. Similarly, fully quantitative analyses are required after step 11. A gene probe is available which may be useful in HACCP protocol development (Paterson, 2006)

6.3.1. Possible patulin control measures

Contamination of the juice could be prevented at stages where rotten or rotting apples could have been rejected from the process, either in the orchard when the fruit is harvested, or during sorting in the factory (FAO, 2003). Post-harvest patulin contamination could have been eliminated, or significantly reduced, by storage at <10 °C and by minimising storage times. Washing, and in particularly pressure spraying, was shown to be effective in removing patulin from apples. Patulin can also be removed from apple juice by filtration, when the patulin bound to solid particles of apple flesh is removed. Inactivation of *P. expansum* spores during pasteurisation at Step 11 reduced the risk of patulin production in the finished juice. A spreadsheet summarising the HACCP plan for patulin in apple juice was developed. The HACCP steps are provided in Table 10.

The HACCP plan was to be audited quarterly, and amended as necessary. Also, it was to be fully documented and appropriate records kept at each CCP. However, no assessment was performed to estimate whether the mycotoxin was lower after these procedures were implemented. It is assumed that they were largely to build confidence in the industries. It appears likely they would have created products with generally lower patulin levels. There is no information as to whether the products fulfilled the ultimate purpose of the HACCP analysis, that of the apple juice having consistently lower patulin concentrations.

7. Conclusion

Mycotoxin contamination of foods in subtle. It is known to cause death and illness on a large scale. The economic consequences are correspondingly high. In developed countries levels tend to be low but at a financial cost. However, in developing countries the problem will be much higher. Some of the issues surrounding mycotoxins are of a political/economical nature, rather than a scientific one. Methods of control are available, with the most effective being the analysis of the particular mycotoxin, and the removal of the contaminated kernels/lots. HACCP will be the most effective way of lowering mycotoxin concentrations in commodities. Finally, the use of mycotoxins as 'bioweapons' in foods requires careful consideration.

Acknowledgements. Russell Paterson is funded by grant SFRH/BPD/14923/2004 from *Fundação para a Ciência e Tecnologia*, Portugal.

8. References

CAST, 2003, *Mycotoxins: Risks in Plant, Animal, and Human Systems* [online]. Council for Agricultural Science and Technology, Ames, Iowa, USA. Available from: http://www.cast-science.org/cast/pub/Mycotoxins.pdf [Accessed 1 Sep 2005].

Coker, R., 1999, *Mycotoxins -The Silent Threat to Human and Animal Health*, Kent, the University of Greenwich, pp. 1-52.

Commission Regulation (EC) No 466/2001 of 8 March. *Aflatoxins in Foodstuffs*. Official Journal of the European Communities. 16/03/2001, L77/1.

Commission Regulation (EC) No 472/2002 of 16 March. *Ochratoxin A in Foodstuffs*. Official Journal of the European Communities. 16/03/2003, L75/18.

Commission Regulation (EC) No 683/2004 of 13 April. *Amending Regulation (EC) No 466/2001 as regards aflatoxins and ochratoxin A in foods for infants and young children*. Official Journal of the European Communities. 14/04/2004, L106/3.

Commission Regulation (EC) No 1425/2003 of 11 August. *Amending Regulation (EC) n° 466/2001 as regards patulin*. Official Journal of the European Communities. 12/08/2003, L203/1.

Commission Regulation (EC) No 123/2005 of 26 January. *Amending Regulation (EC) n° 466/2001 as regards ochratoxin A*. Official Journal of the European Communities. 28/01/2005, L25/3.

Commission Regulation (EC) No 856/2005 of 6 June. *Amending Regulation (EC) n° 466/2001 as regards Fusarium toxins*. Official Journal of the European Communities. 7/06/2005, L143/3.

EMAN, 2004, *European Mycotoxin Awareness Network* [online]. Available from: http://193.132.193.215/eman2/index.asp [Accessed 1 Sep 2005].

FAO, 2003, *Manual on the Application of the HACCP System in Mycotoxin Prevention and Control* [online]. FAO-Food and Nutrition Papers n° 73. Available from: http://www.fao.org/DOCREP/005/Y1390E/y1390e00. htm [Accessed 1 Sep 2005].

Fernandes, A., Venâncio, A., Moura, F., Garrido, J., and Cerdeira, A., 2004, Fate of ochratoxin A during a vinification trial, *Aspects of Applied Biology*, **68**:73.

IARC, 1993, *Monographs on the evaluation of carcinogenic risks to humans, some naturally occurring substances: food items and constituents, heterocyclic aromatic amines and mycotoxins*, IARC (International Agency for Research on Cancer), Lyon, 56, pp. 489-521.

Paterson, R.R.M., Baker, P.S., and van der Stegen, G.H.D., 2001, Ochratoxin A in coffee In: *Coffee Futures. A source book of some critical issues confronting the coffee industry,* Baker, P.S., ed., The Commodities Press, pp. 16-25.

Paterson, R.R.M., 2004a, The isoepoxydon dehydrogenase gene of patulin biosynthesis in cultures and secondary metabolites as candidates PCR inhibitors, *Mycological Research*, **108**:1431.

Paterson, R.R.M., 2004b, Standardized data on 474 metabolites or "extrolites", *Mycologic. Res.* **108**:596.

Paterson, R.R.M., 2004c, Review of: Handbook of Secondary Fungal Metabolites, *Mycologic. Res.* **108**:721.

Paterson, R.R.M., Venâncio, A., and Lima, N., 2004, Solutions to *Penicillium* taxonomy crucial to mycotoxin research and health, *Research in Microbiology*, **155**:507.

Paterson, R.R.M., 2006, Primers from the isoepoxydon dehydrogenase gene of the patulin biosynthetic pathway to indicate critical control points for patulin contamination of apples, *Food Control*, **17**:741-744.

Paterson, R.R.M. and Lima, N., 2005, Fungal Contamination of Drinking Water. In: *Water Encyclopedia: Water quality control*, J. Lehr, J. Keeley, J. Lehr, T. B. Kingery III, John Wiley & Sons. DOI 10.1002/047147844X.wq1516.

Pitt, J.I., 1996, What are mycotoxins?, *Australian Mycotoxin Newsletter*, **7**:1.

Pittet, A., Tornare, D., Huggett, A., and Viani, R., 1996, Liquid chromatography determination of ochratoxin A in pure and adultered soluble coffee using an immunoaffinity column cleanup procedure, *J. Agric. Food Chem.*, **44**:3564.

Sinha, K.K. and Bhatnagar, D., 1998, *Mycotoxins in Agriculture and Food Safety*, Marcel Dekker, Inc, New York Basel, 510p.
Skudamore, K.A. and Banks, J.N., 2004, The fate of mycotoxins during cereal processing, In: *Meeting the Mycotoxin Menace*, D. Barug, H. van Egmond, R. Lopez-Garcia, T. van Osenbruggen, and A. Visconti, (eds), Wageningen Academic Publishers, Wageningen, pp. 131-140.
Smith, J.E. and Solomons, G.L., 1994, *Mycotoxins in Human Nutrition and Health*, Agro-Industrial Research Division, European Union, Directorate General XII, 290 p.
van Egmond, H. and Jonker, M.A., 2003, Current legislation governing mycotoxin limits in food, In: *Mycotoxins in food: Detection and Control*, N. Magan, and M. Olsen (eds.), Woodhead Publishing Limited, pp. 49-68.
van Egmong, H., 1989, *Mycotoxins in Dairy Products*, Elsevier Applied Science Publishers, Barkind, Essex, 272p.

Recommended References

Barug, D., van Egmond, H., Lopez-Garcia, R., van Osenbruggen, T., and Visconti, A., 2004, Meeting the Mycotoxin Menace, Wageningen, Wageningen Academic Publishers, 319 p.
IPCS INCHEM web site. Chemical Safety Information from Intergovernmental Organizations [online]. Available from: http://www.inchem.org [Accessed 1 Sep 2005].
Samson, R.A., Hoekstra, E.S., Frisvad, J.C., and Filtenborg, O., 1996, Introduction to Food-Borne Fungi. Baarn CBS.

3
Preventive Measures for Food Safety
Hygienic Design and Cleaning Validation

GERHARD SCHLEINING *

1. Introduction

Food safety must be considered not only during industrial production but throughout the whole food supply chain (see Figure 1).

Hazard identification is defined by the Codex Alimentarius (1999) as "identification of biological, chemical and physical agents capable of causing adverse health effects".

Biological hazards do not only include the agent capable of causing adverse health effects, but also the consequences of its presence. These hazards are pathogenic micro-organisms and microbial toxins that could be present in various foodstuffs. Micro-organisms of concern in water include bacteria (e.g. pathogenic *Escherichia coli*, *Vibrio cholerae*, *Legionella* spp., *Campylobacter* spp., *Salmonella* spp. incl. *S. typhi/paratyphi*), protozoa (e.g. *Cryptosporidium, Giardia*), viruses (e.g. Enteroviruses, Hepatitis A virus), and toxin producing algae (e.g. *Cyanophyceae*). More detailed information about waterborne hazards can be found in the respective ILSI reports (2001, 2002, 2003).

Chemical hazards include heavy metals, organic compounds, salts, other chemical contaminants, e.g. fertilizers. A detailed reference list of chemical parameters can be found in the EU drinking water directive 98/83/EC (1998).

Physical hazards include any solids and foreign matter from various sources such as material carried with incoming water. More details about hazards related to incoming water and its treatment can be found in the WHO Guidelines (2004).

Hazards are caused not only by human and environment but also by equipment. It is known that the acceptable microbial count of a product does not only depend on the condition of the raw materials and preventive techniques like heat treatment and chilling but also on secondary processes like *cleaning and disinfection* (see Figure 2). The product may, however, also be contaminated with micro-organisms

*Gerhard Schleining, DLWT -Department of Food Science and Technology, BOKU - University of Natural Resources and Applied Life Sciences Vienna, Muthgasse 18, A-1190 Vienna, Austria, http://www.dlwt.boku.ac.at

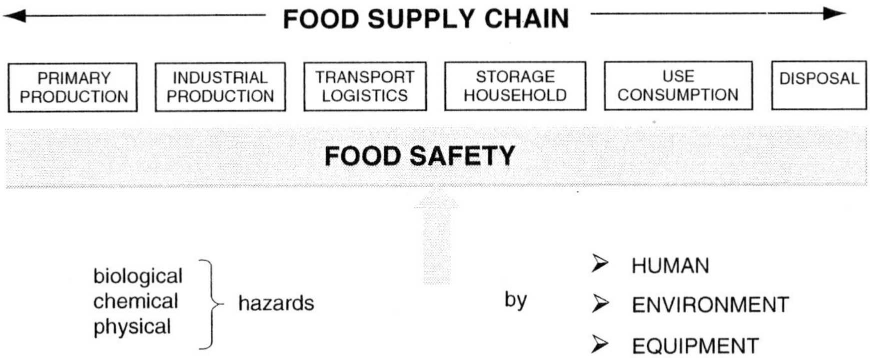

FIGURE 1. Hazards are also caused by equipment and throughout the whole Food Supply Chain.

during processing and packaging. Equipment of poor hygienic design, may be difficult to clean and difficult to free from micro-organisms; the surviving bioburden would then multiply in product residues, in crevices and dead areas thus increasing the risk of post contamination.

Hygiene is vital for the production of a quality of a food product. The product quality depends not only on the capability and safety of the process and preventive measures like GMP, HACCP, maintenance, etc. but also on the quality of the equipment used (see Figure 3). The efficiency of cleaning procedures is also highly dependant on the design and installation of equipment.

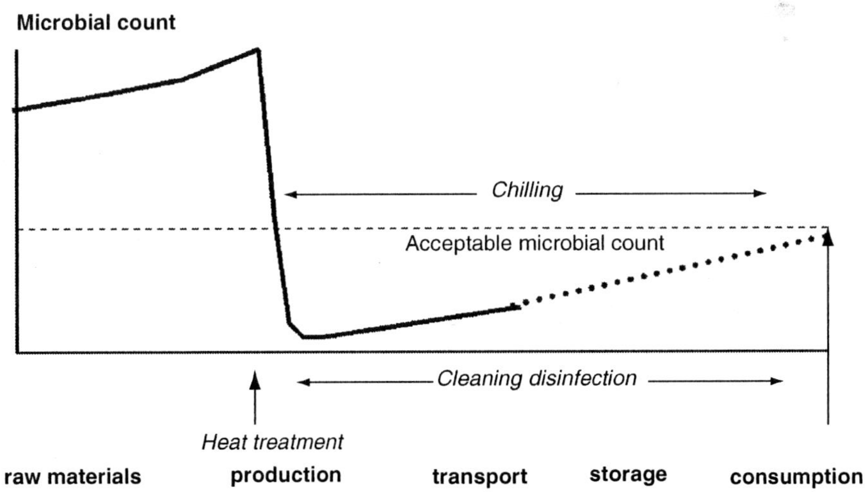

FIGURE 2. Effects on microbial count of a product during its life cycle.

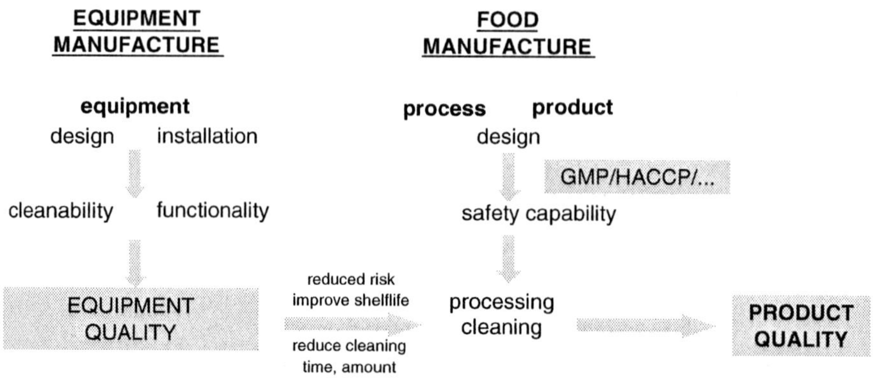

FIGURE 3. The product quality depends also on the quality of the used equipment.

Therefore, when considering food safety, hygienic design and appropriate cleaning are, together with GMP, hygiene, HACCP, maintenance, etc., essential preventive measures.

2. Hygienic Design

A prerequisite for high quality and safe products is correct hygienic design and maintenance of the production systems. A correct hygienic design of process equipment has a tremendous impact on diminishing the risks of contamination of food during production, which also means that the shelf life of products is lengthened. If the process equipment is of poor hygienic design residues (soil) may be retained in crevices and dead areas, allowing the micro-organisms to persist, multiply and then cross-contaminate subsequent batches of product.

Poor hygienic design can affect the equipment manufacturer, the food processing client, government legislative bodies and the final consumer. Many cases of food poisonings are caused by incorrect equipment design. For example the presence of salmonella in milk powder with the evidence of 76 cases of illness and 1 event of death were caused in 1985 in UK by contaminated water. The source of contamination was traced to the fact that birds had access to water as a result of an inadequate closure of a washing water tank, (Holah, 2003).

2.1. Regulations, standards and guidelines

Equipment manufacturers are not always fully aware of the biological, chemical and physical hazards of food systems and often consider operational safety aspects. Food manufacturers, on the other hand, primarily have to take into account potential hazards to the consumer. Potential conflict situations may occur between operational safety and hygiene (see Figure 4). Equipment manufacturers

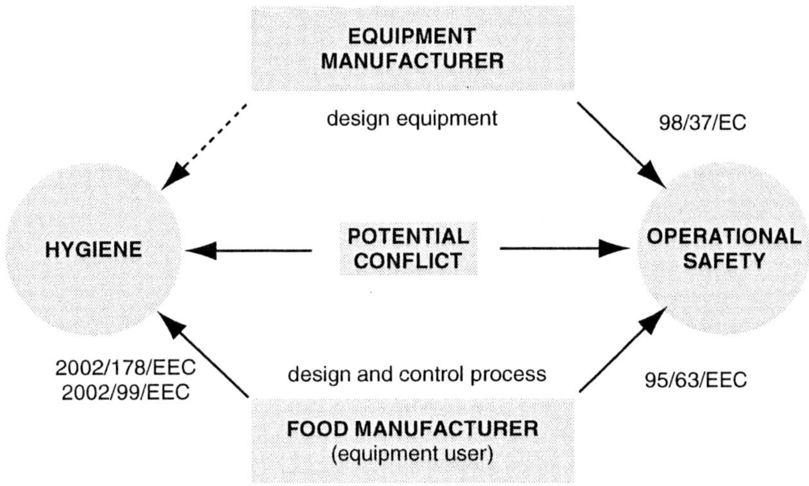

FIGURE 4. Potential conflict between hygiene and operational safety.

usually consult documents such as the Machinery Directive 98/37/EC (1998), EN1672-2 (1997) and EN ISO 14159 (2004); while the food manufacturers have to follow the "hygiene package" documentation that currently consists of 4 regulations and two directives. The two directives in question are the 2002/178/EC (2002) on general principles and requirements of food law and on procedures in matters of food safety and 2002/99/EC (2002) on animal health and the directive 95/63/EEC (1995) on safety and health requirements for the use of equipment by workers. A more exhaustive survey on the legislative in the EC and the USA is given in Lelieveld et al. (2003).

Hygienic design of machines and equipment for the food industry should be based on a well-balanced combination of operational requirements (personnel and process safety) and hygienic requirements (food safety).

During the last decade a series of guidelines for hygienic design and engineering, procedures for evaluation, testing and certification of process equipment and components have been made available to equipment manufacturers who design equipment and to food producers for assessment of hygienic characteristics of equipment. Hygienic design of equipment can thus be verified by standard test procedures. The EHEDG (European Hygienic Equipment and Design Group) is a European association of equipment manufacturers, food industries, research institutes and public health authorities, founded in 1989 with the aim to promote hygiene during the processing and packing of food products. The EHEDG guidelines (2004) cover general design criteria (Doc. 8, 10, 13, 22, 26), specific components like pumps (Doc. 17), valves (Doc. 14, 20), pipes (Doc. 16), seals (Doc. 25), specific issues like welding (in preparation), construction materials (in preparation), passivation of stainless steel (Doc. 18), lubricants Doc. (23), water related aspects (Doc. 27), prevention of *Legionella* spp. (Doc. 24), processes like thermal treatment (Doc. 1, 5, 6, 12), cooling and chilling (in preparation), packing (Doc. 3, 11, 21) as well as procedures for evaluation, testing and certification of process equipment and components like test methods on the assessment of cleanability (Doc. 2), pasteurisation (Doc. 4), sterilisability (Doc. 5), bacteria tightness of equipment (Doc. 7) and bacterial impermeability of membrane filters (Doc. 19).

For process equipment manufacturers it is important to have the design or prototype tested against given requirements. The hygienic design of equipment or parts of it is subsequently assessed and the cleanability of the equipment can be tested. Such tests identify weak points in the hygienic design, including places where product residues may accumulate or where microbes may grow out during prolonged production time.

The procedures for evaluating, testing and certification, described in a validation scheme for process equipment and components (EHEDG, 2004), can be used by authorised test laboratories. Authorisation of test institutes by EHEDG is required before conducting the testing and certification program. Leading food manufacturers in Europe and abroad require conformity of the food process equipment with the EHEDG hygienic design criteria. Certified equipment is listed on http://www.ehedg.org/certequip.htm.

Hygienic design includes building design, equipment design, equipment installation and integration.

2.2. *Building design*

Good hygienic building design should minimise the possibility of infestation by insects, birds, animals, micro-organisms, any accumulation of dust, surface water or condensed water or product. These events are all factors that create the potential for microbial growth and therefore the risk of contamination.

First of all, when a new production facility is planned, risks from neighbouring facilities and activities creating possible contamination sources, such as waste water treatment plants, farms, heavy chemical industries, nearby rivers, canals or other open water areas etc. have to be taken into account.

Building design includes aspects of zoning, construction elements, utility installations, air and water related aspects.

2.2.1. Zoning

To minimise contamination, **dry** and **wet areas** should be separated and washing and drinking facilities should not be placed in production areas. Short **product routing** without crossings together with good equipment location schemes can prevent contamination risks involving raw material, intermediate and end-product handling, waste management, equipment operation and personnel activities.

The classification and separation of areas according to hygienic and cleaning requirements (see Figure 5) is a very effective measure to avoid contamination. In general "Basic" or "Low" hygiene zones are areas with no or closed production.

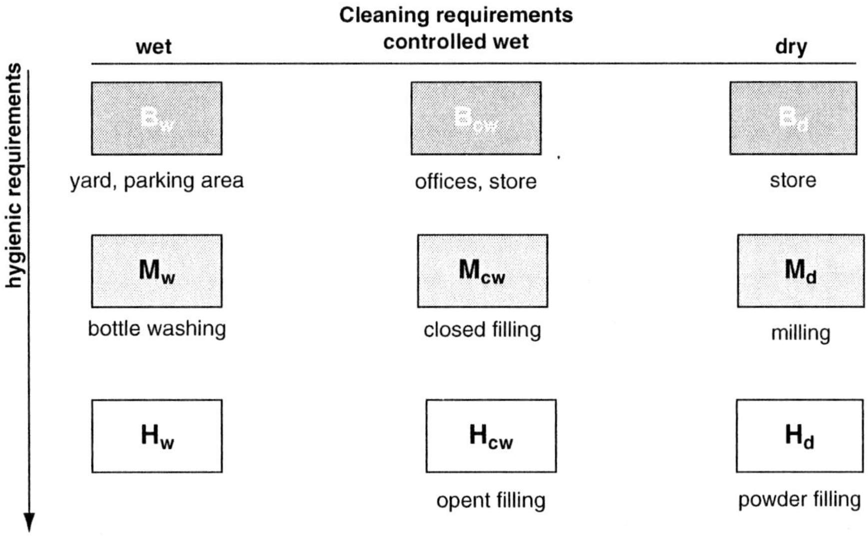

FIGURE 5. Examples for classification of areas according to hygienic and cleaning requirements (Duffey et al., 2003). B ... basic hygiene zones, M ... medium hygiene zones, H ... high hygiene zones, w ... wet cleaning, cw ... controlled wet cleaning, d ... dry cleaning.

Medium hygiene zones are areas with partly open production. High hygiene zones are areas with open production.

Each of these zones can be classified, according to the cleaning requirements, into "**wet**", "**controlled wet**" and "**dry cleaning**". Where "controlled wet" is defined as the use of a controlled amount of water, which is then followed by a drying process.

Zoning means the exclusion and restriction of unwanted items together with the definition of areas and barriers. This implies knowledge of product and process (what is to be prevented). Zoning must be logical and practical for all concerned. Furthermore, zoning must be affordable and the rules must apply for everyone and everything. Training of the personnel is essential. For each zone specific rules for installation, personal behaviour and cleaning procedures should be defined. The different zones should be made clearly visible, especially the perimeters where personnel and traffic will cross. Furthermore, the direction of air, personnel and products must be clearly defined. Zones may temporarily have their classification level changed in cases such as for maintenance or cleaning purposes.

Packaging materials, replacement parts, stockrooms for technical supplies and also maintenance workshops should be located in separate areas, but not in a production area.

Barriers between zones are extremely critical and must be monitored. These barriers may be walls or simply lines on the ground, drains, air filters, windows through which materials may be transferred, etc. Access points for products and personnel, air and utilities, as well as traffic conditions must be systematically reviewed To set appropriate rules to ensure that the appropriate hygienic level requirements are met. Access point includes doors, windows, holes of any size, ventilator openings, drains, etc. For example, material transport from a zone M to a zone H using a conveyer belt should not be carried out with one single belt, but with two. At the boundary the material is transferred from one belt to the other, and a laminar air flow at the boundary prevents the entry of contamination to zone H. A typical example is where bottles and containers are cleaned in a zone M and then transferred to a zone H for filling.

Elevators are classified in the lowest hygienic level due to the non-accessible spaces in the shaft above and below the elevator and air drafts created in the shaft that cause airborne dust movement and can be a major source of contamination. Elevators should never connect different zones if they have not been fitted with additional measures such as airlocks. and the area under the elevators should be monitored very carefully. If different hygienic zones are accessible by stairs or elevators, air locks must be installed.

2.2.2. Construction elements

Interior building element surfaces should be non-electrostatic (stainless steel preferred to polymers), but should be smooth, have round corners and have good accessibility for cleaning. Materials that are resistant to cleaning agents and disinfectants, if required, should be used. The use of mould resistant or disinfectant containing paints is not recommended, because micro-organisms can develop resistance when constantly exposed to disinfectants.

Platforms and *walkways* are difficult to design hygienically and therefore their use should be minimised. All parts should be easily accessible for inspection, maintenance and cleaning. The framework should be constructed from open profiles and all hollow structures, horizontal ledges and pockets should be avoided. If ledges are unavoidable, they should have a slope of 45°. Walkways over open processes should be avoided, because soil and foreign material may be transferred into open equipment located underneath when people walk across the platforms.

Framework should not be mounted directly on to floors that are inevitably uneven, but should be mounted with a rubber seal between the floor and the framework to ensure a tight fit. Double **ceiling** constructions should not be used, as they also accumulate dust and form hollow inaccessible cavities.

Sufficient sloping of **floor** surfaces should be provided for drainage in wet zones To prevent stagnant water formation. Floor coverings have to prevent moisture getting under the floors and should be rounded upwards at the base of the walls. When tiles are applied, all junctions should be made of non-porous material to prevent moisture penetration into the walls.

Any **drainage** must be separated from an open processing area. Drains should always be placed as far as possible from processing equipment. It should be noted that water locks may create air if there is a pressure differential over the water lock. Any bubbling can generate aerosols. Drains and channels should not be present in completely dry areas. When wet cleaning procedures are utilized in a dry production area, the drains should be tightly covered and kept dry during production. The reason being that when there is no water in the drains to form a water lock, there then is direct access to the sewer system.

The number of **windows** should be minimised in processing areas. Window frames should not be made of wood. Windows should not open and have no or 45° sloped sills and ledges. If windows do open to the exterior, they must have insect screens which should be easily accessible, cleanable or replaced regularly.

No untreated standard **glass** should be used for windows, lightings and other fixings in open processing areas due to problems of detection of fragments when this material breaks.. Polymer material like polycarbonate or strengthened glass should preferably be used.. If, however, standard glass were used, it should have a protective film on its surface to prevent shattering on breakage.

As **doors** can be a source of pest infestation and dust accumulation they should be completely flat, without any hollow body and seals around doors should be avoided. An ambient pressure differential would be preferred. If seals are unavoidable, they should be checked regularly, because they deteriorate with time.

Insulations against noise or condensation should be avoided as much as possible, as they can retain moisture and harbour micro-organisms. A better but much more expensive alternative is a so-called hot/cold room concept (see Figure 6), which means that equipment which needs to be heated or chilled is installed in a separate room, where the walls, ceilings and floors of the room provide the insulating boundary. However, if insulation is required, no perforated or electrostatic materials should be used; they should be water tight and must be removable for inspection and cleaning. Insulation should be placed on the side of the wall located in the area of lower hygienic requirements.

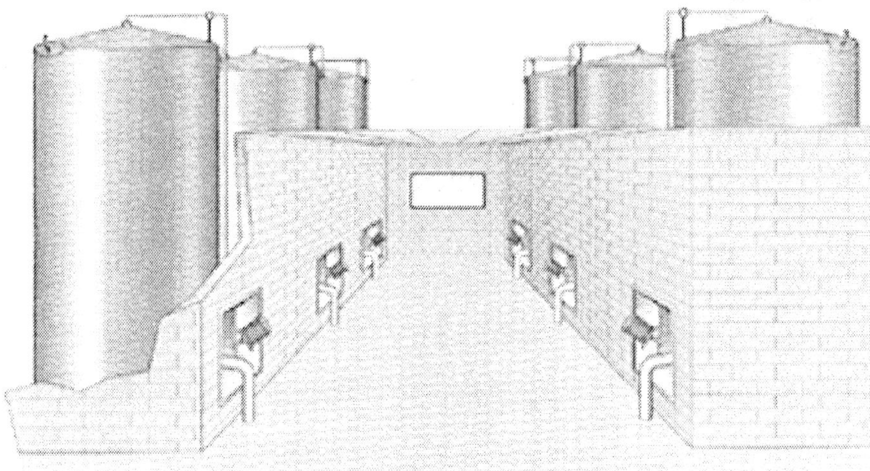

FIGURE 6. Example for a hot/cold room concept.

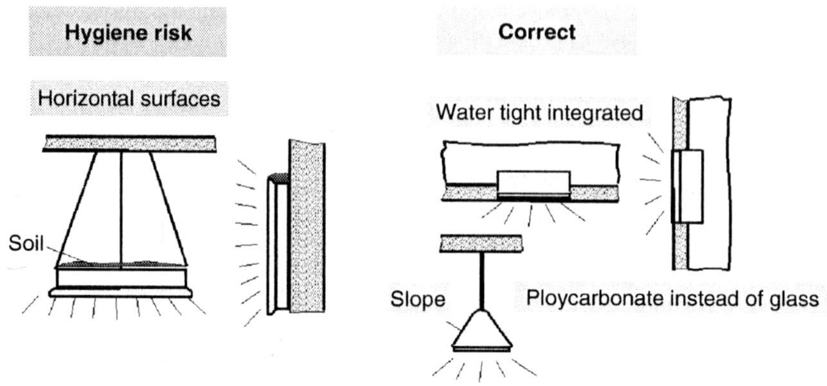

FIGURE 7. Examples of lighting.

Light sources should preferably be watertight and integrated into ceilings and walls in a way that avoids any dust accumulation (see Figure 7). Light sources are not to be placed above open processes, since if damaged, broken fragments will fall into open process equipment.

2.2.3. Utility installations

Inappropriate positioning of utility installations is a major source of contamination by dust and condensation (see Figure 8).

FIGURE 8. Example of inappropriate piping.

Pipes should run in separate and accessible gangways and enter the process area through the ceiling. If this is impossible, open trays that are sealed to walls or to ceilings should be used. These trays should be designed hygienically to minimise presence of horizontal ledges, crevices or gaps where inaccessible dirt can accumulate. Pipe lengths should be as short as possible and should never be installed above open production lines, because drops of condensate may contaminate the product. Exterior surfaces of pipes that traverse walls should have water and air tight contact with the wall, when the wall separates areas with different hygienic requirements. If both sides of the wall are in the same zone, water and air tightness is not essential, but any opening should be large enough for access and cleaning. Special attention must be paid to avoid dead space especially for connections, couplings, seals, valves and sensors.

Trays for **electrical cabling** should be sloped or vertical mounted grids containing only one layer. Cable funnels should be avoided because they are hollow bodies (see Figure 9). If they cannot be avoided for some reasons, the ends should be closed and sealed and able to be easily opened for inspection and cleaning.

2.2.4. Air related aspects

Air contains microscopic particles of dust and micro-organisms, either alone or attached to dust particles. Therefore air is a potential contamination source in food processing. There are different types of air quality to consider as environmental (ambient) air, process air and instrument air.

Ambient air should flow from higher to lower hygiene areas and from lower to higher dust loaded areas. In a high hygienic area, additional measures such as

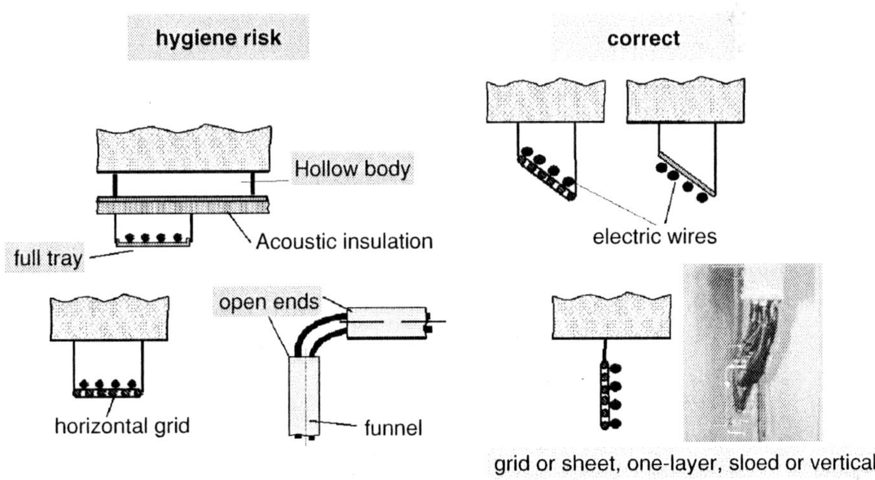

FIGURE 9. Examples for electrical cabling.

air filtering (EN 779, 2003) and operating under conditions slightly above atmospheric pressure are necessary.

Process and transport air should only be drawn in at a single location, being at least 3m above the ground level and at least 10m away from any exhaust discharge point (VDI 6022, 2002).

For **instrument air** the outlet has to be carefully positioned away from open and dry products to avoid formation of dust.

2.2.5. Water related aspects

Water is a vital medium used for many different purposes in the food industry. The safe storage and distribution of water in food factories is described in detail in EHEDG Doc. 27 (2004) while the hygienic re-use of processing water in food plants is referenced in a Codex Alimentarius discussion paper (1999).

Stagnant water should not be allowed to exist on floors, plant and equipment surfaces after wet cleaning. Sufficient **drainage** must be provided if wet cleaning is required.

To prevent contamination risks **water quality control** is very important. Water quality for different use should be specified and quality control procedures should be established. Pipes for different water qualities should never have connections. In general the following water qualities are used: **Product water** is used as an ingredient, for transport purposes, cleaning/disinfection and for rinsing equipment surfaces in direct contact with the product. **Utility water** is used in a secondary process like heating/cooling, cleaning and fire fighting, where no direct contact with the product will occur and **potable water** is for personnel use.

In food factories the proliferation of *Legionella* spp. is of particular concern. This bacterium causes severe respiratory illness (Legionnaires' Disease) when inhaled from contaminated aerosols. Hygienic design to prevent the growth of this bacteria consists of avoiding the presence of stagnant water, the formation of aerosols and the use of the right design and placement of equipment like cooling towers, evaporative condensers, domestic water systems, pressure jetting systems, can/bottle washing systems, emergency showers; fire sprinklers, fountains, garden hoses and sprinklers, spray humidifiers and air washers, machine tool cooling units, conveyor lubrication, etc. where water temperatures in the range of 20-45°C may be found. The prevention, risk assessment and control of *Legionella* spp. bacteria is described in detail in EHEDG Doc. 24 (2004).

2.3. *Equipment design*

Similar to the requirements for buildings, the equipment should be designed without any horizontal surfaces, ledges, hollow bodies, dead spaces, etc. to avoid accumulation of dust, water and product.

In food factories special attention must be drawn to "**product contact surfaces**" which are surfaces that intentionally or unintentionally (e.g. due to splashing) come in contact with the product, or from which product or condensate may drain, drop or be drawn back into the main product or product container, including surfaces (e.g. unsterilised packaging) that may indirectly cross-contaminate product contact surfaces or containers. Product contact surfaces must be inert to the product as well as to detergents, disinfectants or sanitizers under the intended conditions of use. They must not migrate to, or be absorbed by, the food and have to be corrosion resistant, non-toxic, mechanically stable, and their surface finish must not be adversely affected under conditions of use. All surfaces in the product zone must be smooth (roughness Ra $\leq$ 0.8 μm, EHEDG Document No.8, 1993) and non-porous so that tiny particles of food, debris/dirt, bacteria, or insect eggs are not caught in surface crevices and become difficult to dislodge, thus becoming a potential source of contamination.

Dead spaces or other conditions, which trap food and other debris, prevent effective cleaning and/or disinfection and may allow contamination to take place, should be avoided. All surfaces in the product zone must be self-draining without need for dismantling. Noise-reducing materials should not be placed in the food zone.

Like zones, the equipment should also be classified according to hygienic and cleaning requirements (see Figure 6) with specific rules for installation, personal behaviour and cleaning procedures (CIP/non CIP).

Food producers may find it very useful to look for **certified equipment** currently available on the market. Equipment manufacturers may apply to EHEDG for the authorised use of the EHEDG Certification logo for equipment that complies with EHEDG hygienic design criteria (Doc 8). Certification, carried out by authorised organisations may involve cleanability testing of equipment according to the methodology described in EHEDG Doc 2. The use of EHEDG certified equipment and components results in less frequent cleaning, increased life time of equipment, allows for less frequent preventive maintenance, extended shelf life of food products, and last but not least, a lower risk on product (re)contamination during processing and packing.

2.4. Equipment installation and integration

But even hygienically correct designed process and control equipment may cause hazards if it is badly installed and integrated into the process. Problems often occur when line s are moved or added. Sufficient **access space** is required to carry out operational, maintenance and cleaning functions (EHEDG Document No.13, 2004). And clearance between equipment and the civil construction should be adequate for cleaning and inspection. Depending on the cleaning tools used it is generally accepted that small sized equipment should therefore be placed at least 0.3 m above the floor and from walls. For large sized equipment, greater distances apply (at least 0.5m), as it is necessary to be able to walk around such equipment and at least with enough room to facilitate cleaning.

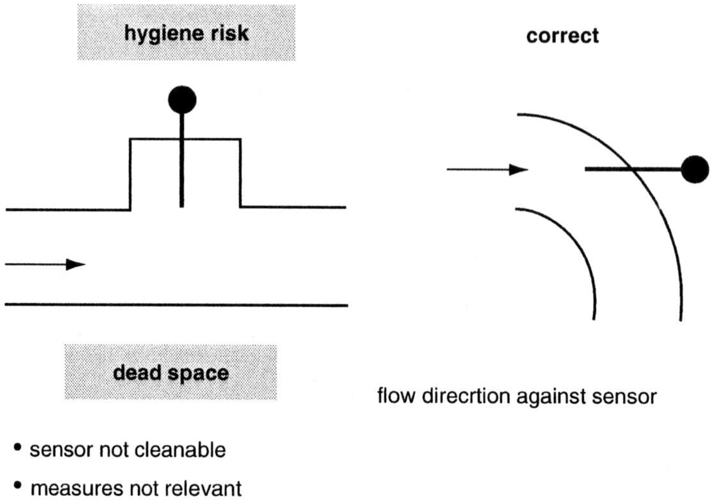

FIGURE 10. Examples for sensor installation.

Hollow bodies should be avoided wherever possible or included in comprehensive monitoring systems. Support structures like foundations need to be well sealed to the floor or wall or ceiling. Care must be taken to avoid any pockets, gaps, cracks or crevices where micro-organisms and insects can survive after wet cleaning operations.

Increasing demands on product quality necessitate an increasing number and variety of **sensors**. All sensors rely on a critical kind of interaction with the product. Therefore, the use of sensors is commonly associated with a certain hygiene risk. The right installation and integration of sensors into the process (see Figure 10) as well as cleaning and maintenance are critical issues.

3. Cleaning Validation

The efficiency of cleaning procedures depends highly on the design and integration of equipment. Insufficient cleaning will compromise Food Safety and on the other hand over-cleaning is wasted money and pollutes the environment. Cleaning procedures must always be developed under consideration of the product requirements.

The validation of cleaning procedures is a well known strategy (FDA, 1993) in the pharmaceutical industry to proof the sufficient removal of residues of previous products, intermediates and cleaning agents as well as to control potential contaminants.

To achieve an effective and efficient cleaning validation it is often found helpful to structure the tasks into steps, beginning from a status evaluation up to the approval procedure and to develop and follow a specified validation program to avoid insufficient and overlapping work. In many cases a structure similar to that shown in Figure 11 has been adopted.

3.1. Status evaluation

To select the equipment and corresponding cleaning procedures which have to be validated and to establish acceptance criteria and analytical methods, products (composition), processes (potential risks), equipment (construction) and cleaning procedures (cleaning agents) should be first evaluated.

Equipment may be grouped into equipment families to take into consideration the product, process, cleaning procedure and construction.. So, if several pieces of equipment are of similar construction and used for products with similar composition and comparable concentration and cleaned by the same procedure, they are treated as belonging to the same equipment family.

To determine **acceptance** criteria for residual products, residual cleaning agents, for process hygiene and for routine monitoring the cleaning procedures are analysed with special emphasis on composition of product and cleaning agents. Due to the fact that a validation process cannot achieve 100% safety, routine monitoring criteria and methods should be designed.. A first choice of parameters could be pH and conductivity for NaOH, Total Organic Carbon (TOC) for organic contaminants and endotoxin in the final rinse.

The **analytical methods** used to detect residues should be specific for the substance to be assayed and provide a sensitivity that reflects the level of cleanliness determined to be acceptable (PICS, 1999). Cleanability can be assessed by test methods described in EHEDG Document No. 2 (2004) and in VDMA (1999).

STEPS	RESULTS	DOCUMENTATION
Status Evaluation	Selected equipment and cleaning procedures	**Cleaning Validation Master Plan**
	Acceptance criteria and analytical methods	
Risk Analysis	Representative equipment, product and sampling locations	
Cleaning Process Studies	Validated analytical and sampling methods	**Cleaning Validation Protocol** for selected representative
	Acceptance limits	Logbook, reports, SOPs
Assessment and Implementation	Corrective actions	Assessment report
Validation of Representative	Approved representative	**Cleaning Validation Report, Approval**
Transfer to other equipment	Approved equipment	Cleaning Validation Protocols, Reports and Approvals

FIGURE 11. Structure of a cleaning validation program.

3.2. Risk analysis

Risk analysis may be used to select representative **equipment** to establish and validate appropriate analytical and sampling methods and to develop and carry out a cleaning validation protocol. The representative equipment must cover all products, processes and equipment of the corresponding equipment family. Worst case conditions, encompassing processing limits and circumstances, within standard operating procedures, which pose the greatest chance of failure (PIC/S, 1999), are analysed for the selection of equipment, product and sampling locations as well as for the determination of acceptance limits. The processed products, potential risks of the process step, the applied cleaning methods (CIP/manual) and the composition of the cleaning agents as well as the functionality, the frequency of usage and the construction of the representative will be carefully examined. Equipment suitable for cleaning and removal of the most stubborn residues should be selected. Sampling locations termed critical areas which are hardest to clean or areas with conditions favourable to reproduction of micro-organisms (e.g. moisture, temperature, crevices and rough surfaces) would be identified. The range of concentration of the critical product component, cleaning procedures with the highest concentration of cleaning agents and cleaning methods with uncertain reproducibility (manual cleaning) will be identified for the determination of acceptance limits during cleaning process studies.

3.3. Cleaning process studies

Cleaning Process Studies are carried out to validate analytical and sampling methods under consideration of worst case by determination of residual products, residual cleaning agents and microbial contaminants before and after cleaning. Special emphasis should be taken on the composition of the product and cleaning agents.

To predefine **acceptance limits** for all acceptance criteria including routine monitoring criteria, the cleaning procedures are analysed on the basis of historical data and/or Cleaning Process Studies. The acceptance limits should be practical, achievable, verifiable and consistent. The limit of quantitation of the corresponding analytical method must be lower than the acceptance limit. The amount of accepted product residue from one batch to the next, based on the amount of the following batch, and the amount of accepted residual cleaning agents depend very much on the kind and the sequence of products. For this reason, a safety factor should be considered and incorporated into the procedure. Acceptance limits for micro-organisms on product contact surfaces depend on the surroundings, where the equipment is located and may not be higher than the acceptance limits of the respective room class.

The **sampling procedure** is most important for the representative validity of the results. When establishing a sampling plan it has to be considered that the contaminants will not be uniformly distributed throughout the system and that a residual contaminant will not be worn off the equipment surface uniformly. The design of the equipment should be carefully examined. Critical areas and different materials of construction should be identified. The sampling locations must be clearly defined in the validation protocol by photographs or drawings and by detailed written descriptions for each location, including a rationale for why certain sampling points will be tested. The act of sampling itself is a cleaning step. Therefore a particular location has only to be sampled once after a particular cleaning run to reflect the true results of the investigated routine cleaning process. After the sampling process the sampling locations have to be cleaned by appropriate methods to remove any contamination introduced during sampling. In addition the equipment has to be cleaned again according the respective SOP (Standard Operation Procedure).

For **direct surface sampling** swabs are used to test flat surface areas. Special focus should be drawn on hard to reach areas such as cracks, crevices, gaskets, seals. Different sampling solvents should be examined; therefore one sampling taking process would have to be designed to cater for different methods. To evaluate the recovery effectiveness and reproducibility of the swab method, which depends on the swabbed material, the sampling solvent, the concentration range of residues and the swab pattern and sequence, recovery tests should be performed on all relevant material types for the relevant concentration range of residues within the validation of the detection method. A known amount of a specific residue should be applied onto the surface and allowed to dry. Then the area will be sampled and the swab is analysed to determine the percentage of recovery. The sufficient number of swabs is estimated with different concentrations of the residues. The predetermined recovery coefficients will be used in the calculation of the residues.

Indirect rinse samples only indicate when sufficient rinsing has been achieved and do not necessarily correlate with residues on the equipment surface. Therefore, wherever possible, both methods should be used, otherwise direct surface sampling should be used as the preferred method. Final rinse samples are be used to test large surface areas, especially inaccessible areas of equipment that cannot be routinely disassembled. The equipment is flushed with a known amount of water and a rinse sample is collected and tested for the required parameters.

3.4. Assessment and implementation

At least before a Cleaning Validation Protocol is approved and executed, it is strongly recommended that assessment of all documents and results should be carried out. Any unacceptable results should be handled as a deviation. In the event that criteria and/or limits are unattainable under practical conditions modifications may be considered. Somebody must be responsible to decide if investigations or corrective actions are necessary under consideration of product risks and/or process capability or if the "**Cleaning Validation Master Plan**" including **SOPs** and **protocols** can be approved and implemented.

The **personnel** performing cleaning and validation must be **qualified** according to This means that persons who perform **sampling** should be trained in the application of validated sampling procedures and a **training plan** containing a time schedule for persons who have to be trained and in which procedures should be issued. Furthermore, a check **list for supervisions** containing this information should be produced..

A **time table for routine monitoring** must be established. An **SOP for Change Control** must be available, to log any changes of product, process, equipment, cleaning and validation procedures. Validation runs should be performed with a minimum of 3 successful consecutive applications of the cleaning procedure.

Revalidation should be performed a) periodically and b) in the case of any change of product, equipment, cleaning procedure, etc., it has to be decided if the change will affect the cleaning procedure and if therefore the cleaning procedure has to be revalidated: The intervals of periodical revalidations have to be defined before approval. Manual cleaning methods should be reassessed more frequently than CIP.

4. Conclusion

A prerequisite for high quality and safe products is correct hygienic design and maintenance of the production systems. The use of appropriate designed and installed equipment and validated cleaning procedures as well will comprehensively contribute to Food Safety and improve shelf life.

The main issues of hygienic design are:

- to avoid contamination by foreign organisms and materials like animals, insects, birds, micro-organism, migrations, construction/equipment parts, lubricants, residues of other products and cleaning agents, etc.
- to avoid conditions which enhance the growth of micro-organism, like accumulation of dust, water, moisture (condensation) and product
- and to improve cleanability

Risks due to poor hygienic design are mainly caused by incorrect positioning of equipment and utility installations, by horizontal surfaces, hollow bodies, dead spaces, bad drainage, insufficient cleanability / accessibility, non-resistant materials etc.

The concepts of zoning and of cleaning validation are well known in the pharmaceutical industry and should be used more frequently in the food industry. There is a lot of knowledge, experience and sometimes simple solutions available through guidelines and experts need to be transferred to equipment manufacturers and to food producers.

5. References

Codex Alimentarius, 1999, Discussion paper on proposed draft guidelines for the hygienic reuse of processing water in food plants.

ILSI Europe Monograph, 2001, Principles of risk assessment of food and drinking water related to human health, ILSI Europe Avenue E. Mounier, 83, Box 6, B-1200 Brussels (http://europe.ilsi.org).

ILSI Europe, 2002, Viruses: An emerging problem.

ILSI Europe, 2003, Foodborne Protozoan Parasites.

98/83/EC,1998, On the quality of water intended for human consumption.

WHO Guidelines, (November 2, 2004); http://www.who.int/water_sanitation_health/dwq/guidelines2/en/, http://www.who.int./water_sanitation_health, http://www.who.int/docstore/water_sanitation_health/GDWQ/Updating/draftguidel/

Holah, J., 2003, Poor Hygienic Design: Affects and Case Histories, ANUGA Foodtec, Cologne.

98/37/EC, 1998, Machinery Directive.

EN 1672-2, 1997, Food processing machinery – Basic concepts – Part 2: Hygienic Requirements. A revised version is discussed (EN 1672-2:2003).

EN ISO 14159, 2004, Safety of machinery – Hygiene requirements for the design of machinery.

2002/178/EC, 2002, General principles and requirements of food law and procedures in matters of food safety.

2002/99/EC, 2002, Animal health.

95/63/EEC, 1995, Minimum safety and health requirements for the use of work equipment by workers at work.

Lelieveld et al., 2003, *Hygiene in Food Processing*, Cambridge: Woodhead Publ.

EHEDG Guidelines (November 3 2004); http://www.ehedg.org/f_guidelines.htm. Also some of them are available as an extended abstract in *Trends in Food Science & Technology*.

Duffey, J.L., Hauser, G., Hutten, H., Mager, K., Masters, K., Meesters, G.M.H., Oosterom, J., Rumpf, W., and Schleining, G., 2003, Hygienic Engineering of Plants for the Processing of Dry Particulate Materials, in: *EHEDG Guidelines*, EHEDG ed., CCFRA Technology Ltd., Chipping Campden GL55 6LD UK, Doc. 26

EN 779, 2003, Particulate air filters for general ventilation.

VDI 6022, 2002, Hygienic standards for ventilation and air-conditioning systems in production facilities and business enterprises. Beuth Verlag GmbH, 10772 Berlin.

FDA, 1993, Guide to Inspections of Validation of Cleaning Processes. http://www.biologicsconsulting.com/docs/1993/FDA199307A.pdf

PICS, 1999, Pharmaceutical Inspection Convention: Recommendations on Cleaning Validation, March 01.

VDMA (Fachgemeinschaft Verfahrenstechnische Maschinen und Anlagen, Fachabteilung Sterile Verfahrenstechnik), 1999, QHD (Qualified Hygienic Design) Hygienische Gestaltung von Komponenten und Maschinen.

4
Packaging

Mona Popa[1] and Nastasia Belc[2]

1. Introduction

Nowadays food packaging is a multifunctional participant in the food industry. The packs have multiple functions; they contain, preserve and protect the product. The outer covering should also to inform the consumer about the product and the design should promote the product. The packaging also has a secondary function, that of loss,, damage and waste reduction for distributor and customer and the facilitatation of storage, handling and other commercial operations.

Packaging technologies bring together a vast range of techniques and materials with two basic objectives: to protect the product and to display items for sale. So packaging has thus progressed from the functional to the expressive as result of some factors: the aim to motivate customers to buy the product and to convey a suitable product image for selling.

According to the FAO report, 50% of agricultural products are destroyed because of the absence of packaging. The causes of this loss are bad weather, physical, chemical and microbiological deteriorations. Industrialization and the consumption of natural resources has accelerated progress so the manufacturer of packaging or manufacturer of packaging machines have to adapt and to anticipate trends and realize that only automation can provide the necessary flexibility to satisfy industrial needs. Progress in the packaging of foodstuffs will prove crucial over the next few years mainly because of new consumer patterns and demands creation and of world population growth which is estimated to 15 billion by 2025.

In this context "Packaging for foodstuffs" became more of a necessity today than yesterday, but not nearly as much as tomorrow (Jean-Paul Pothet, 1996).

At present, foodstuffs represent 50% of the market for packaging, which represents a huge growth of this business field. These changes led to new developments in packaging technology.

[1]AGRAL Programme, University of Agronomical Sciences, and Veterinary Medicine Bucharest,59 Marasti Blvd., 011464, Bucharest 1, Romania
[2]Institute of Food Bioresources Bucharest -Romania

2. Food Packaging Techniques

In the last decade the most important additional function of the packaging method was to prolong the shelf-life of the food product. There are a lot of new methods, used worldwide, in food packaging based on reduced -oxygen atmosphere surrounding the product.

2.1. Vacuum packaging

In this technique the package is evacuated and closed, leaving a very low amount of air, especially oxygen (O_2) in contact with the food.

2.2. Moderate vacuum packaging

The product is stored under a pressure of around 400 mBar at chilling temperature. This can take place in a rigid, airtight container or a plastic pouch. The amount of available O_2 to food, which is about one third of the normal, is slowing down the metabolism of respiring produce and the growth of spoilage microorganisms.

2.3. Active packaging

The composition of the atmosphere in the package can be changed by placing sachets with oxygen scavengers or CO_2 emitters in the package or using another special means. The packaging materials must have a low permeability to different used gases.

2.4. Edible coatings

The technique uses edible coatings or films, which can act to food product as a protective superficial layer i.e. for many years waxing of fruits. Currently, edible films and coatings which protect a food against microbial spoilage as well as loss of quality are developed on the basis of proteins, starches, waxes, lipids, antimicrobial and antioxidant compounds.

2.5. Modified atmosphere packaging

MAP means that an atmosphere with a gas composition different from that of atmospheric air is created in the package.
The properties of the main used gases are the following:
CO_2 – antimicrobial effect;
O_2 – the aim in gas packaging is to reduce the oxygen concentration in the headspace of a package below 1-2 %, even to 0.2 % by replacing oxygen with nitrogen

and /or carbon dioxide. In gas packaging of retailing raw meat, for example, a high oxygen concentration, even 80 % is used in order to extend the duration of oxymyoglobin which gives a cherry red colour to meat surface;
N_2 – inert gas.

In MAP of **non-respiring foods** a high CO_2 content (>20%) is used in most cases with a low O_2 content (<0.5%) and a storage temperature < 5°C is recommended. In MAP of **respiring foods**, i.e. fresh fruits and vegetables, once the atmosphere has been changed to the desired level, the respiration rate of the produce should equal the diffusion of gases across the packaging material in order to achieve an equilibrium atmosphere in the package.

2.6. Aseptic packaging

Normally means that foods after heat processing are transferred to "sterile" and hermetically sealed containers under aseptic conditions, so that no re-infection takes place. The principle is well known for liquid products, e.g. (UHT) milk, fruit juices etc.

The technology called "clean room" is intended to drastically reduce the number of microorganisms in the areas where food products are produced, sliced or packaged, in order increase safety and stability.

3. Packaging Systems

There are eight different types of packaging systems for gas packaging namely: horizontal form-fill-seal machines for rigid and semi-rigid packages, horizontal and vertical form-sill-seal machines for flexible "pillow-pack" pouches, vacuum chambers machines, Gemella packaging system, Fibrelam system, Bag – in – carton systems, Bag-in-box system, Walki-Vent system and Flavaloc.

3.1. Horizontal form-fill-seal machines for rigid and semi-rigid packages (so-called deep-draw machines)

The system consists of 2 webs; the forming and non-forming web (figure 1). The forming web is heated and the film is drawn into a forming die to create the desired pocket by the application of a vacuum or mechanical means. The foodstuff to be packed is then placed in the formed pocket, a vacuum is created, a gas mixture is injected into the package, and the package is sealed. The package usually consists of a semi-rigid base tray (PVC-PE) lidded with a thinner material. Suppliers are e.g. Multivac, Kramer&Grebe, Mahaffy& Harder and Dixie Union. This type of packaging systems is mainly suitable for the retail packaging of meat, poultry, fish, cocked meats, bakery products, cheese and nuts.

FIGURE 1. Different Food Packaging.

3.2. Horizontal and vertical form-sill-seal machines for flexible "pillow-pack" pouches (so-called flow-pack machines)

The package is usually made from a single film and is totally flexible. These machines can also wrap a pre-filled tray of product, but are only capable of a continuous gas flush. Although this allows a good level of purging of the air, mixtures containing high levels of oxygen cannot be used due to the use of hot sealing jaws at the end of the machine. The air from the package is removed only by continuous gas flushing. For certain very porous products (e.g. some bakery goods) substitution of the surrounding atmosphere is still insufficient to achieve the desired results. In such cases, an injection station can be fitted to the machine infeed so that the product itself is impregnated with gas immediately prior to packaging. Suppliers are e.g. Rose Forgrove, Aucoutourier, Fuji and Ilapak. These kinds of packaging systems are mainly suitable for the retail packaging of bakery products, snack foods, coffee, cheese, nuts, meats, fish, salads and fruits.

3.3. Vacuum chambers machines

Preformed plastic bags and trays are used in these machines. Bags have been used for several years for prime meat cuts and cooked meats. Suppliers of machines using bags include Multivac. Trays are more recent development and are becoming more frequently usedl. Examples of such machines include the Dyno or Multivac "Space" machines, which can use pre-made HIPS, HDPE or PET trays.

These types of packaging systems are suitable for both retail and catering packaging of meat, fish, nuts, and prepared meals.

3.4. Gemella packaging system

This system uses either form-sill-seal or preformed composite board-plastic trays. Basically, Gemella packs can be sealed on any appropriate sealing machine with a gas flushing feature. The supplier is Mardon/Smiths in UK. The system can be applied to meat, poultry, fish, cooked meats, fruit, vegetables, cheese, bakery products, salads and ready meals.

3.5. Fibrelam system

The system involves the use of a preformed plastic laminated (polypropylene, polyester or ionomer/polyvinylidenechloride) fibre tray, a compatible plastic lid-forming film and suitable packaging machinery. Two packaging machinery systems have been developed: a special semi-automatic table-top machine for small processors and a modified Tiromat (Kramer&Grebe) vacuum-forming machine for large processors. The fibrelam packages are claimed to be cheaper than plastic trays. The supplier is Keyes Fibre/Maidstone Machine Co. in UK. The system can be applied to meat, poultry, cooked meats, fruit, vegetables, bakery products, salads and ready meals.

3.6. Bag-in-carton systems

The Hermetet system supplied by Akerlund&Rausing is probably one of the best known. The packages are consumer size for dry powders and granules, such as coffee, tea and dried potato powder. The air from the package is removed by a vacuum, after which the package is gas flushed.

3.7. Bag-in-box system

The product is loaded into a barrier or non-barrier bag. It can be in bulk or in conventional EPS trays with film over-wrap. The machine inserts two snorkels into the top of the bag, creates a vacuum and gas flush (figure1). The system has the advantage that a conventional unit pack can be produced and then gas-flushed. The suppliers are e.g. CVP, Corr-Vac, Snorkel-Vac and Bernhardt. Only the Bernhardt machine operates automatically, the others are manually operated.

3.8. Walki-vent system

The system is based on pre-made laminate pouches with a vent and the use of CO_2 pellets, which start gassing as soon as the package is sealed. It is not necessary to feed gas into the package, and normal vacuum packaging equipment can be used. However, a special sealing tool is required to use the Walkivent system on web-fed vacuum packaging machines, since the ventilator is formed in the heat seal. The purpose of the ventilator is to allow the excess pressure formed during the gasification of CO_2 pellets to escape. The system can be applied to raw meat, chicken and fish.

3.9. Flavaloc

Flavaloc is a system that has beeb developed in Australia. It is similar to the conventional gas package except that an intermediate web is introduced. This web is positioned between thermo-formed base and the top lidding web. The Flavaloc machine is similar in design to a standard in-line thermo-forming, filling and lidding packaging machine except that an additional gassing and sealing station is incorporated. This station is also equipped with a stretch-and-flange sealing head that stretches the intermediate web over the pack contents and seals it to a flange located on the side wall of the tray base. The Flavaloc station also evacuates the pack and introduces a predetermined gas mixture into the space between the thermo-formed base and the stretch web before sealing the web onto the base. The second station gas flushes the pack above the stretch web and seals a lidding material onto the packs. Different gas mixtures can be introduced above and below the stretch web. The package is claimed to solve many of the problems, such as the deterioration in the appearance of the lid material during transportation and handling, typically associated with the use of conventional gas packages comprising a thermo-formed tray and a heat-sealed lidding material. It has yet to be widely used commercially.

4. Packaging Materials and Food Applications

To date, packaging materials have been, to a large extend based on non-renewable materials. The only widely used renewable packaging materials are paper and board which are based on cellulose, the most abundant renewable polymer world-wide.

The materials used for packaging today consist of a variety of petroleum-derived plastic polymers, metals, glass, paper and board, or combinations thereof. These materials and polymers are used in various combinations to prepare materials with unique properties which efficiently ensure safety and quality of food products from processing and manufacturing through handling and storage and, finally, to consumer use. Notably, these materials fulfil a very important task as absence of packaging or insufficient packaging would result in fast deterioration of quality and safety, giving way to massive commercial losses of valuable foodstuffs. Individual food products have specific optimum requirements for storage that the packaging materials must be able to provide. When contemplating the concept of food packaging, the entire dynamic interaction between food, packaging material and ambient atmosphere has to be considered.

The biological basis of the starting materials provides the material engineer with a unique opportunity to incorporate a very appealing functionality into the material, that of compostability. This property enables these new materials to degrade upon completion of useful life. Compostability has, so far, been the main focus for applications of biobased packaging materials which is logical consequence for the vast amount of packaging materials used and the waste associated with it.

Design and manufacture of packaging materials is a multi-step process and involves careful and numerous considerations to successfully engineer the final package with all the required properties, The properties to be considered in relation to food distribution are manifold and may include gas water vapour permeability, mechanical properties, sealing capability, thermoforming properties, resistance (towards water, grease, acid, UV, light, etc.), machinability (on the packaging line), transparency, antifogging capacity, printability, availability and, of course costs. Moreover, a consideration of a "cradle to grave" cycle of packaging material is also required; hence, the process of disposal of the package at the end of its useful life must also be taken into consideration.

With regard to the microbiological and sensory shelf-life and quality of gas-packed foodstuffs, appropriate packaging technique also affects in particular the appearance of a product; e.g. a vacuum before gas flushing cannot be used for some products due to crushing and compression, as has already been mentioned. Several factors must be taken into account in determining the combination of packaging material properties required for each specific product and market. Naturally, the basic requirements are the same as generally set for food packaging materials, e.g. the material must be suitable for use in contact with foodstuffs from the health point of view. Particularly with regard to gas packaging, the important factors are the following:

4.1. The gas barrier properties needed

In most gas packaging applications, excluding vegetables and fruits, it is desirable to maintain the atmosphere initially injected into the package for as long a period as possible. The correct atmosphere at the start will not serve for long if the packaging material allows it to change too rapidly. Some of the polymers currently used include PE, PETP, metallic PETP, PP, PS, PVC, PVDC, PA, EVA, and EVOH. Their oxygen and water vapour permeabilities are given in Table 1.

These polymers are normally used as laminated or co-extruded multilayer materials in order to have the barrier properties required. The inner layer is usually polyethylene or its co-polymer which forms the food contact and heat seal medium; polyethylene or ethylene vinyl acetate alone are not suitable for gas packaging because of their high gas permeability.

As a rule, packaging materials with oxygen permeability lower than 100 cm^3/m^2 24 h 101.3 kPa are used in gas packaging. In the literature, on the other hand, very few reported investigations are available on the significance of oxygen permeability between 0-100 cm^3/m^2 24h 101.3 kPa for the quality gas-packed products.

TABLE 1. Barrier properties of selected polymers for 25 μm film thickness

Polymer	Oxygen permeability (cm^3/m^2 24 h 101,3 kPa) 3°C, 75 % RH	Water vapour permeability (g / m^2 24 h) 38°C, 90 % RH
PVDC	0,31	0,31
EVAL	7.5–13.8	21.7–58.9 (40°C)
PA 66	31	93– 155 (40°C)
OPA 6	18,6	155
PA 6	40.3	341
PP	2325	3.9–10.9
PET	74.4– 139.5	27.9–46.5
PVC (Rigid)	77.5–310	14,0–79.1
HDPE	2325	4,7–6.2
LDPE	6510	15.5–23.3
PS	5425	108.5– 170,5

Vegetables and fruits differ from other foodstuffs in that they continue to respire even when placed in a modified atmosphere. Due to the respiration, there is a danger that CO_2 will increase to the levels injurious to the packed commodities. On the other hand, respiration consumes oxygen, and there is a danger of anaerobiosis. A number of special packaging materials intended for vegetables and fruits have been developed, such as smart films, microporous film and microperforated films.

Many foods require specific atmospheric conditions to sustain their freshness and overall quality during storage. Hence, increasing amounts of foods are being in protective atmosphere with a specific mixture of gases ensuring optimum quality and safety of the product in question.

In general, the oxygen permeability and the permeability of other gases of a specific material are closely interrelated and, as a rule of thumb, mineral oil based a fixed ratio between the oxygen and carbon dioxide permeabilities. This relation is also observed for biobased materials. However, for some biobased materials, e.g. PLA and starch, the permeability of carbon dioxide compared to oxygen is much higher than for conventional plastics.

As many of these biobased materials are hydrophilic, their gas barrier properties are very much dependent on the humidity conditions for the measurements and the gas permeability of hydrophilic biobased materials may increase manifold when humidity increased. Notably, this is a phenomenon also seen with conventional polymers. The gas permeability of high gas barrier materials, such as nylon and ethylvinyl alcohol, is likewise affected by increasing humidity. Gas barriers based on PLA and PHA is not expected to be dependent on humidity.

4.2. *Mechanical strength*

In choosing packaging materials for food one has also pay attention to how resistant to mechanical stresses (e.g. puncture), humidity and temperature (frozen or chilled) the material needs to be. If a material is of poor mechanical strength, the mechanical stresses, humidity and low temperature during storage transport and handling can damage the package and cause leakage.

4.3. *Integrity of sealing*

The adequate integrity of the seal is important in order to maintain the correct atmosphere in the package. The seal however must not be too tight; the right balance between tightness and security of the closure and ability to peel back a lidding material must be determined.

4.4. *Type of package*

The type of package to be used; rigid or semi-rigid, lidded tray or flexible film pouch, has to be taken into consideration when choosing packaging materials.

4.5. *Fogging*

In order to improve the appearance of the packages in retail outlets the polyethylene in the packaging laminates can be specially treated to prevent condensation of water, which fogs the package and prevents the consumer examining the product.

4.6. *Microwaveability*

Microwaveability of packaging materials is a factor that should also be considered in gas packaging, particularly in the case of the ready-to-eat food products. For instance, the low melting point of PVC makes the PVC-LDPE-laminate or co-extrusion film much used as a base web material in deep - draw machines unsuitable for microwave oven heating.

4.7. *Biodegradability and recyclability*

These factors are new trends in the packaging business. A major challenge for the materials manufacture is the by nature hydrophilic behaviour of many biobased polymers as a lot of food applications demand materials that are resistant to moist conditions. However, when comparing the water vapour transmittance of various biobased materials to materials based on mineral oil, it becomes clear that it is possible to produce biobased materials with water vapour transmittance rates comparable to the ones provides by some conventional plastics. However, if a high water vapour barrier materials is required, very few biobased materials apply. Notably, developments are currently focusing on this problem and future biobased materials must also to be able to mimic the vapour barriers of the conventional materials known today.

4.8. Thermal and mechanical properties

Most biobased polymer materials perform in a similar fashion to conventional polymers. This indicates that both polystyrene -like polymers (relatively stiff materials with intermediates service temperatures), polyethylene-like polymers (relatively flexible polymers with intermediate service temperatures) and PET-like materials (relatively stiff materials with higher service temperatures) can be found among the available biobased polymers.

The mechanical properties in terms of modulus and stiffness are not very different compared to conventional polymers. In theory, biobased materials can be made having similar strength to the ones we use today.

4.9. Compostability

The compostability of the materials is highly dependent of the materials e.g. the first step of composting is often a hydrolysis or wetting of the material. The rate of this step is very much related to the water transmittance and the water resistance of the material. Hence, the composting rate of the material will be dependent on its properties.

4.10. Manufacturing of food biobased materials

This technology requires knowledge of the processing and material properties of the polymers. If the properties of native polymer are not identical to the required one, or if the polymer by nature is not thermoplastic, a certain modification of the polymer must take place.

Designing and manufacturing of biobased packages and packaging materials require a multi-step approach (figure2):

It should be expected that following requisite processing and product developments of biobased materials resulting properties should equal or better those of the conventional alternatives. Biobased plastic applications are currently targeted

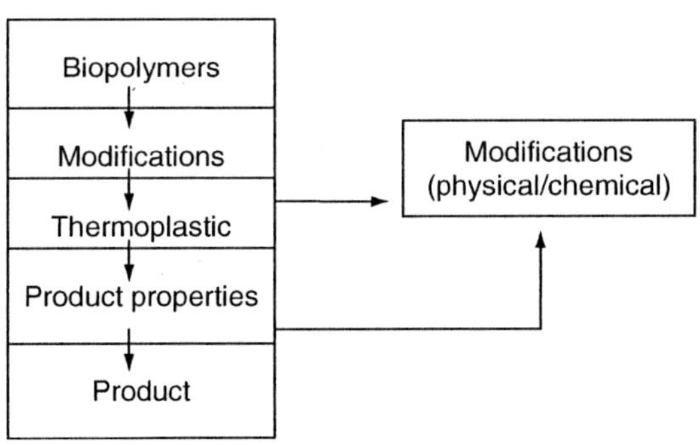

FIGURE 2. Designing and manufacturing of biobased packages and packaging materials.

TABLE 2. The major processing routes to potential biobased products

Processing route	Product examples
(Co-)Extruded film	Packaging film
Cast film	Packaging film
Thermoformed sheets	Trays, cups
Blown films	Packaging film
Injection (blow-)moulding	Salad pots, cutlery, drinking beakers, cups, plates, drink bottles, trays
Fibres and non-wovens	Agricultural products, diapers, certain medical plastics, clothing
Extrusion coating	Laminated paper of films

towards single-use, disposable, short-life packaging materials, service ware items, disposable non-wovens and coating for paper and paperboard applications. In general, the same shapes and types of food packaging can be made from synthetic and biobased resources.

Notably, developments are currently focusing towards the biobased materials which must be able to mimic the water vapour barriers of the conventional materials known today.

4.10.1. Blown (barrier) films

Blown films comprise one of the first product categories to be developed based on mineral oil derived biodegradable polyesters. They have successfully been applied as garbage bags and related applications. Film blowing grades of renewable polymers have been developed based on PLA (polylactic acid). Blown films based on these biopolyesters exhibit excellent transparency and cellophane-like mechanical properties. The sealability depends on the degree of crystallinity and good printability can also be achieved. The possibilities of film blowing PHB/V materials are at this time limited due to their slow crystallization and low melt strength.

Paragon (Avebe,NL) materials which are based on thermoplastic starch can be film co-extrusion set-up with polymers like PLA and PHB/V as coating materials, resulting in a barrier coating which, for example, proved to be successful in the packaging of cheese. In this way, starch-based materials could provide cheap alternatives to presently available gas barrier materials like EVOH and PA6.

The properties (mechanical strength gas and water vapour properties of blown film can be improved by coating of a glass-like ultra thin layer of SiO_x or by producing *nano-composites*.Addition of nano-particles during processing of the film produces composites with improved water and gas barrier properties (Fischer et.al., 2000) and ongoing developments at TNO industry (NL) aims at producing hydrophobic starches based on these composites. A similar approach is use a glass-like ultra-tin coating of SiO_x improving the barriers of the material immensely (Johanson, 2000 and 1997).

4.10.2. Thermoformed containers

A next class of products is thermoformed containers for food packaging. In order to be able to thermoform a polymer it should be possible to process this material from the melt (extrusion) into sheets and consequently thermoforming these sheets above the Tg or Tm of the material. Again, it is possible to produce thermoformed articles from laminates based on Paragon as well as other thermoplastically processable biopolymers.

4.10.3. Foamed products

Starch-based foams for loose fill applications (Novamont, National Starch (USAa.o.) have been commercially introduced with success from some years ago and the market for these products is still growing. Foamed products like trays and clamshells based on starch for food packaging have not yet been introduced commercially.

Foamed products based totally on PLA are still in a developmental phase. In order to be able to use these starch-foamed products in food contact applications coatings should be applied on the starch-based foams. Adhesion between the foam and the coating is of importance. Paraffin and other oligomer based coatings are proposed next to PLA and PHB/V based coatings. Protein and medium chain length PHA based coating (ATO, 2000) are close to market introduction.

4.10.4. Coated paper

Paper and board have excellent mechanical properties; however, the gas permeabilities are too high for many food applications. The hydrophilic nature of the paper based materials is a major challenge of these materials when packaging moist foods. To date, the paper-based materials have been coated with a thin layer of synthetic plastic which has water resistance.

Paper-based materials coated with PE are readily repulpable as the hygrophobic PE is easy removed in the pulping process. Hence, paper-based materials coated with biobased, hygrophobic polymeric materials are, likewise, going to be repulpable.

4.10.5. Conclusions and perspectives

Biological derived polymers may be used for the production of all types of packaging (trays, cups, bottles, films, etc.) using same equipment and technology used for conventional materials. Comparing the properties of biobased polymeric materials with the conventional synthetic petroleum-derived polymers shows a major potential of these polymers for the production of well-performing food packaging.

However, when using proteins and polysaccharides in the materials their sensitivity towards relative humidity must be overcome. The biobased materials have an inherent potential of being compostable which may help the commercialization of these materials. Similar to the synthetic materials used today it will be necessary to use several polymeric materials in multi-layers or composites tailoring the properties of the packaging to meet the demands of specific foodstuffs.

To be able to produce a 100% biobased packaging development of biobased additives is needed. Additives used in the production of packaging are plasticizers, UV-stabilizers, adhesives, inks and paints, natural pigments and colorants. So far, few developments have been made in this field and it is suggested to direct research to this area.

4.11. Biobased food packaging materials

These materials may be divided into three main categories based on their origin and production:

Category 1: Polymers directly extracted/removed from biomass. Examples are polysaccharides such as starch and cellulose and proteins like casein and gluten.

All this are, by nature, hydrophilic and somewhat crystalline-factors causing processing and performance problems, especially in relation to packaging of moist products. On the other hand, these polymers make materials with excellent gas barriers.

Category 2: Polymers produced by classical chemical synthesis using renewable biobased monomers.

A good example is polylactic acid (PLA), a biopolyester polymerised from lactic acid monomers. The monomers themselves may be produced via fermentation of carbohydrate feedstock. PLA may be plasticized with its monomers or, alternatively, oligomeric lactic acid. PLA may be formed into blown film, injected mould objects and coating all together explaining why PLA is first novel biobased material produced on a major scale.

Category 3: Polymers produced by microorganisms or genetically modified bacteria. To date, this group of biobased polymers consists mainly of the polyhydroxyalkanoates, but developments with bacterial cellulose are in progress.

4.11.1. Starch and derivatives

As a packaging material, starch alone does not form films with adequate mechanical properties, unless it is first treated by either plasticization, blending with other materials, genetic or chemical modification of the above approaches. Corn is the primary source of starch, although considerable amounts of starch are produced from potato, wheat and rice starch in Europe, the Orient and United States.

Common plasticizers for hydrophilic polymers, such as starch, are glycerol and other low-molecular-weight-polyhydroxy-compounds, polyethers and urea. These plasticizers lower the water activity thereby limiting microbial growth. Therefore the blending of such materials with more hydrophobic polymers produces formulations that are suitable for injection moulding and blowing films.

Starch-based thermoplastic materials have been commercialized during the last few years and are to day dominating the market of biobased, compostable materials.

4.11.2. Cellulose and derivatives

Cellulose, an almost linear polymer of anhydroglucose, is the most abundantly occurring natural polymer on earth.. Due to its regular structure and array of hydroxyl groups, it tends to form strongly hydrogen bonded crystalline microfibrils and fibres and is most familiar in the form of paper and cardboard in the packaging context. Waxes or polyethylene coating are used in some areas of primary food packaging, however the bulk of paper is used for secondary packaging. Cellulose is a cheap raw materials, but difficult to use because of its hydrophilic nature, insolubility and crystalline structure. The cellophane produced is very hydrophilic and, therefore, moisture sensitive, but it has good mechanic properties. Cellophane is often coated with nitrocellulose wax or polyvinylidine chloride (PVdC) to improve barrier properties and in such form it is used for packaging of baked goods, processed meat, cheese and candies.

A number of cellulose derivatives are produced commercially, most commonly carboxy-methyl cellulose, methyl cellulose, ethyl cellulose, hydroxyethyl cellulose, hydroxypropyl cellulose and cellulose acetate (CA). Of these derivatives, only cellulose acetate (CA) is widely used in food packaging (baked goods and fresh produce). CA has relatively low gas and moisture barrier properties. Many cellulose derivatives posses excellent film-forming properties, but they are simply too expensive for bulk use. Research is required to develop efficient processing technologies for production of cellulose derivatives.

4.11.4. Chitin/Chitosan

Chitin is a naturally occurring macromolecule present in the exoskelton of invertebrates and represents the second most abundant polysaccharide resource after cellulose.

Chitosan is a polymer derived from chitin. In general, chitosan has numerous uses: flocculant, clarifier, thickener, gas-selective membrane, plant disease resistance promoter, wound healing promoting agent and antimicrobial agent. Chitosan also readily forms films and, in general, produces materials with very high gas barrier, and it has been widely used for production of edible coating. Furthermore, chitosan may very likely be used as coating for other biobased polymers lacking gas properties.

However, as with other polysaccharides-based polymers, care must be taken moist conditions. The cationic property may further be used for incorporation and/or slow release of active components, adding to the possibilities for the manufacturer to tailor the properties. Another interesting property of chitosan and chitin in relation to food packaging are their antimicrobial properties and their ability to absorb heavy metal ions. So far, the major interest for chitosan as a packaging material has been in edible coating.

The biodegradable laminate consisting of chitosan-cellulose and polycaprolactone can be used in modified atmosphere packaging of fresh produce.

4.11.5. Biobased monomers

Biobased monomers may be obtained from biobased feed stocks. This may be prepared using chemical and biotechnological routes, or a combination of both. Oleochemicals, such as the unsaturated fatty acids oleic and ricinoleic acid, are derived from feedstock such as coconut and castor beans and have long been recognised as useful chemical precursors in preparing polymeric materials.

Carbohydrate sources such as woody material, molasses and maize give to a rich array of chemical and biotechnological transformations leading to a wide spectrum of potentially interesting chemicals. Fermentation of carbohydrate materials using selected microorganisms has led efficient pathways to the formation of multifunctional acids such as succinic acid.

At present, biobased monomers may not be directly commercially attractive, however, biobased monomers derived by biotechnological pathways present promising alternatives to petrochemical polymer routes.

4.11.6. Protein engineering

Protein engineering is a field of growing interest for production of synthetic analogues to nature's polymers. Other developments include the possible production of biodegradable polymers currently derived from petroleum sources from biobased feedstock.

4.11.7. Polyhydroxyalkanoates (PHAs)

Poly(hydroxybutyrate)(PHB) is most common of PHA, a large number of bacteria contain it as energy and carbon reserves. Due to their biodegradability and biocompatability these biopolyesters may easily find industrial applications. The properties of PHAs are dependent on their monomer composition, and it is therefore, of great interest that recent research has revealed that, in addition to PHB, a large variety of PHAs can be synthesized by microbial fermentation. The monomer composition of PHAs depends on the nature of the carbon source and microorganisms used. A very interesting property of PHAs with respect to food packaging applications is their low water vapour permeability which is close to that of LDPE. Recent application developments based on medium chain length PHAs range from high solid alkyd-like paints to pressure sensitive adhesives, biodegradable rubbers.

4.11.8. Bacterial cellulose

This material is a useful polymeric material that has remained rather unexploited. Bacterial strains are able to produce an almost pure form of cellulose with a chemical and physical structure that is identical to the cellulose formed in plants.Production costs of bacterial cellulose are high due to the low efficiency of the bacterial growth process; approximately 10% of the cellulose used in the process is incorporated in the cellulose. Several high-added-value specialty applications have been developed. The material has been used as an artificial skin, as a food-grade non-digestible fibre, as an acoustic membrane, and as a separation membrane.

4.11.9. Polysaccharides

The polysaccharides are produced by fungi and bacteria: xanthan; curdlan; pullan and hyaluronic acid, will receive more interest in the future.

5. Safety Aspects on Active and Modified Atmosphere Packaging

5.1. General packaging safety considerations

Packaging has often been thought of as a source of risk for foods and seldom as a technology that could be used to enhance food safety (Wolf, 1992).

It is obviously safety of food may be compromised through different ways and main causes are two, namely when the integrity of the pack is lost and some pathogenic microorganisms can contaminate the product or some package components migrate to food and some chemical contaminants can occur. There are a lot of examples concerning the safety problems some are listed in table 3 (Hotchkiss, 1995).

5.2. Conventional packaging safety considerations

5.2.1. Packs as barrier to contamination

The most safety packaging system taking into account the barrier properties between food and environment has been demonstrated to be packaging in glass and metal, for both environmental chemicals and pathogenic microorganisms contamination.

In the last decades the polymeric materials have been developed because of the economical and functional advantages in comparison with glass and metal. A lot of research and experimental data are available regarding the barrier properties of these package materials and now polymers with high oxygen and water vapour barrier properties as well as aroma/flavour barriers are available on the market.

Post packaging microbial contamination of foods is now not only a function of closure integrity but material integrity too (Downes et al., 1985).

Two aspects are very important regarding the integrity of the package, namely the strength and the completeness. The strength means the seal is strong enough to be able to bear up the food handling in the distribution chain without failure and the completeness means the packages are free of holes in the material or seal. The polymeric films are more capable to the pin-holes than glass or metal packs even during the shipping. Through these pin-holes can enter some microorganisms although there are no signs of leakage on the external surface of the packs. The strength and integrity are very important issues nowadays

TABLE 3. Safety problems in food packaging

Examples	Consequences
Microbial contamination	
Loss of integrity	Seal rupture, film puncture, leaking cans, incomplete glass finishes – allow pathogenic and spoilage microorganisms contamination
Anaerobiosis	Low oxygen environment resulting from product or microbial respiration that can be in favour of some toxicogenic microorganisms (anaerobic ones) growth.
Chemical contamination	
Migration	Transfer of chemical compounds from package to food product
Environmental contamination	Environmental toxicants can permeate the films
Recycled packaging	Contamination of post-consumer packaging is transferred to foods after recycling
Insect contamination	Some insects can bore through many common packaging materials
Foreign objects	Glass shards, metal pieces
Injury	
Exploding pressurized containers	Soft drinks, beer in glass etc.
Broken containers	Cuts, lacerations
Environmental impact	Disposal, recycling, CFCs
Loss of nutritional and sensory quality	Aroma and nutrient sorption by polymers
Tamper evidency	Malicious and innocuous
Inadequate processing	
Conventional	Underprocessing can lead to food poisoning
Aseptic	Loss of integrity or insufficient sterilization of packaging can lead to food poisoning

because the flexible packed foods are transported further and stored for an extended period and are more often used than rigid packaging systems. There is a permanent need for research to improve the testing of integrity of polymer-based packages.

There is a big concern regarding the chemical potential of contamination because the polymers are permeable to organic vapours and foods are very susceptible to absorption of environmental contaminants. In addition, many environmental organic compounds, which permeate films, can transmit undesirable odours to food products.

The general trend to use recycled materials lead to an increased concern regarding the microbiological risk of fibre based packaging materials (Klungness et al., 1990). In the technological process of paper making the microbial cells can be a potential risk for pathogenic spores to be transferred to foods from recycled materials.

5.2.2. Packs as source of migration

A major safety concern of packaging is the potential of packaging material to transfer the chemical contaminants to foods. A lot of research has been developed on this subject:

Migrants consist of inorganic toxicants like lead from soldered cans as well as organic toxicants such as vinyl chloride monomer which is a known human carcinogen.

New concerns over migration have been arisen recently because of the use of the recycled materials or refillable containers for food and beverage packaging (Begley and Hollifield, 1993).

One solution to the post-consumer contamination of recyclable plastics is the use of "sniffers" that are designated to sample the air inside the container and determine if volatile compounds are present prior to refilling.

Another solution is to reform the basic polymer after chemically break down the existent polymeric structure or to put a functional barrier between recycled polymer and the product in order to prevent the migration of contaminants from the recycled polymers to product.

A new safety concern regarding migration of contaminants to food has been recently resulted from the use of microwaveable polymer-based packs. In this type of migration there is an influence first of all of the rising temperature after the Arrhenius kinetics law and secondly if a polymer degradation can release some toxic additive and transmit it to food product.

5.3. Unconventional packaging safety considerations

5.3.1. Active elements interactions

The concept to incorporate substances which absorb or emit some functional gases inside of pack to control microbial growth or psychological status of the product is more desirable than active sachets because of accidental ingestion risk reducing.

These absorbers or emitters have the same safety implications like conventional MAP.

For those active packaging systems which indirectly add components to foods, the governmental regulatory and health issues will be similar to those related to migration of residual monomers or other polymer components (Crosby, 1981). In some cases, the use of functional barriers to prevent migration of the active components will be required (Hotchkiss, 1995).

Another safety effect seems to be the change in the microbial ecology of the packed food through the inclusion of antimicrobial agents, scavengers or emitters which change the environmental conditions for the existing microflora. This change can enhance the safety in some cases when CO_2 is added to high pH food products or can be compromised like in the fresh mushroom vacuum packaging where *Clostridium botulinum* can growth. From this point of view it is necessary to make a laborious study to establish an adequate packaging technology for each type of food in order to obtain better results for food shelf-life and to maintain food safety status.

A good example for food safety is that one concerning the safety/spoilage indicators as time temperature indicators, for example, which can show the historical data regarding the parameters of the storage conditions for a certain product in a certain period. It is well known that temperature abuse is main factor in food spoilage and pathogens growth. Now are available biosensors for certain toxin detection in food pack and new immunologically based sensor for food adulteration detection.

5.3.2. Modified atmosphere packaging

One of the most effective alternative methods used for fresh food shelf-life prolonging is modified atmosphere packaging (MAP).

MAP is a technique that involves first to make vacuum and second to flush the pack with a certain gas mixture with functional role. Selection of the used gases depends on the role of each gas, the food product type, film permeability, storage conditions after packaging and so on. Carbon dioxide (CO_2) acts as an anti-mould agent, nitrogen (N_2) slows chemical reactions and oxygen (O_2) prevents discoloration. Nitrogen, an inert gas, is also valuable as a "filler" to cushion the product and protect the integrity of the package during handling and storage.

It is clear that traditional post processing inspection of perishable packaged food is not sufficient to protect public health.

The risk analysis for the ingredients, product and packaging materials incorporate into the HACCP flow system for Vacuum/MAP products gives us the potential hazards and the necessary corrective actions (see table 4). CAP/MAP technologies represent much that is potentially both good and bad in food science. Errors can have dire effects, but when detailed safety protocols are followed, these

TABLE 4. Hazards in Vacuum/MAP food products

Control point	Potential hazard	How monitored	Action to take for deviation
Raw material inspection	Microbial	Establish specifications	Notify vendor; reject if not within specification
		Vendor certification and warranty	Audit quality
		Inspect shipping vehicle	
		Measure/record temperature of product when received	
		Visual examination of product	Rework/reject
		Note physical condition of packaging material	As per PQC/TQC program
Package integrity	Microbial	Seal integrity check	Rework
		Vacuum and MAP check	Rework
		Date code check	Rework

technologies can provide consumers with the safe, attractive, tasty products they deserve, and at an affordable price (Andrew Ebert, 1994).

6. References

Ahvenainen, R., 1989, *Gas packaging of chilled meat products and ready-to-eat food,* Technical Research Centre of Finland, VTT, Espoo, Finland.

Andrew Ebert, G., 1994, *Toxicology and Food Safety of CAP/MAP: HACCP,* Modified Atmosphere Packaging, Edited by Aaron L. Brody, Institute of Packaging professional, Virginia, USA, 195-205.

Begley, T.H. and Hollifield, H.C., 1993, Recycled polymers in food packaging: Migration considerations. *Food Technology,* **47**(11):109-112.

Brody, A.L. (editor), 1994, *Modified Atmosphere Packaging,* Institute of Packaging Professionals, Virginia, USA.

Bureau, G. and Multon, J.L. (editors), 1995, *Food Packaging Technology, Volume I and II,* Wiley – VCH, Inc., USA.

Crosby, N.T., 1981, *Food Packaging Materials, Aspects of Analysisand Migration of Contaminants,* Applied Science Publishers Ltd, London.

Downes, T.W., Arndt, G., Goff, J.W., et al., 1985, *Factors affecting Seal Integrity of Aseptic Paperboard/Foil Packages.* Aseptipak`85: Proceedings of the Third International Conference and Exhibition on Aseptic Packaging, pp. 363-401.

Farber, J.M. and Dodds, K.L. (editors), 1995, *Principles of modified – atmosphere and sous vide product packaging,* TECHNOMIC PUBLISHING CO. INC, Lancaster – Basel, USA.

Hotchkiss, J.H., 1983, *Tamper evident packaging for foods: Current technology.* Proceedings Prepared Foods, **152**(10):66-67.

Hotchkiss, J.H., 1995, *Safety considerations in active packaging,* Active Food Packaging, Blackie Academic & Professional, edited by M.L.Rooney, CSIRO, Australia, 238-255.

Klungness, J.H., Lin, C.H., and Rowlands, R.E., 1990, *Contaminant removal from recycled wastepaper pulps.* Pulping Conference proceedings, 1.8-12.

Weber, C.J. (editor), 2000, *The Food Biopack Conference Proceedings,* EU-concerted action project sponsored by EU DG XII, Denmark.

Weber, C.J. (editor), 2000, Biobased Packaging Materials for the Food Industry – Status and Perspectives, EU-concerted action, Denmark.

Wolf, I.D., 1992, *Critical issues in food safety,* 1991-2000. Food Technology, **46**(1):64-70.

Food Consumption and Packaging Usage Factors, Summary of Workshop organised by ILSI Europe, Brussels, 1996.

Part II
Benefits and Risks of Microorganisms

5
HACCP in the Cheese Manufacturing Process, a Case Study

ELEFTHERIOS H. DROSINOS AND PANAGIOTA SIANA*

1. Introduction

HACCP (Hazard Analysis Critical Control Points) is a system that analyzes the potential hazards related to food production, identifies critical control points at which the potential hazard can be controlled and ensure the prevention of foodborne diseases. It is a management tool that dovetails with quality management systems (ISO 9000 and BRC Global Standard) and total quality management (TQM) strategies. The HACCP constitutes a preventive as well as a proactive intervention system in food production that changes according to the occurrence of emergent hazards. It is also dependent on prerequisite programmes or general hygienic rules without which it cannot be operational (Corlett, 1998; Mortimore and Wallace, 1998).

2. Prerequisite Programmes

Prerequisite programmes serve as the basis of any HACCP system, without such programmes in place food safety cannot be assured. These systems establish procedures that address operational conditions and provide assurance to public health. These procedures include good manufacturing practices (GMPs), sanitation standard operating procedures (SSOPs), water safety control, receiving, storage and shipping control, pest and waste control, supplier control, trace and recall programmes, equipment calibration and employee training. Development of an effective HACCP system depends on the implementation of these prerequisite programmes, which typically are not embraced in a HACCP system and rarely characterised as critical control points (CCPs).

* Eleftherios H. Drosinos, Laboratory of Food Quality Control and Hygiene, Department of Food Science and Technology, Agricultural University of Athens, 75, Iera Odos, Str., Votanikos, Hellas (Greece), EL-118 55. Panagiota Siana, Hellenic Ministry of Rural Development and Food, Veterinary Laboratory of Tripolis, Pelagos, Arcadia, Hellas (Greece), EL-221 00.

2.1. *Building location and environment*

The location of the dairy plant should be such that it has a constant and reliable supply of potable water. Moreover, it should ensure that the processing and removal of waste from the location is easily managed. Additionally, the climate of the area should be taken into consideration: strong winds or high temperatures, would undoubtedly affect the quality of the product. The surroundings of the facility should be clean and pleasant for the personnel. In general, the surrounding space should be tidy, without useless accumulation of items such as boxes, equipment or machinery where it would be easy for pests to find shelter (Zerfiridis and Litopoulou-Tzanetaki, 1998).

2.2. *Sanitary design and engineering*

The interior of the plant should be designed to prevent contamination and assure the safety of food products. It should facilitate cleaning and sanitation procedures. The floors of the dairy processing plants should be made of acid tolerant material without being slippery. The walls should be covered with tiles. Equipment should not be in contact with the walls or the floor. Engineering and machinery should be made of stainless steel without sharp edges. It should be designed and located to provide easy and effective cleaning and sanitation.

Proper ventilation of the production plants provides a healthy environment for the personnel. Furthermore, it must be designed to prevent contamination. In processing areas, positive pressure should be used, to avoid airflow from raw material areas to finished products. Additionally, adequate lighting of processing, receiving and storage areas should be provided.

There must be rest rooms for the personnel while hand-washing stations should be adequate and accessible. Water taps should be foot or electronically activated to prevent hand recontamination. Waste pipes of the manufacturing plants should be isolated from human sewerage. The processing plant floors should be inclined to allow the drainage of the waste and proper cleaning of the plants.

Finally, preventive maintenance of equipment should be performed, including those items of equipment used for thermal treatment of milk (pasteurization). Production facilities should be designed to allow a thorough supervision of the whole process (Zerfiridis and Litopoulou-Tzanetaki, 1988).

2.3. *Good manufacture practices*

Good manufacturing practices are procedures designed to prevent contamination of food during processing. These prerequisites of safe food production impinge on the whole manufacturing process and involve all concerned.

Food handlers should be healthy, have received a certified medical examination and be dressed in a way that contamination of food is prevented (appropriate outer garments, gloves, hair restraints, limitations on jewellery, etc). Eating, drinking or smoking is forbidden in food processing plants.

All the incoming raw materials and ingredients as well as the finished products should be stored in areas segregated from processing and packaging ones. The containers should be cleaned thoroughly. Waste should be handled properly. Finally, procedures that ensure the exclusion of extraneous to food material should be established.

Good manufacturing practices also include the maintenance of a water safety programme. Water used during food processing should be potable; hence, a water safety programme should be designed, installed and regularly maintained with all procedures performed kept on record.

2.4. *Cleaning and sanitation*

Cleaning and sanitation are of major concern to the dairy industry as milk can easily be contaminated. Effective cleaning can only be achieved through a specifically designed programme that should be implemented as part of the every day procedures. The personnel responsible for application and supervision of the cleaning and sanitation procedures should be trained. Furthermore, the programme and personnel should be regularly checked for efficacy and efficiency. Finally, the cleaning procedures should be recorded.

2.5. *Receiving, storage and shipping control*

Food should be handled in a manner that promotes its safety not only during the reception but also during storage of raw materials and distribution of final products as well. The reception of raw materials should be checked for high quality and hygiene and the enclosed documents should be recorded and filed. Recorded information about incoming material should include the date, supplier, temperature, condition of product and certificates of analysis. Storage of raw materials as well as final products should take place in a thoroughly cleaned, pest-free and tidy area, the environmental conditions of which, such as temperature or humidity, should be controlled.

2.6. *Pest control*

A pest control programme should be included among the prerequisites of a food manufacturing process. Documentation referring to the capability and liability of the pest control operator, procedures and chemicals applied as well as the pesticides' location are necessary.

2.7. *Waste control*

The dairy industry must implement a waste control programme. Waste control will eliminate environmental and water pollution caused by rubbish or dangerous chemical wastes.

2.8. *Chemical control*

The chemical control programme ensures the elimination of hazardous chemical substances from food. Chemical hazards, which can be found in a dairy industry, include sanitizers, pesticides, mycotoxins, allergens, refrigerants, solvents, and acids. These materials may interact with the raw materials and the ingredients or the additives and the equipment producing untoward effects. Certain chemicals are also used for food processing, cleaning, sanitation and pest control. Generally, chemical control procedures are related to receiving, storage, usage and disposal of chemicals.

2.9. *Supplier control*

The supplier control programme provides the company with a tool to evaluate the suppliers of raw materials, ingredients and services. Without an effective supplier control programme, the HACCP system can easily fail to provide safety assurance. The description of the industry requirements from the suppliers is of primary concern. These include the HACCP programme of the supplier, certificates of analysis, and records of applied prerequisite programmes.

2.10. *Trace and recall*

The trace and recall programme is used to control the movement of food products from the reception of raw materials and ingredients to the distribution of final products. This programme provides the company with the ability to withdraw products defective or dangerous to human health. The establishment of an effective withdrawal programme includes recorded recall procedures, in which comprehensive information about the process and the responsible personnel are retained. This would include information regarding, raw materials, finished products coding, customer complaints and appropriate handling of recalled products.

2.11. *Equipment calibration*

Equipment used during food processing should be standardized and calibrated. Equipment such as thermometers and pH meters, whose functionality affects food safety, must be calibrated on a regular basis. Certificates provided by national or international organizations are a means of ensuring reliability.

3. Preliminary Tasks

3.1. *Assembling the HACCP team*

Assembling a HACCP team is the first of the five preliminary tasks that should be completed before the HACCP principles are applied to a specific product. The executive manager has to ensure that there is the necessary budget and resource to effect implementation and maintenance of the HACCP system. The team, on the other hand, should clearly define the goals of the programme together with when it is expected to be operational (Surak, 2002). The HACCP team leader is responsible for coordination of the other team members and the provision of an effective HACCP system. Balanced representation from all plant departments in the HACCP team, is also important. A HACCP team should include scientists with different technical or scientific backgrounds as well as personnel with technical and practical knowledge of all stages of the manufacturing process.

3.2. Description of the product

Since HACCP system design is product specific, the second preliminary task is to describe this product (e.g. intrinsic factors of the product such as pH or a_w). 'Feta' cheese is a traditional white brined cheese, which has been produced in Greece since ancient times. It is made exclusively from ewe's milk or from a mixture of ewe and goat's milk and it is ripened in brine for two months. Feta cheese can only be produced under strict product specifications in certain areas of Greece. Feta is registered as a Protected Designation of Origin (PDO) product.

3.3. Description of the intended use and end user

Feta cheese is consumed primarily as it is, or less commonly, cooked. It has high nutritional value and is consumed by all groups of the population (except infants).

3.4. Developing the process flow diagram

During the development of the process flow diagram, the HACCP team should describe each step and operation that takes place during the manufacturing of the specific food. The flow diagram is a very important tool required by the HACCP team for the development and implementation of an effective HACCP plan. The flow diagram is also utilised by the auditors of the competent authorities called in to scrutinise the programme; these individuals, must understand all aspects of the manufacturing process during the evaluation of the HACCP programme.

The flow diagram should depict the processing steps controlled by the company. Furthermore, it may include steps that precede or follow the processing operation, such as receiving of raw materials, storage and possible pre-treatment of them, handling, packaging, storage, distribution or usage of the end-product. The flow diagram should be detailed and contain a variety of data and information on raw materials, ingredients, and packaging materials. It should also provide information on the layout of the factory, equipment, relationship between temperature and time for every step of the processing operation from receiving and storage of raw materials to handling of intermediate and end-products. Furthermore, the flow diagram may contain information about the storage and distribution of the end-product. The flow diagram for the production of Feta cheese is illustrated in Fig. 1. An outline of the Feta cheese flow diagram is given below.

COLLECTION OF MILK
↓
CENTRIFUGAL SEPARATION-STANDARDIZATION
↓
PASTEURIZATION
(68°C for 10 min)
↓
STORAGE AT 34-36°C
↓
ADDITION OF STARTER CULTURES
↓
ADDITION OF RENNET AND $CaCl_2$
(clotting at 35-36°C for 50 min)
↓
CUTTING OF THE CURD/GEL INTO CUBES
↓
MOULDING
↓
FIRST SALTING
(12 hours)
↓
SECOND SALTING
(2 days at 15-16°C)
↓
WASHING WITH BRINE OR COLD WATER
↓
PACKAGING INTO BARRELS - ADDING OF PASTEURIZED BRINE
↓
PRE-RIPENING
(at 15-16°C for 15-20 days)
↓
RIPENING – STORAGE
(at 2-3°C for 2 months)

FIGURE 1. Flow diagram of Feta cheese.

3.4.1. Milk collection and reception

Milk is collected and transported from the farms to the dairy industry in refrigerated milk cans (churns). On reception at the dairy plant, the milk is firstly examined and weighed on a platform balance. Routine examination, by experienced personnel, of the content of each milk churn includes odour, colour, texture and temperature. In addition, the milk should be tested for the presence of antibiotics and acidity. The milk, which has been certified as appropriate for cheese manufacturing is then stored at 4°C (in milk tanks) for a time that should not exceed 24 hours (Walstra et al., 1999).

3.4.2. Centrifugal separation

The aim of this procedure is to remove foreign objects and dirty particles from the milk. Another purpose is to separate the fat content of the milk. In the next step, the fat will be mixed again with skimmed milk to standardize it to a desired fat content. There are several types of milk separators currently commercially available (Abd El-Salam et al., 1999).

3.4.3. Milk standardization

The standardization of milk refers to the adjustment of its fat content to a legally determined percentage. The rationale used for the standardization of the milk for the manufacturing of Feta cheese, is the ratio of its fat content to its crude protein (casein). A fat to crude protein (casein) content of 1.2/1 ensures production of high quality Feta cheese.

3.4.4. Pasteurization

During this step, the milk is treated at 68°C for 10 min. Pasteurization destroys pathogenic microorganisms and enzymes harmful for the ripening. This thermal treatment is sufficient to inactivate alkaline phosphatase but not superoxide dismutase. A more intense thermal pasteurization above 78-80°C would inactivate superoxide dismutase but this would cause denaturation of milk proteins, through the formation of κ-casein and serum proteins complexes. As a result the rennetability and the syneresis of the curd are decreased (Zerfiridis, 2001).

3.4.5. Addition of the starter culture

An appropriate starter culture is essential for the quality of the product. The lactic acid bacteria used for the production of Feta cheese are *Streptococcus thermophilus, Lactococcus lactis* subsp. *lactis* and *Lactobacillus delbrueckii* subsp. *bulgaricus*. The starter cultures should be of excellent quality. It is applied at 1% v/v and the inoculated milk is incubated for 30 min.

3.4.6. Addition of rennet and $CaCl_2$

Cheese is produced by milk clotting through the action of chymosin that is the main component of rennet. This step is primarily essential for the development of a uniform gel and finally Feta with firm texture. Factors that affect the clotting with rennet are milk acidity, temperature and the amount of rennet added. This amount is usually 1.0-3.0 g rennet, of rennet strength 1: 100.000, for 100 litre of milk. The milk should be at rest during clotting for 30-60 min at 30-35°C.

Additionally to rennet, an aquatic dilution of calcium chloride (40%) is added with a maximum amount equal to 200 cm^3/ 1000 l milk.

3.4.7. Cutting of the coagulum and moulding

After coagulation caused by rennet has been completed, the curd/gel formed is cut in 2-3 cm cubes, ladled in thin layers into perforated metal moulds and kept at rest for 5-6 hours for syneresis to occur. A very important factor for this essential stage of Feta production is temperature. The temperature of the operation plants, where syneresis as well as salting takes place should be 16°C. At this temperature a product of better texture and flavour is produced. When curd is firm enough to remove the moulds, it is cut into four sub-blocks of 23 x 12 x 6 cm.

3.4.8. Salting

Salting of Feta cheese takes place in two stages. The first stage includes dry salting and takes place on a salting table, where the blocks of curd are placed. Coarse salt is sprinkled on the surface of the cheese to allow its slow penetration into the curd. Dry salting is repeated two or three times every 12 hours. During the second stage of salting, Feta is kept immersed in a concentrated solution of NaCl (brine). The final concentration of salt in Feta cheese should be 3-4%. Salt combined with acidity (pH<5.1) is essential for the quality and safety of the final product.

3.4.9. Ripening

Maturation of Feta cheese actually begins before the curd making is finished, and can be separated into two phases. The first one takes place simultaneously with dry salting in the ripening plants, at 18°C, while the second one occurs during the storage of Feta in refrigeration. The duration of the first phase or pre-ripening is approximately 15 days. During this time, syneresis is completed while a wide range of biochemical reactions takes place.

3.4.10. Storage and packaging

During the final stage of the Feta cheese flow diagram, Feta is packed in either wooden barrels or tin cans. Containers should be thoroughly cleaned. Following packaging, the containers should be filled with brine. The brine added should cover the upper surface of the cheese in the container but it must also allow a free space of 2 cm. Then, it is kept at 14-16°C until a pH of 4.6 and moisture content less than 56% are reached. At this time, the containers are sealed and stored at 4-5°C. The relative humidity of the plant where maturation of Feta cheese takes place should be 95-100%. Care should be taken so that the surface of the cheese is completely covered with brine (Zerfiridis, 2001).

3.5. Verifying the accuracy of the process flow diagram

After the flow diagram of the manufacturing of the specific food has been developed, the HACCP team should regularly inspect the plants of the industry to verify its implementation and accuracy. The inspection of the manufacturing processes is essential for the validation of how clearly the flow diagram depicts the process. The person responsible for the verification of the flow diagram is the quality and safety assurance manager. Moreover, the HACCP team must evaluate all the data concerning the production. Consultations or observations should take place and cases of deviations recorded. The HACCP team should conduct inspections every month without notice, at any time during working hours.

4. Principles of HACCP

4.1. Hazard analysis

Hazard analysis is considered the foundation of a HACCP plan. The process of hazard analysis includes the identification of all potential hazards of a food and the probability of their occurrence (Surak, 2002). In a food production plant the HACCP system identifies three types of hazards relevant to food. Based on their origin they are classified in microbiological, chemical or physical.

Microbiological hazards include pathogens and microbial toxins, which are responsible for human foodborne illnesses. In the dairy products *Brucella spp., Clostridium botulinum, Listeria monocytogenes, Salmonella spp., Escherichia coli* O157:H7, *Yersinia enterocolitica, Campylobacter jejuni, Staphylococcus aureus,* toxins and parasites are considered as microbiological hazards. Potential chemical hazards include chemical contaminants of food, either naturally occurring or added during the production process. Any object or material that is either naturally occurring in the food but must be removed during the process, or is a foreign object that accidentally was introduced into the product during its handling (such as metal, pieces of glass, hard plastic), is defined as a physical hazard.

Following identification of hazards related to the food, a hazard analysis of the production process should be developed. This includes identification of the stages in the flow diagram where there is a risk for the safety of the final product. Elimination, avoidance or restriction of the above hazards into definite boundaries would ensure the hygienic quality of food and consumer safety.

During the identification of microbial hazards, it is important to record all potential pathogen or toxins, responsible for food born human illnesses. Public regulatory agencies regularly publish epidemiological summaries that establish the hazard profiles of different products. The probability of finding these factors in the raw materials used during the manufacturing of Feta cheese should be evaluated. The pathogens usually detected in milk are shown in Table 1.

In a next step, a definition of how the production process of Feta cheese affects the above pathogens has to be established. Finally, the HACCP team should include the microbiological hazards of every pathogen or toxin, which could be responsible for product recontamination during processing. Hazard analysis of Feta cheese manufacturing steps is given below and summarized in Form 1.

TABLE 1. Pathogens transmitted to human via consumption of milk[a]

Mycobacterium spp. (*Mycobacterium bovis, M. tuberculosis*)
Brucella abortus
Salmonella spp.
Listeria monocytogenes
Bacillus anthracis
Yersinia enterocolitica
Shigella spp.
Escherichia coli (enteropathogenic)
E. coli O157:H7 (verotoxigenic, enterohaemorragic)
Streptococcus pyogenes
Campylobacter jejuni
Staphylococcus aureus
Clostridium botulinum
Bacillus cereus
Clostridium perfringens
Coxiella burnetii
Pseudomonas aeruginosa
Vibrio spp.
Aeromonas hydrophila

[a]Based on Mantis, 1993; Teuber, 2000

4.1.1. Milk collection

4.1.1a. Microbiological hazards.

Sheep and goat's milk comprises the major and basic ingredient for the manufacturing of Feta cheese. Quality and microbiological safety of Feta is primarily dependant on the quality of milk as well as the microbial population as milk contains a wide range of nutrients, which provide an excellent culture medium for many pathogens. Only milk from healthy animals should be collected and received by the dairy industry. Unhealthy animals may provide milk with pathogenic organisms such as *Listeria monocytogenes* or microorganisms responsible for milk spoilage and deterioration of milk or both. The most significant pathogens are shown in Table 1 (Mantis, 1993).

Mycobacterium spp. Both *Mycobacterium tuberculosis and Mycobacterium bovis* are pathogenic to humans. Although *Mycobacterium* is killed by low temperature milk pasteurization, highly contaminated milk may need more intense thermal treatment. *M. paratuberculosis* may be associated with Crohn's disease (ulcerative colitis).

Brucella spp. Strains of *Br. melitensis* and *Br. abortus* are pathogenic to humans (causing illness resembling Malta fever in man). The transmission from infected animals to humans takes place through contaminated food and especially dairy products. The pathogen is killed by pasteurization. In the acidic environment of matured Feta cheese, it can survive for as long as a month.

Salmonella spp. As low temperature pasteurization is adequate to kill these pathogens, special care should be taken to avoid post-pasteurization

FORM 1. Hazard Analysis, identification of hazards and control measures for production of Feta cheese

Ingredients & process step	Potential hazards	Justification	Hazard to be addressed in plan? Y/N	Control measures
Reception of milk (kept cool)	**M**: pathogens (e.g. *Listeria monocytogenes*, *Escherichia coli*, *Staphylococcus aureus*, *Salmonella*, *Mycobacterium*, *Brucella*, *Yersinia enterocolitica*, *Coxiella burnetii*) and viruses	Coming from: unhealthy animals, farm environment, milk containers, and inadequate refrigeration. Constitute public health hazard, responsible for foodborne illnesses and poisoning	Yes	Good Manufacturing Practices (GMP) during milk collection and reception, Low storage temperatures Adequate thermal treatment (pasteurization)
	C: antibiotics, hormones, toxins, heavy metals, sanitizers, additives		Yes	Farmer and animal records, raw materials testing, GMPs.
	P: metal, pieces of glass, foreign objects		No	Centrifugal separation-Filtering
Starter culture	**M:** contamination of starter culture	Abnormal fermentation		Starter culture analysis and GHPs
Rennet	**M:** recontamination microorganisms	Abnormal fermentation Foodborne illnesses		Rennet analysis, rennet reception records
Salt	**M:** microorganisms (pathogens or spoilage psychrotrophs) **C:** element residues (copper, iron, lead)	Affects the microbial cultures, enzymes and biochemical reactions	Yes	Brine pasteurization Brine reception records
Pasteurization	**M:** survival of heat-tolerant spores and toxins (or/and survival of pathogens – inadequate thermal treatment)	Presence of heat tolerant spores or toxins in milk	Yes	Farmer and animal records, raw materials testing , GMPs Temperature control and recording, Cheese maturation
Process steps after pasteurization	**M:** contamination with pathogens **P:** foreign objects, dust	Poor hygiene, source: personnel, equipment	Yes	GMP, prerequisite programs
Ripening	**M:** contamination, insects **C:** mycotoxins **P:** foreign objects, dust	Inadequate maturation, finished product spoilage	Yes	GMP, temperature control and recording, pH

Key: M: microbiological, C: chemical, P: physical

recontamination of milk. Implementation of prerequisite programmes such as good hygiene practice would eliminate this risk.

Listeria monocytogenes. It has been associated with several outbreaks in humans caused by dairy products. It grows at temperatures as low as 5°C and is also quite resistant during thermal treatment. It is killed by pasteurization. During maturation of Feta cheese, it appears that the population of *Listeria* increases for the first 10-14 days but then decreases slowly.

Escherichia coli. Enteropathogenic species of *E. coli* may be responsible for foodborne infection and poisoning if their population exceeds 10^7-10^8 cells/g of food. Low temperature pasteurization is capable of destroying this pathogen population in milk.

Staphylococcus aureus. *Staph. aureus*, present in numbers of 10^5 cfu/ml or g of food, is responsible for the production of a heat resistant staphylococcal enterotoxin, which has been implicated in human foodborne illness. Incidents related to consumption of cheese are associated with milk containing enterotoxin prior to its pasteurization. Pasteurization destroys the pathogen but not its toxin.

Yersinia enterocolitica. Is a psychrotroph pathogen able to grow at 4°C. In Feta cheese, it can survive as long as 5-30 days, depending on the initial population and the acidity of the product.

Coxiella burnetii. High temperature pasteurization (72°C for 15 sec) destroys the pathogen, whereas low temperature pasteurization (63°C for 30 min) is not always effective when microbial population is too high.

Furthermore, several potentially pathogenic microorganisms of less importance could be found in milk, but they are killed or destroyed during manufacturing of Feta cheese (thermal treatment of milk, ripening, acidity, storage at refrigerator temperatures). Additionally, implementation of the prerequisite programmes such as good manufacturing practices (GMPs) eliminates the possibility of milk contamination with various microbiological hazards.

4.1.1b. Chemical hazards.

The milk collected must be free from chemical contaminants that are able to inhibit microbial activity responsible for cheese production. Potential chemical hazards include antibiotics, pesticides, cleaners and sanitizers, heavy metals and toxins. Since presence of antibiotics is the most common chemical problem for the dairy industry, milk should be tested at the point of collection. Another potential chemical hazard is mycotoxins, which are commonly associated with animal feed of low quality. Mould growth, especially *Aspergillus* spp., is related with the production of heat resistant aflatoxins, which are transmitted, through the animals, to the milk.

4.1.2. Pasteurization

The aim of performing pasteurization is to kill or inactivate milk pathogens. Pasteurization is not capable of destroying microbial spores (such as *Bacillus cereus or Clostridium perfringens* spores) and heat resistant toxins (such as

Staphylococcus aureus enterotoxin). As viruses are sensitive to heat treatment, pasteurization is adequate for their elimination. Among milk pathogens *Mycobacterium tuberculosis* and *Coxiella burnetii* are considered the most heat-resistant (Mantis, 1993).

4.1.3. Post pasteurization process

The implementation of good manufacturing practices (GMP) could result in the exclusion of post-pasteurization milk recontamination. Salt, rennet or other substances added may also contain microorganisms. Additionally, microbial hazards may be caused by equipment, as well as by the personnel. Microorganisms usually implicated with post-pasteurization milk recontamination during processing are coliforms, *Pseudomonas, Achromobacter, Proteus*, moulds and yeasts. Poor hygienic conditions during production could even cause recontamination with pathogens, such as *Salmonella, Shigella* and *Staphylococcus* (Mantis, 1993).

Finally, salt quality and hygiene must be taken into consideration. The dry salt used should be free from harmful microorganisms while the brine is usually pasteurized. The concentration of salt in insoluble matter shall not be exceeded 0.03% and it may not contain contaminants in amounts that may be harmful for the health of the consumer. The maximum limits that shall not be exceeded for copper, iron and lead are 0.002%, 0.01% and 0.0005%, respectively, as they affect biochemical reactions during cheese manufacturing (CAC, 2000; Anifantakis, 2004).

4.2. Determination of critical control points

A critical control point (CCP) is defined as "a point, step, or procedures at which control can be applied and which is essential to prevent, eliminate, or reduce a hazard to an acceptable level" (NACMCF, 1998). Loss of control of a critical control point provides unsafe food to the consumer. All hazards identified and evaluated during the hazard analysis must be controlled in a specific step of the production process, from receiving of the raw material to shipping of the products.

Critical control point determination is usually based on a decision tree. The one proposed by Codex Alimentarius (CAC, 2001), is one of the most widely used (Fig. 2). For every hazard identified in hazard analysis a series of questions should be answered as shown in Form 2.

4.3. Establishment of critical limits

The National Advisory Committee on Microbiological Criteria for Foods (NACMCF) defines a critical limit as "a maximum and/or minimum value to which a biological, chemical, or physical parameter must be controlled at a CCP to prevent, eliminate, or reduce to an acceptable level the occurrences of a food safety hazard" (NACMCF, 1998).

Q1 Do preventative control measures exist?
↓ ↓ ↑
Yes No Modify steps in the process or product
↓ ↓ ↑
Is control at this step necessary for safety? → Yes
↓
No → Not a CCP → Stop*

Q2 Is the step specifically designed to eliminate or reduce the likely
occurrence of a hazard to an acceptable level? → Yes
↓ ↓
No
↓

Q3 Could contamination with identified hazard(s) occurs in excess
of acceptable level(s) or could this increase to unacceptable levels?**
↓ ↓
Yes No → Not a CCP → Stop*
↓

Q4: Will a subsequent step eliminate identified hazard(s)
or reduce likely occurrence to acceptable level(s)?**
↓ ↓
Yes No → **Critical Control Point (CPC)**
↓
Not a CPC → Stop*

FIGURE 2. Codex Alimentarius decision tree.

Establishment of critical limits is based on different sources of information and scientific knowledge. The major sources used to establish critical limits are literature, Government and Federal Regulations or specialized scientists (Surak, 2002).

In order to establish the critical limits, critical factors/parameters concerning every CCP should be determined, as well as the boundaries out of which, the product can be characterized as dangerous and unsafe. A factor is critical if loss of its control leads to the loss of food safety control. Food safety assurance may require control of several factors for every CCP. Based on the hazard nature, three categories of critical limits can be distinguished: (i) critical limits of microbiological hazards, (ii) critical limits of chemical hazards, and (iii) critical limits of physical hazards.

Safety assurance of microbiological hazards involves the establishment of critical limits, based on microbiological analysis. However, the use of traditional cultural techniques is not appropriate for controlling these hazards, because of their time consuming character. Additionally, low level of pathogen population is difficult to identify using traditional techniques, while a more extensive examination increases the cost. Instead of traditional microbiological analysis, measurements of physical or chemical factors indicative of the microbiological safety of the product are used. In this case, the following Critical Limits for every CCP identified in Feta cheese production were established.

FORM 2. Determination of critical control points (CCPs) for the production of Feta cheese

Process step	Possible Hazards	Q1:	Q2:	Q3:	Q4:	CCP
Reception of milk	**M:** pathogens, viruses					
	C: Antibiotics, hormones, toxins, heavy metals, sanitizers, additives	Yes	No	Yes	No	CCP 1
	P: metal, pieces of glass, foreign objects	Yes	No	Yes	Yes / centrifugal separation -filtering	-
Pasteurization	**M:** survival of heat-tolerant spores and toxins (or/and survival of pathogens – inadequate thermal treatment	Yes	Yes	-	-	CCP 2
Addition of LAB, rennet, $CaCl_2$	**M:** growth of contamination microorganisms	Yes	No	Yes	Yes / ripening	-
Curd handling	**M:** growth of contamination microorganisms	Yes	No	Yes	Yes / ripening	-
Salting	**M:** growth of contamination microorganisms	Yes	No	Yes	Yes / ripening	-
Ripening	**M:** growth of contamination microorganisms	Yes	Yes	-	-	CCP 3

Q1. Do control measures exist for the identified hazard? (no: not a CCP, yes: Q2).
Q2. Is this process step specifically designed to eliminate or reduce the likely occurrence of this identified hazard to an acceptable level? (No: Q3, Yes: CCP).
Q3. Could contamination with the identified hazard(s) occurs in excess of acceptable level(s) or increase to unacceptable level(s)? (No: not a CCP, Yes: Q4).
Q4. Will a subsequent step eliminate the identified hazard(s) or reduce its likely occurrence to an acceptable level(s)? (No: CCP, Yes: Not a CCP).
Key: M: microbiological, C: chemical, P: physical

4.3.1. Milk reception (CCP 1)

Milk supplied to the dairy must be tested for: i) milk acidity (pH) should be 6.65-6.45, ii) presence of antibiotics (absence), iii) temperature (<4°C), iv) periodically, bacterial count (at 30°C Total Viable Count (TVC) should be $\leq 10^6$ cells/ml of milk).

4.3.2. Pasteurization (CCP 2)

Thermal treatment at 68°C for 10 min, periodical bacterial counts of milk samples after pasteurization should be determined (at 30°C TVC should be $\leq 10^5$ cells/ml of milk).

4.3.3. Ripening (CCP 3)

Pre-ripening: takes place at 17-18°C for 5-15 days. After this time, the pH of the product must be 4.6-4.7. ii) Ripening during storage occurs at refrigeration temperature (4°C). It is legally defined that this period must be at least 2 months. The microbiological standards of the final product, as established by national and European legislation, are illustrated in Table 2.

4.4. Establishment of monitoring procedures

As defined by the NACMCF, monitoring is a planned sequence of observations or measurements to assess whether a CCP is under control and to produce an accurate record for future use in verification (NACMCF, 1998). Ideally monitoring of a CCP should be on a continuous basis. When this is not possible, monitoring frequency should be statistically established.

Monitoring procedures must ensure absolute control of CCPs and should include either observation of procedures or measurement techniques. The first one characterizes control based on quality while the second one on quantity. The choice of which one would be preferred depends upon established critical limits, methods and time available, as well as costs. Effective measurement techniques

TABLE 2. Microbiological criteria for soft cheese (made from heat-treated milk) in European Union legislation[a]

	Sampling plan[b]		Limits	
Microorganisms	N	c	m	M
Listeria monocytogenes	5	0	absence in 25 g	
Salmonella spp.	5	0	absence in 1 g	
Staphylococcus aureus	5	2	100 cfu/g	1000 cfu/g
Esherichia coli	5	2	100 cfu/g	1000 cfu/g
Coliforms	5	2	10^4 cfu/g	10^5 cfu/g

[a] EEC 1992. [b] N = number of units comprising the sample; c = number of sample units giving values between m and M

require properly calibrated equipment, qualified personnel and sufficient data recording procedures.

4.4.1. Milk reception (CCP 1)

During milk reception each container should be separately examined and data recorded: a) milk lot coding and farm code, b) pH, c) specific gravity, d) fat content, e) solids-not-fat-content, f) presence of antibiotics, g) temperature, h) storage time, i) adulteration, j) periodically measurement of bacterial and coliform counts and k) periodically sampling for pathogen presence.

4.4.2. Pasteurization (CCP 2)

Data recorded during pasteurization of milk must include: a) temperature of heat exchanger, b) time of thermal treatment, c) phosphatase activity test, d) superoxidase activity test and, d) periodically, measurement of bacterial and coliform counts.

4.4.3. Ripening (CCP 3)

The parameters below should be monitored and appropriately recorded at the end of the first week, before storage at 4°C and at the end of the two-month ripening at refrigeration temperature. The parameters measured are: a) temperature of maturation plants, b) time of ripening, c) cheese pH, d) water content of cheese, e) fat content of cheese, f) relative humidity of the maturation plants and, g) periodically, measurement of bacterial, coliform and pathogen count in cheese samples.

Monitoring procedures as described above refer to critical control points of the production process. Besides these, a series of additional points should be monitored and recorded:

1. Centrifugal separator cleaning (monitored daily by observation and periodically by luminometry).
2. General equipment cleaning. This should include cleaning of thermal exchangers, milk containers, pipes and any other secondary equipment used during processing. Checking of hygiene and effective cleaning and sanitation takes place after cleaning but before sanitation processes.
3. Sampling and microbiological analysis of water used in processing plants, monthly.
4. Testing of starter culture activity, periodically.
5. Rennet microbiological analysis.
6. Lot coding and microbiological analysis of salt.
7. Concentration of salt -Baumé (°Bé of brine).
8. Lot coding of $CaCl_2$.
9. Lot coding of rennet.
10. Temperature/ time of brine pasteurization.
11. Barrel code.

12. Dairy plants temperature.
13. Microbiological analysis of the air.

4.5. Establishment of corrective actions

The aim of a HACCP system is to assure the production of safe foods, to identify possible heath hazards and establish the measures that will eliminate the risk of loss control. An ideal implementation of a HACCP plan would not include the establishment of corrective actions. However, when the results of monitoring a CCP indicate a loss of control, corrective actions and measures must be taken to control the hazard and to prevent consumers from consuming unhealthy and hazardous food (Surak, 2002). Personnel responsible for corrective actions must be well informed on the HACCP plan, and possess a thorough understanding of the production process.

The aim of corrective actions is to (i) provide the necessary corrections to a product, manufactured under low safety, (ii) correct the cause of the deviation from critical limits, and thus to ensure that CCP is under control. Every corrective action should be documented and evaluated during HACCP system verification.

In HACCP plan for Feta cheese manufacturing, when loss of the critical control points control occurs, the following corrective actions must be taken.

4.5.1. Milk reception (CCP 1)

When the pH of the milk is out of the critical limit or antibiotics presence is detected, the lot must be rejected. In case of significant temperature deviations during transporting above the critical limit of 4°C are detected, the milk lot is marked and its microbiological quality is reconsidered after thermal treatment.

4.5.2. Pasteurization (CCP 2)

In case of deviation from the desired temperature/time ratio or alkaline phosphatase has not been inactivated after pasteurization, the thermal treatment must be repeated. Additionally, it is necessary to determine the reasons caused these deviations (e.g. improper maintenance of equipment).

4.5.3. Ripening (CCP 3)

When parameters such as pH and temperature, checked during monitoring procedures are out of the defined critical limits, evaluation of the product microbiological safety must be conducted by sampling and application of microbiological techniques. Temperature is adjusted to the appropriate degree and maturation time may, if necessary, be prolonged. Finally, a thorough optical observation must take place. If optical observation and microbiological analysis reveals a danger to consumer health, the foodstuff has to be rejected.

4.6. Establishment of record keeping and documentation procedures

Establishment of effective record keeping and documentation procedures is essential for the implementation and maintenance of a HACCP plan. This sixth principle of HACCP provides the documents for the implementation of the HACCP plan, available for internal or external audits. These documents should be filed and be available for a specific period. Record keeping includes all documents about chemical, physical and microbiological monitoring tests of critical control points, deviations from critical limits and corrective actions applied. Documents of a HACCP system should include:

1. Title and date of the document.
2. Product description (code, production date, weight).
3. Ingredients used and equipment.
4. Processes, which took place.
5. Critical limits.
6. Corrective actions-responsible personnel.
7. Signature of employee and supervisor.

The records of a HACCP system that are of primary importance concern raw materials, CCPs, critical limits, monitoring of CCPs, deviations of critical limits and corrective actions taken. Important documents also include packaging and storage records, as well as validation and verification documents and documents describing the HACCP system.

4.7. Establishment of verification procedures

Verification procedures are essential for the implementation of an effective HACCP programme. The aim of verification procedures are a) to confirm that HACCP programme is working in accordance with established HACCP plan, and b) to verify that HACCP system is functional and effective (Surak, 2002). Verification procedures include various methods and activities such as procedures monitoring, records checking, or random sampling and analyzing of final products, raw materials or intermediate products. Verification procedures may include:

1. Checking of prerequisite programmes applied.
2. Validation of HACCP plan.
3. Confirmation of critical limits maintenance for each CCP.
4. Validation of corrective actions taken when deviations from critical limits occurred.
5. Validation of record keeping and documentation procedures.
6. On-site inspection of the manufacturing process.
7. Audit report.

The HACCP team is responsible for the development, establishment and implementation of the verification procedures and activities as well as the time scheduled for them. Verification procedures are implemented by qualified industry personnel (or external consultants). Additionally, relevant governmental agencies have the regulatory responsibility to perform regular external validations of industry's HACCP plan and its effective implementation.

5. Concluding Remarks

The present chapter has presented basic information on the occurrence of hazards in and an indicative HACCP plan for Feta cheese manufacture. An outline of GMP's that should be implemented as the prerequisite of HACCP system has been given. The owners of such establishments should try to comply with the new EU food legislation, namely Regulation 178/2002 (EU, 2002), the Food Hygiene Regulation 852/2004 and the specific Regulation on Hygiene of foods of animal origin (EU, 2004a; b) that determine that HACCP is a compulsory system for safety assurance in food business.

6. References

Abd El-Salam, M.H., Alichanidis, E., and Zerfiridis, G.K., 1999, Domiati and Feta type cheeses, in: *Cheese: Chemistry, Physics and Microbiology*, Vol. 2, P. F. Fox, ed., 2nd ed. Aspen Publishers, Inc, Gaithersburg, pp. 301-335.

Anifantakis, E.M., 2004, [*Cheese making: Chemistry, Physico-Chemistry, Microbiology*], 2nd ed. Stamoulis Publications, Athens, Greece. (in Greek).

Codex Alimentarius Commission, 2001, Basic Texts on Food Hygiene, 2nd ed., Food and Agriculture Organization of the United Nations -World Health Organization, Codex Alimentarius -Joint FAO/WHO Food Standards Programme, Rome.

Codex Alimentarius Commission, 2000, *Codex Alimentarius*, Vol 1A, *General Requirements*, 2nd ed., Food and Agriculture Organization of the United Nations -World Health Organization, Codex Alimentarius-Joint FAO/WHO Food Standards Programme, Rome.

Corlett, D.A., 1998, *HACCP User's Manual,* Aspen Publishers, Inc, Gaithersburg, Maryland.

EEC, 1992, Council Directive 92/46/EEC of 16 June1992 laying down the health rules for the production and placing on the market of raw milk, heat-treated milk and milk-based products. Official Journal of the European Communities, L 268, 1-32.

EU, 2002, Regulation (EC) No 178/2002 of the European Parliament and of the Council of 28 January 2002 laying down the general principles and requirements of food law, establishing the European Food Safety Authority and laying down procedures in matters of food safety. Official Journal of the European Communities L 31, 1-24.

EU, 2004a, Regulation (EC) No 852/2004 of the European Parliament and of the Council of 29 April 2004 on the hygiene of foodstuffs. Official Journal of the European Union L 139, 1-54.

EU, 2004b, Regulation (EC) No 853/2004 of the European Parliament and of the Council of 29 April 2004 laying down specific hygiene rules for food of animal origin. Official Journal of the European Union L 139, 55-205.

Mantis, A.G., 1993, [*Hygiene and Technology of Milk and Dairy Products*], 2nd ed., Kyriakidis Brothers S.A. Publishing House, Thessaloniki, (in Greek).

Mortimore, S. and Wallace, C., 1998, *HACCP, A Practical Approach,* Aspen Publishers, Inc, Gaithersburg.

National Advisory Committee on Microbiological Criteria for Foods, 1998, Hazard Analysis and Critical Control Point Principles and Application Guidelines, Journal of Food Protection, **61**:1246-1259.

Surak, J.G., ed., 2002, The Quality Auditor's HACCP Handbook. ASQ Food, Drug, and Cosmetic Division. American Society for Quality. Quality Press, Wisconsin, USA.

Teuber, M., 2000, Fermented milk products. In: *The Microbiological Safety and Quality of Food*, B.M. Lund, T.C. Baird-Parker & G.W. Gould, eds, Aspen Publishers, Inc., Gaithersburg, Maryland pp. 535-589.

Walstra, P., Geurts, T.J., Noomen, A., Jellema A., and Van Boekel, M.A.J.S., 1999, Dairy technology. Principles of Milk Properties and Processes, Marcel Dekker Inc, New York.

Zerfiridis, G.K., 2001, [Dairy technology - Cheese making], Giahoudis-Giapoulis' Publications, Thessaloniki, (in Greek).

Zerfiridis, G.K. and Litopoulou-Tzanetaki, E., 1988, [*Hygiene of Dairy Industry*], Greek National Dairy Committee, Athens. (in Greek).

6 Genetically Modified Organisms and Food Safety

DEZIDER TOTH[1]

1. Genetically Modified Organisms

Directive 2001/18/EC of the European Parliament and of the Council of 12 March 2001 on the deliberate release into the environment of genetically modified organisms defines GMO as an organism, with the exception of human beings, in which the genetic material has been altered in a way that does not occur naturally by mating and/or natural recombination.

Within the terms of this definition techniques of genetic modification are:

i. recombinant nucleic acid techniques involving the formation of new combinations of genetic material by the insertion of nucleic acid molecules produced by whatever means outside an organism, into any virus, bacterial plasmid or other vector system and their incorporation into a host organism in which they do not occur but in which they are capable of continued propagation
ii. methods involving the direct introduction into an organism of heritable material prepared outside the organism including micro-injection, macro-injection and micro-encapsulation
iii. cell fusion (including protoplast fusion) or hybridisation techniques where live cells with new combinations of heritable genetic material are formed through the fusion of two or more cells by means of methods that do not occur naturally.

In the Cartagena Protocol on Biosafety to the Convention on Biological Diversity instead of GMO (genetically modified organism) the term Living Modified Organism (LMO) was introduced. This is defined as any living organism that possesses a novel combination of genetic material obtained through the use of modern biotechnology. The modern biotechnology means the application of:

[1]Dezider Tóth, PhD., DSc, Slovak Agricultural University of Nitra, Tr.A.Hlinku 2, SK-94976 NITRA, Slovak Republic, E-mail: dezider.toth@uniag.sk

i. in vitro nucleic acid techniques, including recombinant DNA and direct injection of nucleic acid into cells or organelles, or
ii. fusion of cells beyond the taxonomic family, that overcome natural physiological reproductive or recombinant barriers and that are not techniques used in traditional breeding and selection.

Both definitions are fairly similar in meaning, even though the approach is different. To consider the potential impact of genetically modified organisms on environment and human health, a sound understanding of the biology of an organism is required, as well as its relationship and interactions with the environment and other organisms. The most important information focused on those biological attributes used for evaluation of GMOs safety are compiled by OECD in the so called Consensus Documents on the biology of several crop plant species, trees and microorganisms (such as OECD 2002). These documents deal with many important issues in assessing safety, including genetic characteristics, reproductive biology, data concerning the origin and biodiversity, characteristics on pests, diseases and ecology.

Genetically modified organisms have already entered the food stream in many parts of the world (FAO, 2005). Various species of microorganisms, mostly bacteria and fungi, have been modified for increased production of proteins, amino acids and commercial chemicals (Tourte, 2003). Early work in this area relied primarily on discovery of naturally occurring or by mutagenesis-induced variant microbial strains. Often these variant genotypes were blocked in specific metabolic pathways, or they expressed higher levels of a key rate-limiting enzyme, with the result that their metabolic output was being channelled into the desired product. Such mutants provided valuable biological tools for researchers, and for the fermentation industry they also represented a key commercial asset.

However, the current generation of GMOs consists mostly of plants modified for a limited number of traits. With the expected increase in the availability of genomic information for many species in the next years, the floodgates of genetic modifications could open and release an unprecedented variety of genetically modified products onto the market. In parallel with this rapid market penetration, there is increasing concern about the use of genetic engineering for food production, particularly about possible deleterious effects on human health and about the possible impacts of the widespread deployment of GMOs in the environment.

Plants play a critical role in global life as they are the base of the food chain. Most animals consume plant material as food, although some animals are carnivores, dependent on plant-eating animals. Microorganisms can also utilize either living or decaying plants as source of essential substrates. Pathogenic microorganisms may induce a hypersensitive reaction in plants or a systematically acquired resistance to pathogens. Along with complex interactions inside of plant kingdom (competition for light, space and nutrients) there are additional interactive relations between plants and pests, herbivores, pollinating insects or seed-dispersing animals.

It is generally accepted that the plants play a number of roles in the environment and may affect other organisms through different direct and indirect mechanisms. The risk/safety assessor, in considering the safety of transgenic plants, should evaluate the potential for disruption of these interactions. Applications of crop genetic engineering could be subdivided into several groups:

i. Transformation for insect resistance – This is important in field and horticultural crops as well as plantation trees. The most common modification involves use of genes for protein toxins from the soil bacterium *Bacillus thuringiensis* (Bt). The Bt plants are insecticidal to a limited range of specific caterpillars and beetles like European corn borer (Bt maize), bollworm and budworm (Bt cotton) and Colorado potato beetles (Bt potatoes). Another example involves the transformation of peas with a kidney bean amylase inhibitor gene to develop a source of resistance to pea weevils. A cowpea gene has also been used in strawberries to confer weevil resistance.
ii. Transformation with plant quality genes – This is important for both food and feed production. The first quality-enhanced product on the market was tomatoes with a reduced level of the polygalacturonase gene product. The lower activity of this enzyme is responsible for slowing down the long pectin chains cleaving in the cell walls of the fruit so that the ripe fruit does not soften so quickly and remains for firm for a longer time. Another case is canola with a thio-esterase transgene from a bay laurel tree prevents the synthesis of longer chain fatty acids and thus enabling formation of high lauric acid in the oil (an useful raw material in detergent manufacture). The third example that applies to a range of horticultural crops (potatoes, apple, lettuce, bananas, grapes, lettuce, pineapple), is prevention of damage-induced browning by blocking polyphenoloxidase genes.
iii. Transformation for disease resistance –The transformation for virus resistance is well developed. Transgenic plants containing various parts of a viral genome can be protected against a virus. For example, the expression of viral coat protein genes in the target plant, as in tomato and tobacco plants resistant to mosaic virus, potatoes resistant to potato leaf roll virus (PLRV) and white clover resistant to alfalfa mosaic virus.
iv. Transformation for herbicide tolerance – The development of crop resistance to a herbicide is an important economical and environmental achievement of weed control. Commercially-used transgenic herbicide-resistant crops are maize, soybeans, cotton and canola. These are resistant to the broad-spectrum herbicides such as glyphosate (Roundup®) and glufosinate (Basta®). Herbicide-tolerant pasture crops, such as subterranean clover, enable farmers to control broadleaf pasture weeds earlier in the growing season with low levels of herbicide.
v. Transformation for nutritional purposes – Examples of other transformations include enhanced lysine content in maize or enhanced vitamin A content in canola. Edible vaccines, for viral and diarrhoeal diseases, using proteins expressed in transgenic plants is extremely important for developing countries. The vaccine is expressed in the fruit or vegetable and even multiple vaccines could be produced in one plant.

1.1. GMO: new biotechnology products

Biotechnology, broadly defined, includes any technique that uses living organisms, or their parts, to make or modify products, to improve plant or animals, or to develop microorganisms for specific use. It ranges from traditional biotechnology to the most advanced modern biotechnology (Doyle and Persley, 1996). Commercial biotechnology consists of an expanding range of interrelated techniques, procedures and processes for practical applications in the health care, agricultural and industrial sectors. The background of biotechnology is formed by a mixture of scientific disciplines – biochemistry, cell biology, embryology, genetics, microbiology and molecular biology combined with practical disciplines such as chemical engineering, information technology and robotics. From the synergy of these building blocks are born the desired results – new thoughts, new products. Biotechnology should be seen as an integration of new techniques emerging from modern biotechnology with well established approaches of traditional biotechnology, such as plant and animal breeding, food production, fermentation process and production of pharmaceuticals, biopesticides and fertilizers.

Taking into account the historical progressive aspects, biotechnology consists of a gradient of technologies, ranging from the long-established and widely used techniques and traditional processes like fermentation to novel techniques of modern biotechnology. As shown by Persley (1990), these technologies started with the oldest microbial fermentations; these were then followed by plant tissue cultures, embryo transfer in animals, monoclonal antibody production up to recombinant DNA technology used for genetic engineering of microorganisms, plants and animals.

One of the major issues relating to the role and application of biotechnology in agriculture is the safety of organisms with novel traits and the appropriate regulatory measures for research and development, field testing, marketing of organisms with novel traits and their use as food or feed. There is fear that uncontrolled introduction of genetically modified organisms might cause undesirable effects on ecological or genetic relationships in some communities or potential harm to some consumers in the food chain. Therefore organisms with novel traits should be carefully designed and their introduction into the environment or the market allowed only after a proper testing.

Both the traditional and the modern plant breeding have the potential to alter the nutritional value of plants and their products or lead to unexpected or unintended changes in concentration of various natural toxicants or antinutrients (Table.1).

It is important to evaluate all new varieties in order to reduce the likelihood that unexpected changes will produce adverse health effects. Unintended changes in levels of nutrients can theoretically arise in several ways:

i. Insertion of genetic material could disrupt or alter the expression of normally expressed plant genes.
ii. Expression of the introduced gene resulting in a protein synthesis might reduce the availability of amino acids used for synthesis of normal plant compounds.

TABLE 1. Plant toxins and antinutrients in some common crop plants

Crop	Toxin/antinutrient
Rape (Brassica napus, B. rapa)	Glucosinonales
	Erucic acid
	Phytate
Maize (Zea mays)	Phytate
Tomato (Lycopersicon esculentum)	Alpha-tomatine
	Solanine
	Chaconine
	Lectins
	Oxalate
Potato (Solanum ruberosum)	Solanine
	Chaconine
	Protease inhibitors
	Phenols
Soybean (Glycine max)	Protease inhibitors
	Lectins
	Isoflavones
	Phytate

iii. Production of normal plant compounds might also be affected if the expressed protein diverted substrates from other important metabolic pathways.
iv. Finally, either the expressed protein or altered levels of other proteins might have antinutritional effects. These possible concerns are related to the randomness of DNA insertion. However, changes in gene expression can also occur when traditional breeding method are used; such changes may be less frequent in transgenic plants since only a limited number of genes are transferred during the genetic modification.

Food safety should consider the potential for any change in nutritional composition, especially in key elements that have a significant impact on the diet, as well as the potential for any change in the bioavailability of key nutritional components. Where additional assurance of safety is needed, analytical methods traditionally applied in the evaluation of food constituents such as total protein, fat, ash, fibre and micronutrients may be supplemented with additional analyses to identify unexpected effects.

Because of the potential for broad changes in nutrient levels and interactions with other nutrients and unexpected effects, it may be necessary in certain instances to undertake feeding studies with animals to determine outcomes that result from changes in nutrient profiles and nutrient bioavailability. Nutritional modifications which are within the normal range of nutrient variation might require a less extensive evaluation than those outside normal ranges.

In GM plants that were not developed to have intentionally altered nutritional value, the aim_ of the nutritional evaluation is to demonstrate that there have been no unintentional changes in the levels of key nutrients, natural toxicants or antinutrients, or in the bioavailability of the nutrients. In this case, food substitution using products from the genetically engineered plant should not adversely affect the health or nutritional status of the consumer. Implications for the population as a whole and for specific subgroups like children and/or the elderly should be considered.

2. Familiarity and Substantial Equivalence

The concept of familiarity is a key approach used in identifying and evaluating environmental risks (hazards) that may be associated with the release of a GM plant and also in the information management practices that may be needed to manage recognized risks. Familiarity considers the biology of the plant species, the trait, and the production practices (Bannon et al., 2004).

To conduct a safety assessment of a transgenic plant is advisable to elaborate a detailed monograph describing the biology of the species under review. From this, the species-specific characteristics that may be affected by the novel trait, such as the tendency to become weedy, invade natural habitats, or be otherwise harmful to the environment can be identified. It can also provide details on significant interactions between the plant and other life-forms that must be evaluated in the impact analysis. The core of such document is represented by the following components:

- Taxonomic description
- Consumption and uses of the crop plant
- Regional/national breeding, seed production, and agronomic practices
- Reproductive biology of the crop plant, details on pollination, mechanisms for dispersal of pollen and seed, and any other means of gene escape
- Occurrence and viability of intraspecific, interspecific, and intergeneric hybrids
- Details on the centres of origin and genetic diversity for the plant species
- Details on the ploidy of the cultivated crop, its progenitors and any sexually compatible species
- Distribution and ecology of related species or feral biotypes, including any evidence of weediness
- Common diseases and pests
- Potential interactions with other organisms such as pollinators, mycorrhizal fungi, animal browsers, birds, soil microbes and soil insects.

Although much experience has been gained in the environmental risk/safety assessment of the first-generation of GMOs, new environmental safety questions will arise as the volume and types of genetically products on the global market increase. This is true especially in the case genetically modified food and/or feed. The range of organisms, the number of traits and the geographical locations of deliberate releases will inevitably increase in the near future (NASS, 2002). This will lead to challenges in the application of currently accepted concepts such as familiarity. Therefore, the risk/safety assessors and the relevant institutions should keep the risk/safety assessment methods under review and should continue to exchange experiences and to collaborate towards a further improvement in risk/safety assessment and monitoring methods. One perfect tool is available to fulfil the presented task – namely the global Clearing House of the Convention on Biological Diversity (CBD, 2005), where all national and regional regulations, experiences, case studies and all information on genetically modified organisms will be continuously cumulated in the subunit named The Biosafety Clearing House.

Transfer of knowledge obtained by those countries with experience in risk/safety assessment to all those who need it, seems to be an additional challenge, which is under progress. Along with the above mentioned CBD Clearing House, which was established primarily for the Parties of the Convention, there are many other biosafety information exchange mechanisms and databases, e.g. those of OECD, UNEP, UNDP, FAO and others (such as OECD 2005).

In performing the environmental safety assessment of GM plants, the concept of familiarity is coupled in some countries with that of substantial equivalence. The latter is based on the principle that novel plants can be compared with their non-transformed counterparts that have an established history of safe use (Kuiper et al., 2002). The objective is to determine if the novel plant presents any new or greater risk in comparison with its traditional counterpart, or whether it can be used interchangeably with its traditional counterpart without negatively affecting the environment in which it is grown. The goal is to establish a relative level of safety, so that there is a reasonable certainty that no undue risk to the environment will result from the cultivation of the novel plant under anticipated conditions of production.

The OECD and FAO recommended that substantial equivalence could be taken as an important component in the safety assessment of foods and food ingredients derived from genetically modified plants intended for human consumption (OECD, 1993; FAO, 1996). This concept embodies a science-based approach in which a genetically modified food is compared with its existing, appropriate counterpart. The approach is not intended to establish absolute safety, which is an unattainable goal for any food. Rather, the aim is to ensure that the food or any substance that has been introduced into the food as a result of genetic modification is as safe as its traditional counterpart.

In most countries there are common safety concerns that must be addressed on a case-by-case basis prior to commercialization of a novel plant. As summarized by OECD and FAO these include:

- stability of the genetic modification
- gene transfer to related plants
- gene transfer to unrelated organisms
- potential for weed growth
- secondary and non-target adverse effects.

The safety assessment of a genetically modified food is directed by the results of a comparison between the GM food and its conventional counterpart. It follows a stepwise process aided by a series of structure questions. Factors taken into account in the safety assessment include:

i. identity
ii. source
iii. composition
iv. effects of processing/cooking
v. transformation process
vi. the recombinant DNA (stability of insertion, potential for gene transfer)
vii. protein expression product of the novel DNA (effects on function, potential toxicity, potential allergenicity)
viii. possible secondary effects from gene expression or the disruption of the host DNA or metabolic pathways, including composition of critical macronutrients, micronutrients, antinutrients, endogenous toxicants, allergens, physiologically active substances
ix. potential intake and dietary impact of the introduction of the genetically modified food.

The above factors are particularly pertinent to the assessment of foods derived from GM plants. When assessing the safety of foods derived from GM animals and microorganisms, other factors may need to be taken into account on a case-by-case basis.

It was noted that very little is known about the potential long term effects of growing or eating any food, including traditional ones, let alone GM foods (OECD, 1996). In many cases the wide genetic variability in the human population inevitably means that some individuals may have a greater predisposition to suffer from food-related effects. Therefore it should be accepted that for GM foods, the pre-marketing safety assessment already gives assurance that the food is as safe as its conventional counterpart.

FAO and WHO (FAO, 1996) are of the opinion that there are presently no available alternative strategies that would provide better assurance of the safety of GM foods than the appropriate use of the concept of substantial equivalence. Nevertheless, some aspects of the steps in the safety assessment process could be refined to keep abreast of developments in genetic modification technology. New methodologies, such as profiling techniques, offer the means of providing a more detailed analytical comparison (WHO, 2000).

Substantial equivalence acknowledges that the goal of the assessment is not to establish absolute safety but to consider whether the genetically modified food is as safe as its traditional counterpart, where such counterpart exists. Consumers have a right to be informed about the products they buy. However, whether or not the labelling of genetically modified foods is the most appropriate way to enable consumers to make informed choices about the food products is the subject of an active and ongoing debate in number of countries (Hoban, 1997; Codex, 1999; IFT, 2000). There is substantial variation in the labelling policies and procedures for GMOs that have been adopted by various governments.

3. Novel Foods Based on GMOS

3.1. Principles and practice of GMOs and novel food safety assessment

Firstly it is important to stress that all foods contain DNA, which is consumed and subsequently ingested in significant quantities. Dietary intake of RNA and DNA is usually in the range 0.1 – 1.0 g per man per day (Doerfler and Schubbert, 1997). With respect to a genetically modified food consumed in human diet, it must be realized /noted that this novel DNA would represent less than 1/250 000 of the total amount of DNA consumed. In view of this and the digestibility of dietary DNA, the probability of transfer of genes from GM plants to mammalian cells is extremely low. However, this possibility should be examined and the consequences of such transfer, if it were to occur, evaluated.

The transfer of plant DNA into microbial or mammalian cells under normal dietary conditions would require occurrence of a set of the following events:

i. the relevant genes in the plant DNA would have to be released, probably as linear fragments
ii. the genes would have to avoid attack of nucleases in the plant and in the gastrointestinal tract
iii. the genes would have to compete for uptake with dietary DNA
iv. the recipient bacteria or mammalian cells would have to be competent for transformation and the genes would have to keep their integrity after processing by their restriction enzymes
v. the genes would have to be inserted into the host DNA by rare repair or recombination events.

Numerous experiments have been aimed at evaluating the possibility of transfer of plant DNA to microbes and mammalian cells (Schubbert et al., 1994; Nielsen et al., 1998; Beever and Kemp, 2000). To date, there are no data on transfer of marker genes of plant DNA to the cells of these organisms.

There are many unanswered questions concerning the risks of GMOs in terms of food safety and environmental impact. Regulatory agencies claim they are formulating their standards according to science-based assessments of risk (Royal Society of Canada, 2001; Nap et al., 2003). If so, why are there such differences between the standards applied in Europe and in the USA? It seems that decision-making based on science is the only objective way to set policy in a world of diverse opinion, values and interests.

Risk analysis is a process consisting of three components:

- risk assessment
- risk management
- risk communication.

3.2. Risk assessment

Risk assessment is a process of gathering diverse data to identify possible risks in research and development involving genetically modified microorganisms, plants, animals. Risk assessment should focus on the characteristics of the product itself but at the same time take into account the techniques that are used to produce it.

There are some differences in the approach to the risk assessment framework in the USA and Europe. In the framework suggested by the U.S. National Academy of Sciences (Essential Biosafety, 2002), the following factors are recognized to be important in assessing risks of organisms with novel characteristics:

i. For the environment:

- properties of the organism and of the environment into which it may be introduced
- possibility of containing and controlling the organism
- probable effect on the environment should the organism or genetic trait persist longer than intended or spread to non-target environments
- risks to human health and the environment that are associated with introduction of organisms with novel traits.

ii. For the organism with novel trait, taking into account:

- the recipient host or parental organism that receives the new trait
- the donor organism from which the trait is derived
- the vector used to transfer the trait from donor to recipient
- the inserted or introduced trait, including potential toxicity of a gene product or its metabolites
- empirical data on the novel organism
- the intended application (contained use, release to environment or planned production)
- the potential receiving environment.

Another way of looking at risk assessment is preferred in Europe (Conner et al., 2003). In this approach hazard and exposure are distinguished as separate parameters. Hazard assessment means evaluating whether an organism can be harmful and assessing whether it is a pest or a pathogen or if it will introduce new pests or pathogens or enhance existing ones.

Exposure assessment involves evaluating the extent to which the environment or humans might be exposed to organisms with novel traits. The degree of exposure depends on the following parameters (Kuiper et al., 2002; Kleter and Kuiper, 2002):

- the route of introduction
- the survival and reproductive potential of the organisms with novel traits
- the mode and rate of dispersal beyond the site of introduction (by wind, water, insects)
- the location and size of any receptive or susceptible population.

Risk in the context of safety includes two elements:

i. hazard, an intrinsic factor (e.g. a biological, chemical or physical agent in, or condition of, food, with the potential to cause an adverse effect on health) that indicates the damage if the event occurs
ii. the probability or chance that the event will occur.

Thus, in relation to chemicals, risk is taken to be the product of hazard and chance of exposure.

In relation to quarantine, it is the product of potential damage by the pest and chance of introduction.

It can be concluded, the risk assessment is a scientifically based process consisting of the following steps:

i. hazard identification
ii. hazard characterization
iii. exposure assessment
iv. risk characterization.

Hazards, and the chance of those hazards occurring, are thereby studied and models constructed to predict the risk. The two components of risk both contain a measure of uncertainty, and it is this measure of uncertainty that is the subjected of many discussions. For example, there is some doubt as to whether risk estimation methodologies used for related purposes (e.g. pesticide residues in food and pest introduction) have sufficient predictive value for GMOs. In particular, the hazard component of risk analysis is subject to close scrutiny (Kuiper et al., 2001).

3.3. Risk management

Risk assessment is the process of weighing policy alternatives in consultation with interested parties, considering risk assessment and other factors relevant for the protection on consumers' health and for the promotion of fair trade practices as well as, if necessary, selecting appropriate prevention and control options (FAO 2001).

Environmental hazard is surely less easy to quantify than health hazard. There have been many attempts to monitor some consequences of the effects of GMOs but this task is too complicated, as in most cases the potential change (especially in the environment) can be seen only after very long period of time (Conner et al., 2003). Moreover, only long-term experience can show if both risk assessment and risk management have been successful. When a sound risk management strategy is applied to environmental problems, as distinct from safety problems, it will begin by describing a problem and the goals, objectives and values to be pursued by solving that problem. An analysis of alternatives is then carried out to consider as many solutions as possible. Rather than narrowing the analysis, this allows the creation of new options or combinations of options. When the benefits and drawbacks of a wider range of solution scenarios can be compared, fuller participation by the concerned society can be better assured.

The type of risk management for contained use and/or planned introductions of genetically modified organisms depend on the organism involved and on the intended application. The process involves reviewing alternatives and selecting the most appropriate regulatory actions based on the findings of the risk assessment. Measures to be taken to minimize risk include physical and biological containment. Doyle and Persley (1996) suggest posing the following questions:

i. What are the risks?
ii. How probable is it that they will occur?
iii. How serious is the damage if they occur?
iv. What can be done to minimize the risks and contain the damage?
v. Do the benefits outweigh the risks?

The term containment is used to describe safe methods for maintaining control over the distribution of GMOs in the laboratory and in the environment into which they are introduced. The purpose of containments is to minimize unnecessary exposure of laboratory workers and the environment to potentially hazardous organisms.

Biological containment of microorganisms principally involves the use of specific combinations of vector and host in such a way that the probability of the transfer of a vector to an unintended host and the subsequent survival of the host-vector combination in the environment is limited. The growth of plants which require special environmental conditions for their survival, like biological containment, can be achieved in either in either the greenhouse or field. Similar results can be obtained with studies using contained animal facilities (Kleter and, Kuiper, 2002).

Physical containment involves physical constraints on the movement of organisms of uncertain risk or potential hazard. The aim of physical containment is to prevent inappropriate exposure of humans and the environment to organisms. Physical containment is achieved by following the principles of good laboratory practice, occupational safety, and hygiene – together with involvement of well-qualified and competent personnel who follow safe, standard procedures and with keeping a working place designed to prevent the unintended spread to the environment of genetically modified organisms.

Risk management is an attempt to select the most appropriate regulatory strategy or action to prevent or at least to minimize the potential harmful effect of GMOs. It integrates the results of risk assessment with technical, social, economic, and political concerns. Carried out by regulatory agencies under legislative mandates, risk management is a decision-making process that determines reasonable control costs by requiring value judgments that compare potential risks and benefits (Dale et al., 2002). Risk management of the products of biotechnology involves the comparison of their benefits and risks with those associated with the products they replace. The process of risk management follows the process of risk assessment at each stage of development and use of a genetically modified organism and its products (Dale et al., 2002).

3.4. Risk communication

Risk communication is the interactive exchange of information and opinions among assessors, risk managers, consumers, industry, the academic community and other interested parties throughout the risk analysis process. The information exchange concerns risk-related factors and risk perceptions, including the explanation of risk assessment findings and the basis of risk management decisions. It is vitally important that risk communication with the public comes from credible and trusted sources.

In the framework of the Cartagena Protocol on Biosafety, a Biosafety Clearing House was established as a part of the clearing-house mechanisms under the Convention of Biological Diversity (CBD-BCH 2002). The principal tasks of the Biosafety Clearing House (BCH) are as follows:

i. Facilitate the exchange of scientific, technical, environmental and legal information on, and experience with, genetically modified organisms.
ii. Assist the Parties (to the Protocol) to implement the Cartagena Protocol on Biosafety, taking into account the special needs of developing countries as well as countries that are centres of origin and centres of genetic diversity.

The Biosafety Clearing House will serve as a means through which information relevant to the implementation of the Protocol is made available. It will also provide access to other international biosafety information exchange mechanisms. Each Party to the Protocol will make available to the BCH information relevant to the biosafety issues such as:

i. Existing laws, regulations and guidelines for implementation of the Protocol, as well as information required for the advance informed agreement procedure.
ii. Bilateral, regional and multilateral agreements and arrangements.
iii. Summaries of its risk assessments or environmental reviews of GMOs generated by its regulatory process including relevant information regarding products and processed materials that are of GMO origin, containing detectable novel combinations of replicable genetic material obtained through the use of modern biotechnology.
iv. Final decisions regarding the importation or release of GMOs.
v. Reports submitted on monitoring of GMOs interactions with environment and human health.

The Biosafety Clearing House will use electronic and other systems for the exchange of information relevant to the Cartagena Protocol on Biosafety. The BCH may draw on a wide variety of existing mechanisms as those of the Organization for Economic Cooperation and Development -OECD (Biotrack http://www.oecd.org/biotrack), United Nations Industrial Development Organization -UNIDO - Biosafety Information Network and Advisory Service (BINAS http://binas.unido. org/binas/), International Centre for Genetic Engineering and Biotechnology -ICGEB (Biosafety Bibliographic Database http://www.icgeb.org/-bsfdata1.htm), United Nations Environment Programme – UNEP, Microbial Strain Data Network

(MSDN), Information Resource for the Release of Organisms (IRRO) http://panizzi.shef.ac.uk/msdn/.

4. Ethical Questions Related to Safety Issues

Many ethical questions about modern biotechnology are related to the potential consequences of applying GMOs, or other new technologies, to intensify food production and achieve greater food security (FAO 2001). As GMOs enter the food and feed supply chains, they will be increasingly released into ecosystems, including agro-ecosystems with the potential for unforeseen consequences.

Some ethical aspects of GMOs fall within the context of the right to adequate food, which is derived from the Universal Declaration of Human Rights. FAO and other international organizations have also declared in several documents (World Food Summit, Rome Declaration on World Food Security), the right of anyone to an adequate food supply. Other important human rights principles that could bear upon the assessment of GMOs, include the rights to informed choice and to democratic participation.

4.1. Modified foods on market

Foods are complex mixtures of compounds characterized by a wide variation in composition and nutritional value. Although priorities vary, food safety is a concern among consumers in all countries. They would like assurances that genetically modified products reaching the market have been adequately tested and that these products are being monitored to ensure safety and to identify problems as soon as they emerge. The nature and complexity of food products implies / means that research on the safety of genetically modified foods is still thought to be more difficult to carry out than studies on components such as pesticides, pharmaceuticals, industrial chemicals and food additives (FAO 2001). Through the Convention on Biological Diversity, Codex Alimentarius Commission and other international fora, countries discuss standards for GMOs and ways to ensure their safety. One approach, which is being used widely in assessing the risks of GMOs, derives from the concept of substantial equivalence (Kuiper et al., 2002). Numerous foods based on modern biotechnology and derived from microorganisms, plants and/or animals are entering the market in several countries.

The assessment of genetically modified food with enhanced nutritional properties should focus on the simultaneous characterization of inherent toxicological risks and nutritional benefits (Kuiper et al., 2001). This requires an integrated multidisciplinary approach, incorporating molecular biology, toxicology, nutrition and genetics (Figure. 1).

As shown in Figure.1, the issues to be addressed are:

i. evidence for nutritional/health claims and target population
ii. toxicological and beneficial dose ranges of selected compounds
iii. impact on overall dietary intake and associated effects on consumers

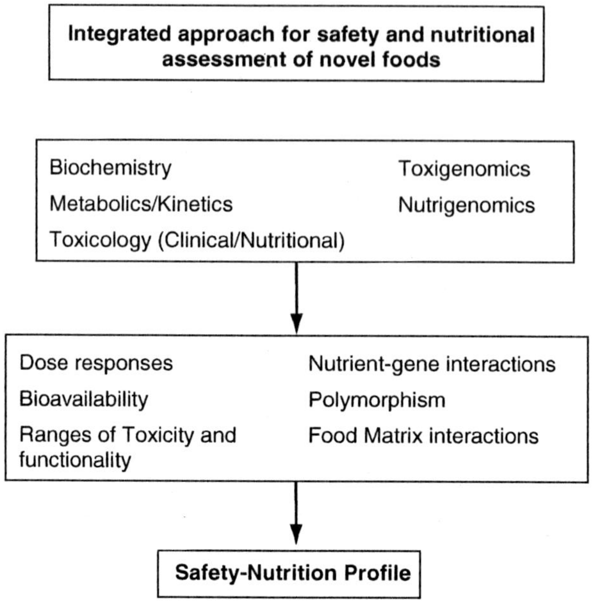

FIGURE 1. Integrated approach for safety evaluation of genetically modified food (Kuiper et al., 2001).

iv. interaction between food constituents and food matrix effects
v. possibilities for effective post-market surveillance.

5. MON 810, A Case Study on Genetically Modified Maize

The European Corn Borer (Ostrinia nubilalis; ECB) causes massive damage every year to cultivated maize (Essential Biosafety 2001). As reported by some countries (e.g. in Switzerland) the ECB is controlled on approx. 5-20% of maize fields through the release of the parasitic wasp Trichogramma, in other countries (like Germany) are as a countermeasure in addition to the release of Trichogramma applied as well an insecticide (Decis®), to combat the ECB. As permission to use insecticides will in future be limited, it is necessary to find new ways to stop the negative effects of the ECB. One possibility is seen in using insect-resistant maize lines (Essential Biosafety, 2002).

Toxins produced by the bacteria *Bacillus thuringiensis*, are effective in killing the ECB by provoking death through lysis of the gut lining. The genes from *B. thuringiensis* have now been cloned and inserted directly into the corn genome for expression in various plant parts, to protect the plant against the ECB. The commercialized transgenic lines vary in their expression of the toxin: NOVARTIS Bt 176 maize expresses the Bt toxin in the kernels, leaves and pollen. NOVARTIS Bt 11 expresses toxin in the kernels and MONSANTO's MON 810 expresses toxin in all parts, except in the pollen (Essential Biosafety 2002). Bt toxins are, depending on the strain of *B. thuringiensis* from which they originate, very specific in attacking primarily Lepidoptera and Coleoptera.

Under the trade name YieldGard®, a maize line MON 810 was developed through a specific genetic modification securing resistance to attack by the European corn borer (ECB, Ostrinia nubilalis), a major insect pest of maize in agriculture. The novel variety produces a truncated version of the insecticidal protein, Cry1Ab, derived from *Bacillus thuringiensis*. Delta-endotoxins, such as the Cry1Ab protein expressed in MON 810, act by selectively binding to specific sites localized on the brush border midgut epithelium of susceptible insect species. Following binding, cation-specific pores are formed that disrupt midgut ion flow and thereby cause paralysis and death. Cry1Ab is insecticidal only to lepidopteran insects, and its specificity of action is directly attributed to the presence of specific binding sites in the target insects. There are no binding sites for delta-endotoxins of *B. thuringiensis* on the surface of mammalian intestinal cells, therefore, livestock animals and humans are not susceptible to these proteins.

This GM line of maize received its first regulatory approval in the USA in 1995 and has since been approved for environmental release and use in livestock feed and/or human food from a number of countries in America, Asia, Australia and Europe (Table 2).

Maize, taken generally, is the third most important crop in the world after rice and wheat; with an annual production over 600 million tones (FAOSTAT Database Records,). It has been grown as a commercial crop in many countries worldwide, starting some 8000 years ago in Mexico and Central America (Mexico, 1994; OECD, 2002). In Europe it has been known for 500 years.

Maize is naturally cross-pollinated and until 1925 mainly open pollinated varieties were cultivated. To produce hybrid seed the tassels are removed from the plants prior to pollen shedding, so that only one sort of pollen will be received by the silks (OECD, 2002). The hybrid plants grown from this seed give more vigorous growth and higher yields. The two leading maize producers are USA and China.

TABLE 2. Regulatory approval status of insect-resistance maize MON 810

Country	Approved for Environment	Approved for food/feed
Argentina	1998	1998
Australia		2000
Bulgaria	2000	2000
Canada	1997	1997
European Union	1998	1998
Japan	1996	1997
Russia		2000
South Africa	1999	1999
Switzerland		2000
United Kingdom		1997
United States of America	1995	1996

The behaviour and characteristics of MON 810 have been studied in a range of field experiments in USA and Europe since 1992 (Essential Bisafety, 2001). The assessment of the interaction of this maize line with the environment has included studies on:

- susceptibility to insects and diseases
- survival capacity (volunteers)
- seed multiplication capacity (yields)
- Cry1Ab protein expression in leaves and grain
- seed composition analysis
- safety for birds
- safety for mammals.

Observation and field trials have demonstrated that the mode and reproduction of MON 810 are typical of other maize varieties (Essential Biosafety, 2002). Expression levels of the Cry1Ab, CP 4 EPSPS, GOX and NPTII proteins are shown in Table 3.

Disease and pest susceptibilities of MON 810 were tested and no differences in agronomic quality, disease, or insect susceptibility other than European corn borer control were detected between MON 810 and non-transgenic plants (Essential Biosafety 2002). Diseases observed included northern leaf blight (*Excerohilum turcicum*), southern leaf blight (*Bipolaris maydis*), bacterial leaf blight (*Erwinia stewartii*), common corn snut (*Ustilago maydis*), maize stripe virus and common maize rust (*Puccinia sorghi*).

In 1990, the FAO and WHO agreed that the comparison of a final product with one having an acceptable standard of safety provides an important element of safety assessment (OECD, 2002). The OECD further elaborated this concept and advocated the approach to safety assessment based on substantial equivalence as being the most practical one to address the safety of foods and food components derived through modern biotechnology. It was concluded that the safety assessment of genetically modified foods requires an integrated and stepwise, case-by-case

TABLE 3. Summary of specific protein levels measured in tissues of MON 810

PROTEIN	Protein concentration in µg/g of fresh weight			
	LEAF	GRAIN	WHOLE PLANT	POLLEN
Cry1Ab	9.35	0.31	4.15	0.09
CP4 EPSPS	nd	nd	nd	na
GOX	nd	nd	nd	na
NPTII	na	na	na	na

Abbreviations:
Cry1Ab – insecticidal protein of the delta-endotoxin group, product of cry1Ab gene
CP4 EPSPS – product of the epsps marker gene (plant selectable marker enabling the identification of corn cells with introduced cry1Ab gene)
GOX – product of glyphosate oxidase encoding gene gox (another selectable marker)
NPTII – enzyme neomycin phosphotransferase II, product of gene nptII, the bacterial selectable marker conferring resistance to kanamycin or neomycin
nd – not detected
na – not analyzed

approach, which can be aided by a structured series of questions. A comparative approach focusing on the determination of similarities and differences between the genetically modified food and its conventional counterpart aids in the identification of potential safety and nutritional issues and is considered the most appropriate strategy for the safety and nutritional quality of genetically modified foods. It should be seen as a key step in the safety assessment process although it is not a safety assessment in itself, it does not characterize hazard.

An important element in the determination of substantial equivalence is elaboration of compositional comparison, where a comparison of critical components can be carried out at the level of the food source (i.e. species) or the specific food product (FAO, 1996). Critical components are determined by identifying key nutrients and key toxicants and antinutrients for the food source in question. The comparison of critical components should be between the modified variety and non-modified comparators with an appropriate history of safe use. The data for the non-modified comparator can be the natural ranges published in the literature for commercial varieties or those measured levels in parental or other edible varieties of the species. The comparator used to detect unintended effects for all critical components should ideally be the near isogenic parental line grown under identical conditions.

Yield comparisons for MON 810 and representative controls were assessed in 9 different locations of USA (Essential Biosafety, 2001; Essential Biosafety, 2002). A comparison of a non-transgenic hybrid with the same hybrid in which one parent was a backcross-derived MON 810 line, showed no significant difference between these hybrids in yield.

6. Future Trends

One of the difficulties in discussing the benefits and concerns that attend any technology is consideration of the rapid and extensive advances. A few generations ago, most of our population lived on farms or in small towns. Nearly all of our food was grown at home, or nearby, and processed in our neighbourhood. The technologies were simple and available to everybody; therefore all had confidence in it. The present technology is no longer simple and familiar to all. It is complex and, to most consumers, unknown. Connected with this, is the current complex debate concerning the benefits and concerns of using unfamiliar technologies including, of course, modern biotechnology. Effective progress can be made through education in close cooperation with the science.

The role of science and new genetic technologies in food production is an area of considerable controversy and concern to many people across the world. It is an area in which there are strongly conflicting views and opinions, in which scientific progress and individual morals and beliefs are often opposed and in which the science itself is sometimes uncertain and open to interpretation. However, food security is also one of the major challenges facing humanity, and new genetic technologies have real potential to ameliorate the current situation – a world in which some 900 million people lack access to sufficient nutritious food at affordable prices (FAO, 2004). Hence, the scientific community has a duty to develop responsibly and to explain its research in this area. Presently the scientists and researchers and teachers in this branch of modern biotechnology are not informing the public in a manner commensurate with the volume and quality of the scientific data and analysis available. The scientific community could play a more active and better organized role in raising public awareness about emerging modern biotechnology and what bit means for different societies, in terms of choices, risks and benefits (ICSU, 2003).

Much data has been generated over the past decade on the behaviour of genetically modified organisms in various environments. It would be helpful to guide future regulatory decisions if more of this data was made available publicly. Additional, publicly funded research that addresses key gaps in present knowledge would be valuable to inform the debate about the use of genetic engineering. The value of this research would be increased if the key questions are framed in an environment that reflects the concerns of the public, policy makers, politicians, regulators, and all of this at both national and international levels.

In the regulatory area, additional research is necessary to assist in the continued development of regulatory approaches that keeps abreast of new scientific developments. For example, there is a need for the continued development of food safety assessment methods to deal with emerging products such as nutritionally enhanced foods and other complex traits controlled by multiple genes. There is also a need for the development of internationally agreed standards for the assessments of environmental risks and benefits of genetically modified organisms.

A genetically modified product may be the subject of both a food safety assessment and an environmental safety assessment. In many countries these two different procedures are the responsibility of different ministries or agencies. In many cases the same or similar information is used in both types of assessment, such as information on the biology of the host organism, the molecular characterization of the inserted gene and the properties of the modified organism. Even though the different ministries and agencies perform their assessments independently, a better understanding of the way in which similar considerations can be used in the two assessments may lead to an increase of the effectiveness of the overall risk assessment of the product.

The broad range of modern biotechnology applications in agriculture could contribute more towards improving the efficiency and sustainability of agriculture in emerging economies. Currently available applications of new biotechnology could

- improve the efficiency of plant breeding
- develop new diagnostics and vaccines for the control of pests, parasites and diseases in crops, trees, livestock and fish
- generate disease-free planting material with substantial increases in productivity.

Genetically modified crops also offer the promise of contributing more towards food security and poverty reduction. New varieties of crops with useful traits may result from public or private investments. However, as outlined by ICSU (2003), the successful deployment of new products will require:

- public acceptance of new products
- an enabling policy and regulatory environment, including safety assessments and intellectual property management
- investment in research and development
- local private sector development for distribution and marketing of seeds and other genetically modified products.

Regulatory differences are hampering the move towards global norms in the field of genetically modified organisms and products. The latent public demand for strict regulatory control in synergy with the constant process of maintaining and improving the quality of the environment and human health by avoiding even potential injury by the new generations of GMOs could lead to a regulatory solution that is accepted globally.

7. References

Bannon, G., Astwood, J., Goodman, R., Hefle, S., and Taylor, S., 2004, Allergy assessment for food biotechnology, in: Bhalgat, M.K., Ridley, W.P., Felsot, A.S., Seiber, J.N.(eds.), *Agricultural Biotechnology. Challenges and prospects.* ACS Symp. Series 866.Oxford University Press 2004, pp. 151-163.

Beever, D.E. and Kemp, C.F., 2000, Safety issues associated with the DNA in animal feed derived from genetically modified crops. *Nutrition Abstract Reviews, series Livestock Feeds and Feeding* **70**:175-182 CBD 2005: www.biodiv.org/chm.

CEC (Commission of the European Communities) 2000, Communication from the Commission on the Precautionary principle. Brussels, Belgium (http://europa.eu.int/comm/dgs/health_consumer/ library/pub/pub07_en.pdf)

Codex, 1999, Report on the Twenty seventh session of the Codex Alimentarius Committee on food labelling. ALINORM 99/22A. Codex Alimentarius Commission, Rome.

Conner, A.J., Glare, T.R., and Nap, J-P., 2003, The release of genetically modified crops into the environment. Part II. Overview of ecological risk assessment. *The Plant Journal* **33**:19-46

Dale, Ph.J., Clarke, B., and Fontes, E.M.G., 2002, Potential for the environmental impact of transgenic crops. *Nature Biotechnology* **20**:567-586.

Doerfler, W. and Schubbert, R., 1997, *Fremde DNA im Säugersystem. Deutsches Ärzteblatt* **94**:51-52.

Doyle, J.J. and Persley, G.J. (eds.), 1996, Enabling the safe use of biotechnology. Principles and practice. Environmentally Sustainable development Studies and Monaograph Series No. 10, The World Bank, Washington, D.C., 75 pp.

Essential Biosafety, 2001, The latest scientific and regulatory information for genetically modified and other novel crops and foods. AGBIOS, Agriculture and Biotechnology Strategies, Inc., Merrickville, Canada.

Essential Biosafety, 2002, The latest scientific and regulatory information for genetically modified and other novel crops and foods. AGBIOS, Agriculture and Biotechnology Strategies, Inc., Merrickville, Canada.

FAO, 1996, Biotechnology and food safety. Report of joint FAO/WHO consultation. FAO Food and Nutrition Paper 61, FAO Rome.

FAO, 2001, Genetically modified organisms, consumers, food safety and the environment. FAO Ethic Series 2, FAO Rome.

FAO, 2004, The state of food insecurity in the world. FAO Rome.

FAO, 2005, Summary Document to Conference 12 of the FAO Biotechnology Forum. Feb. 2005: http://www.fao.org/biotech/logs/C12/summary.htm.

FAOSTAT, Database Records, http://apps.fao.org./page/form?collection

Hoban, T.J., 1997, Consumer acceptance of biotechnology: An international perspective. Nature Biotechnology **15**:232-234.

ICSU, (International Council for Science) 2003, New genetics, food and agriculture: Scientific discoveries – Societal dilemmas. ICS, 56pp.

IFT, 2000, Institute of Food Technologists Expert Report on Biotechnolog and Foods, Food Technology vol. 54, No. 9.

Kleter, G.A. and Kuiper, H.A., 2002, Cosiderations for the assesment of the safety of genetically modified animals used for human food or animal feed. *Livestock production Science* **74**:275-285.

Kuiper, H.A., Kleter, G.A., Noteborn, H.P.J.M., and Kok, E.J., 2001, Assessment of the food safety issues related to genetically modified foods. *The Plant Journal* **27**:503-528.

Kuiper, H.A., Kleter, G.A., Noteborn, P.J.M., and Kok, E.J., 2002, Substantial equivalence – an appropriate paradigm for the safety assessment of genetically modified foods? *Toxicology* **181-182**:427-431.

Mexico, D.F., 1994, Maize seed industries revisited: emerging roles of the public and private sectors. World Maize Facts and Trends 1993/1994 CIMMYT.

Nap, J-P., Metz, P.L.J., Escaler, M., and Conner, A.J., 2003, The release of genetically modified crops into the environment. Part I. Overview of current status and regulations. *The Plant Journal* **33**:1-18.

NASS, 2002, National Agricultural Statistics Service of USDA: Prospecive plantings. March 2002. http://usda.mannlib.cornell.edu/reports/nassr/field/pcp-bbp/pspl0302.pdf

Nielsen, K.M., Bones, A.M., Smalla, K., and Elsas, J.D., 1998, Horizontal gene transfer from transgenic plants to terrestrial bacteria – a rare event? *FEMS Microbiology Reviews* **22**:79-103.

OECD, 1993, Safety considerations of foods derived by modern biotechnology. Concept and principles. OECD Paris.

OECD, 1996, Food safety evaluation. Workshop report, Oxford, UK, ISBN 92-64148671.

OECD, 2002, Consensus Document on compositional considerations for new varieties of maize (Zea mays): Key food and feed nutrients, antinutrients and secondary plant metabolites. Series on the Safety of Novel Foods and Fedds, OECD Environmental Health and Safety Publication No. 6, OECD Paris.

OECD, 2005, www.icgeb.org/biosafety/

Persley, G.J., 1990, Beyond Mendel's Garden: Biotechnology in the service of world agriculture. CAB International, Wallingford.

Royal Society of Canada, 2001, Elements of precaution: Recommendations for the regulation of food biotechnology in Canada. www.bio-scope.org

Schubbert, R., Lettman, C., and Doerfler, W., 1994, Ingested foreign (phage M13) DNA survives transiently in the gastrointestinal tract and enters the blodstream of mice. *Mol.Gen.Genet.* **242**:495-504.

Tourte, Y., 2003, Genetically Modified Organisms. Transgenesis in Plants. Science Publishers, Inc. Enfield, NH, USA.

WHO, 2000, Safety aspects of genetically modified foods of plant origin. Report of a Joint FAO/WHO Expert Consultation on Foods Derived from Biotechnology. WHO Geneva.

7
Nutritional Strategies Targeting the Beneficial Modulation of the Intestinal Microflora with Relevance to Food Safety: the Role of Probiotics and Prebiotics

KONSTANTINOS C. MOUNTZOURIS*

1. Introduction

The concern and fear raised by recent zoonoses and food poisoning cases (e.g. bovine spongiform encephalopathy, Creutzfeldt-Jacob disease, dioxins in poultry *E. coli* 0157:H7 outbreaks) and the increasing consumer awareness and sensitivity on food safety and quality issues has led the agri-food industry to seek new standards and more effective measures in order to safeguard the quality of food and feed throughout the food chain.

Worldwide, most foods are produced geographically far away from the place of consumption. In addition, there is a significant lag time between production and consumption of these foods. To prevent outbreaks of foodborne illness, the microbiological safety of foods is therefore of primary concern.

Thus, certain principles and control measures need to be observed in order to ensure safety of the raw materials, their proper processing and handling by the food industry (manufacturers, distributors, retailers) as well as by the consumers. In this sense, the implementation of the HACCP (Hazard Analysis Critical Control Point) concept, backed up by good manufacturing practice (GMP) and the use of rapid detection methods, is enhancing the safety assurance of foods by reducing the risk of microbial spread from farm to fork (Jay, 1992).

However food borne illnesses do occur, and in developed countries it is usually pathogenic bacteria or their toxins that contaminate food and cause host infection

*Department of Animal Nutrition, Agricultural University of Athens, Iera Odos 75, 118 55, Athens, Greece

and disease. Among the most common pathogenic microorganisms responsible for food poisoning cases are: *Staphylococcus aureus*, *Bacillus cereus*, *Clostridium botulinum*, *Clostridium perfringens*, *Salmonella* spp., *Campylobacter jejuni*, *Escherichia coli*, and *Listeria monocytogenes* (Southgate, 2000).

Apart from acute gastroenteritis cases reported, some microorganisms are thought to be behind the aetiology of many chronic diseases of the digestive system such as gastritis (i.e. *Helicobacter pylori*), duodenal ulceration, irritable bowel syndrome and ulcerative colitis (Marteau et al., 2002; Madden and Hunter, 2002).

Most known bacterial pathogens have some mechanism of attachment to host cells in order to establish at a body site, a condition which may then proceed to infection and ultimately disease. Microbial establishments at any site of the body that do not cause infection are characterised by the term "colonisation". Colonisation in nature is evidenced by a profound interaction between various microorganisms and humans or animals. (Salyers and Whitt, 1994).

Vast numbers of prokaryotic and eukaryotic microbial cells, estimated in the order of 10^{14} cells, colonise the various body surfaces and gastrointestinal (GI) tracts of humans and animals. In fact, the GI tract is the most heavily colonised organ of the body. From the various anatomical regions of the gastrointestinal tract, extending from the oral cavity to the anus, the large intestine (i.e. colon) is the most diversely populated region of the intestine. The colon contains bacteria from well over 400 different species and in concentrations ranging from 10^{10} to 10^{12} microbial cells per g of intestinal contents (Berg 1996). The type and number of microbial species that persist and colonise different intestinal sites is thought to depend on a number of: (a) environmental and dietary factors, (b) microbial associated factors and (c) host related factors including genetic factors. Generally, physicochemical conditions in the digestive system such as gut motility, pH, redox potential, nutrient supplies and host secretions (e.g. hydrochloric acid, digestive enzymes, bile and mucous) play an important role in the gut colonisation process (Savage, 1977; Mackie et al., 1999; Sanderson, 1999).

As mammalian cells of the adult human are approximately 10^{13}, they are outnumbered by a factor of around 10 by colonising microbial cells (Savage, 1977). Therefore, a profound influence of the latter on the metabolic, nutritional, immunological and physiological processes of the host can be postulated.

In the rapidly growing sector of functional foods a lot of scientific and commercial effort is being directed at the elucidation of the physiological effects that foods and various other bioactive ingredients have on the GI tract. The potential beneficial metabolic (i.e. systemic) effects that result from their consumption are a keen area of interest.

The role of intestinal microflora in health and disease is rapidly gaining recognition (Macfarlane and McBain, 1999). There is evidence that the composition and activities of GI microflora may have beneficial and pathogenic outcomes for the host.

Recent scientific data support the view that the composition and activities of the intestinal microflora can be modulated through diet (Gibson and

Roberfroid, 1995; Salminen et al., 1998). For example, there is accumulating evidence that dietary modulation using functional foods (e.g. probiotics and prebiotics) could be beneficial for the host by effecting a health-promoting modification of the composition and activities of the intestinal microflora (Salminen et al., 1998).

In this sense, foods that contain ingredients that promote a good gastrointestinal environment, that is perceived as beneficial and health promoting for the consumer, hold much promise for the development of "safer" foods. From a food safety perspective, such products could contribute to reduction of gastrointestinal infections through the fortification of the beneficial members (e.g. lactic acid bacteria) of GI microflora.

This chapter starts with a presentation of the composition and metabolic activities of the GI microflora that are of relevance in health and disease. Later, the chapter describes probiotics and prebiotics as nutritional approaches that can be used to improve gastrointestinal condition and health and therefore can be considered as relevant for an overall food safety perspective. Finally the application of probiotics and prebiotics is considered under the context of emerging functional foods.

2. Intestinal Microflora Composition and Metabolic Activities

The complex microbial ecosystem of the GI tract of humans and animals consists of at least 50 bacterial genera that are represented by more than 400 different species (Berg, 1996). These numbers should be considered as indicative only as there is evidence of many more organisms that cannot be cultured and identified by the traditional culture methods and techniques. It is expected that with the advent of high fidelity techniques in molecular biology, our current knowledge of the overall microbial ecology and diversity occurring in the GI tract will rapidly advance.

The development of the microbial ecology and diversity in the GI tract is a complex and long term process which depends on many factors that have to do with the environment, host physiology, nutrition, genetics and the microbes themselves.

Shortly after birth, the previously sterile infant gut starts being colonised by an array of bacteria that belong to facultative anaerobes and strict anaerobes (Mountzouris et al., 2002a) given in Table 1.

The introduction of solid foods in the infant diet triggers a period of change in their intestinal microflora that gradually becomes more complex and starts to approximate that of the adult human (Table 2) by the second year of life (Benno, 1984; Mountzouris et al., 2002a).

Despite the fact that the intestinal microflora in the human large intestine could be characterised as generally stable, it is clear from Table 2 that there exists a large diversity between different individuals.

TABLE 1. Major bacterial genera and species colonising the GI tract of human infants

Facultative anaerobes	Strict anaerobes
Escherichia (E. coli)	*Bifidobacterium (B.breve, B. longum, B. adolescentis, B. bifidum, B. infantis)*
Staphylococcus (S. aureus, S. epidermidis)	*Bacteroides (B. fragilis, B. distasonis B. vulgatus, B. ovatus, B. thetaiotaomicron B. uniformis)*
Streptococcus (S. faecalis, S. faecium)	*Clostridium (C. perfringens, C. difficile, C. butyricum, C. tertium, C. paraputrificum)*
Enterobacter (E. cloacae)	*Lactobacillus (L. acidophilus, L. fermentum, L. brevis, L. salivarious, L. plantarum)*
Klebsiella (K. pneumoniae)	*Eubacterium (E. aerofaciens, E. lentum, E. rectale)*
Proteus (P. mirabilis)	*Veillonellae (V. parvula)*
Citrobacter (C. freundii)	*Peptococcus (P. saccharolyticus)*
Pseudomonas (Ps. Aeruginosa)	*Peptostreptococcus (P. productus, P. anaerobius)*

TABLE 2. Predominant bacterial genera in the human large intestine (From Salminen et al., 1998)

			Concentration ($\log_{10}$ /g dry weight of faeces)	
Bacteria	Gram reaction	Morphology	Mean	Range
Bacteroides	Gram –	Rods	11.3	9.2 – 13.5
Eubacteria	Gram +	Rods	10.7	5.0 – 13.3
Bifidobacteria	Gram +	Rods	10.2	4.9 – 13.4
Clostridia	Gram +	Rods	9.8	3.3 – 13.1
Lactobacilli	Gram +	Rods	9.6	3.6 – 12.5
Ruminococci	Gram +	Cocci	10.2	4.6 – 12.8
Peptostreptococci	Gram +	Cocci	10.1	3.8 – 12.6
Peptococci	Gram +	Cocci	10.0	5.1 – 12.9
Methanobrevibacter	Gram +	Cocco bacilli	8.8	7.0 – 10.5
Desulfovibrios	Gram –	Rods	8.4	5.2 – 10.9
Propionibacteria	Gram +	Rods	9.4	4.3 – 12.0
Actinomyces	Gram +	Rods	9.2	5.7 – 11.1
Streptococci	Gram +	Cocci	8.9	3.9 – 12.9
Fusobacteria	Gram –	Rods	8.4	5.1 – 11.0
Escherichia	Gram –	Rods	8.6	3.9 – 12.3

There is evidence that the composition and activities of the gastrointestinal microflora can have beneficial and pathogenic outcomes for the host. Gibson and Roberfroid (1995) have generally split the predominant human faecal bacteria into groups that can exert harmful or beneficial effects on the host or both.

Beneficial health-promoting effects could arise from: (a) energy salvage from the fermentation of dietary carbohydrates and proteins reaching the colon (Cummings and Macfarlane 1991), (b) the synthesis of vitamins, primarily of the B and K group (Tamura, 1983; Berg, 1996), (c) production of short chain fatty acids (SCFA) as bacterial metabolic end products. SCFA can exert an antipathogenic effect by lowering the pH of the intestinal lumen thereby facilitating water absorption by the colon [Tamura 1983; Gibson and Roberfroid, 1995), (d) production of antimicrobial compounds (Kim, 1993; Yildirim and Johnson, 1998) and (e) enhancement of the gut barrier function [Berg, 1996; Cunningham-Rundles and Lin, 1998; Cebra, 1999) via the mechanisms shown in Figure 1.

However, the intestinal microflora can also have detrimental effects upon health of the host. Such pathogenic effects could arise from: (a) a conversion of non-carcinogens or pre-carcinogens to carcinogens and the production of toxic end products of protein metabolism such as ammonia, phenolic compounds and amines, (b) intestinal overgrowth following disruption of the gastrointestinal microecology by oral antibiotics, shock or other conditions and (c) opportunistic infections by passage through the mucosal barrier to the mesenteric lymph nodes and other extra intestinal sites (known as bacterial translocation) especially in immuno-compromised or traumatised hosts (Cummings and Macfarlane, 1991; Gibson and Roberfroid, 1995).

Generally, bacteria belonging to the genera of *Lactobacillus* and *Bifidobacterium* are perceived as beneficial bacteria and are usually regarded as probiotics (see later). On the contrary, bacteria thought to exert harmful effects on the host usually belong to the genera of *Clostridium*, *Staphylococcus*, *Pseudomonas*, *Proteus* and *Veillonella*. Finally, the genera of *Bacteroides*, *Eubacterium*, *Escherichia*, and *Enterobacter* are thought to include bacteria that may have either beneficial or harmful effects on the host (Gibson and Roberfroid, 1995).

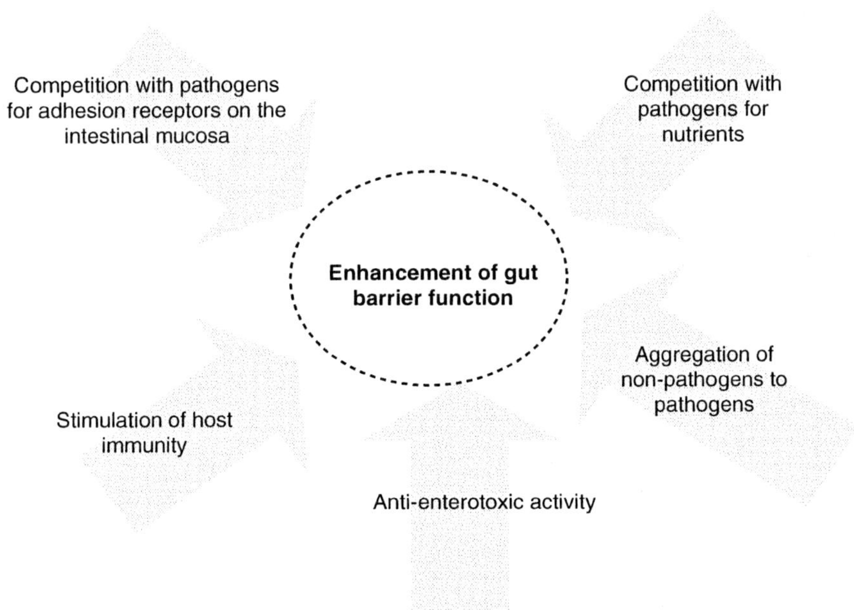

FIGURE 1. Mechanisms involved in the enhancement of gut barrier function by the host intestinal microflora.

The more our knowledge on gut microbial ecology expands the more likely it is that the beneficial or pathogenic nature of bacteria at species and even strain level will be identified and therefore the above classification should be regarded as indicative only.

3. Probiotics

Probiotics represent a nutritional approach whereby maintenance of a healthy GI environment and improved intestinal function is pursued through the consumption of beneficial microorganisms. Dietary consumption of microorganisms dates many centuries ago. Many bacteria, fungi and yeasts have a well documented history of use in human food. Traditionally microorganisms have been used in fermentation of milk, wine, meat, cheese, bread, and vegetables. Such fermented foods have been globally consumed for a very long time without indications considered to be harmful to health of consumers.

The health benefit of consuming live microbes was given a scientific basis almost a century ago by Metchnikof (1907), who attributed the longevity of Bulgarian farmers to the regular consumption of yoghurt and the elevated numbers of microbes contained therein. Since that historical association of microbes, later known to be lactic acid bacteria (LAB), in fermented milk and longevity, accumulating scientific knowledge on the health benefits linked with the consumption of live microorganisms gave birth to the development of the contemporary probiotic concept.

The term probiotic is derived from the Greek language and means for-life (probiotic). It was firstly introduced by Lilly and Stillwell (1965), to describe substances produced by one micro organism that stimulated the growth of other microorganisms. Later the definition was linked with the effects of microbes on health and well being of humans and animals. According to Fuller (1989), probiotics are live microbial feed supplements, which beneficially affect the host animal by improving its intestinal microbial balance. More recently, a joint FAO/WHO working group on drafting guidelines for the evaluation of probiotics in food defined probiotics as "live microorganisms which when administered in adequate amounts confer a health benefit for the host" (FAO/WHO, 2002).

The most popular probiotic microorganisms belong to the genera *Lactobacillus* and *Bifidobacterium*. Most probiotic strains are considered commensal microorganisms (Table 2) with little or no pathogenic potential (Salminen et al., 1998). These include (Collins and Gibson, 1999):

- Lactobacilli: *L. acidophilus*, *L. casei*, *L. delbrueckii* subsp. *bulgaricus*, *L. reuteri*, *L. brevis*, *L. cellobiosus*, *L. curvatus*, *L. fermentum*, *L. plantarum*.
- Gram-positive cocci: *Lactococcus lactis* subsp *cremonis*, *Streptococcus salivarius* subsp *thermophilus*, *Enterococcus faecium*, *Staphylococcus diaacecetylactis*, *S. intermedius* and
- Bifidobacteria: *B. bifidum*, *B. adolescentis*, *B. animalis*, *B. infantis*, *B. longum*, *B. thermophilum*.

In addition to LAB, *Bacillus* spp., fungi such as *Aspergillus* spp. and yeasts such as *Saccharomyces boulardii* have been used as probiotics.

3.1. Selection criteria for probiotics

To be able to exert its beneficial effects, a probiotic microorganism should be able to survive a range of harsh physicochemical factors during transit through the stomach and small intestine, and finally it should be able to persist in the gut. For these reasons, main selection criteria for probiotic strains include the following characteristics (Ouwehand et al., 1999; FAO/WHO, 2002):

- Non-pathogenic and non-toxic.
- Resistance to gastric acidity and bile acids.
- Adherence to mucus and/or human epithelial cells and cell lines.
- Colonisation of GI tract.
- Production of antimicrobial substances.
- Competitive exclusion and antagonism against pathogenic bacteria.
- Ability to stimulate host's immune response.
- Good technological properties (e.g. long self life, high viability, stability and functionality).
- Clinically validated and documented health effects.

3.2. Legislation, safety considerations and claims for the use of probiotics in foods

Although LAB have a long term history of safe use in human food (Adams and Marteau, 1995), it is generally accepted that zero risk can never be applied to the ingestion of live microbes whether these are probiotics, starters or components of natural microbiota of fresh foods e.g. fruits and vegetables (Saarela et al., 2002).

Therefore, even among a group of bacteria that is generally recognised as safe (GRAS) as for example *Lactobacillus* spp., it is recommended that probiotic stains be characterised at a minimum with the following tests (FAO/WHO, 2002):

1. Determination of antibiotic resistance patterns
2. Assessment of certain metabolic activities (e.g. deconjugation of bile salts)
3. Testing for toxin production.
4. Haemolytic potential.
5. Lack of infectivity by a probiotic strain in immunocompromised animals.
6. Assessment of side-effects during human studies (i.e. systemic infections, deleterious metabolic activities, excessive immune stimulation in susceptible individuals and gene transfer).
7. Epidemiological surveillance of adverse incidents in consumers (post market).

Careful assessment is also needed in the case of immunocompromised persons or those under antibiotic treatment as probiotic supplementation could possibly create complications (Pletincx et al. 1995). However, it should be noted that documented correlations between systemic infections and probiotic consumption are few and all occurred in patients with serious underlying medical conditions such as immunocompromised persons (FAO/WHO, 2002; Saarela et al., 2002).

General health claims allowed for foods containing probiotics fall under the guidelines for the use of nutrition claims adopted by the Codex Alimentarius since 1997. FAO/WHO (2002), has recommended that specific health claims such as enhanced function and/or disease prevention claims on foods containing probiotics should be allowed only in cases where sufficient scientific evidence is available. In the European Union such evidence has to be assessed by the European Food Safety Authority, prior to its use in the labelling, presentation and advertising of foods, according to the European Commission 2003/0165 (COD) proposal on nutrition and health claims made on foods.

3.3. Health benefits – established effects

In most cases, the health benefits attributed to probiotic consumption depend on the probiotic strain used. In addition, other factors such as host age, health status, stress and overall diet will most probably have various effects on the efficacy of the probiotics consumed. Since probiotics are beneficial microbes, their health promoting effects generally fall under the beneficial for health scheme such as the enhancement of gut barrier function (Fig. 1), reported earlier for the GI microflora.

Generally, health benefits claimed for probiotics are summarised in Figure 2 and refer to improved states of: a) intestinal condition and health (e.g. relieve of constipation, alleviation of lactose intolerance, recovery from viral and antibiotic induced diarrhoea, detoxification effects – antimutagenic/anticarcinogenic effects, alleviation of colitis symptoms), b) improved host nutrition (e.g. synthesis of vitamins, lowering of serum cholesterol), c) immune modulation (e.g. stimulation of immune response, enhancement of gut barrier function). In addition, ongoing research studies investigate the potential of probiotics towards the prevention of colo-rectal cancer (Saarela et al., 2002).

From the health benefits claimed for probiotics, the most well established effects in humans refer to the alleviation of symptoms of lactose intolerance and food allergies as well as the preventative and therapeutic effects against diarrhoea (Rowland, 1999). Therefore, it is obvious that probiotic fortification of foods is of relevance to food safety since their consumption could improve gut function and resistance to gastrointestinal infection.

For most probiotic microorganisms efficacy is demonstrated with daily consumption of $10^8 - 10^{11}$ organisms per day (Scheinbach, 1998). Thus, in order to provide consumers with useful information, it is recommended that food manufacturers apart from designating on the labels of their products the probiotic stain(s) used, they should also point the suggested serving size needed to deliver the effective dose of probiotics related to the health claim.

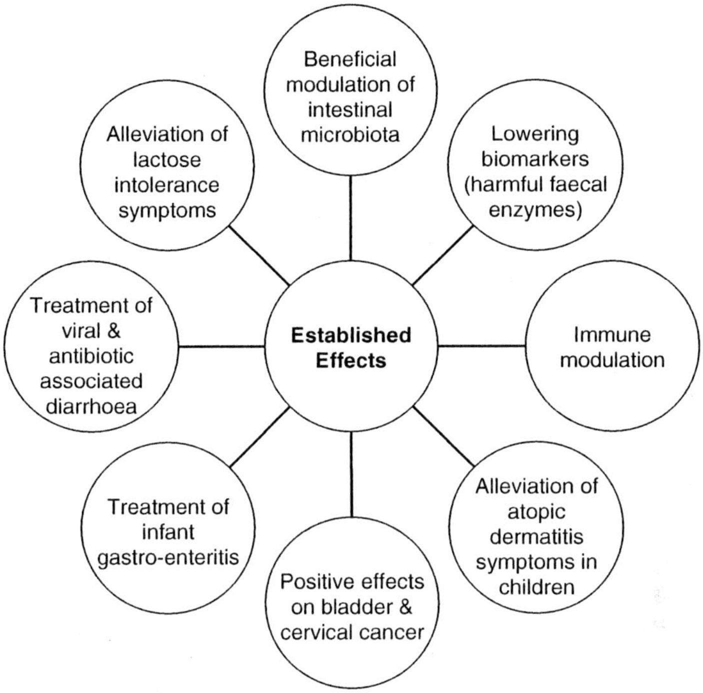

FIGURE 2. Beneficial health effects claimed for probiotics.

4. Prebiotics: Role as Bioactive Nutrients

The essential role of carbohydrates in human and animal nutrition is well established. Further to their numerous biological functions in the body, carbohydrates are the principal nutritional components in the diet that are used metabolically by the host for the generation of maintenance energy, growth and development. Most of the dietary carbohydrates ranging from oligosaccharides to polysaccharides get hydrolysed in the small intestine to their monomeric sugar moieties (e.g. glucose, fructose etc) prior to their uptake by the intestinal enterocytes.

Dietary components that totally, or even partially, escape digestion by the host's enzymatic system will arrive in the hindgut where they are then subject to metabolic activities of the colonic microflora. Carbohydrates that resist the digestive process in the small intestine are characterised as non-digestible or dietary fibre (Roberfroid, 1993).

Non digestible carbohydrates or dietary fibre can be either insoluble or soluble. Insoluble dietary fibre (e.g. cellulose) reduces gut transit time and increases faecal volume, while it might get slightly metabolised by the intestinal microflora. On the contrary, soluble dietary fibre gets highly metabolised by the gut microflora and results in: a) acidification of the intestinal content, b) production of short chain fatty acids (SCFA) and gases and c) increase in faecal bacterial mass and faecal volume (Roberfroid, 1993).

Accumulating scientific evidence suggests that from the non-digestible carbohydrates specific oligosaccharides and polysaccharides can be selectively metabolised by the "beneficial" bacteria belonging to the genera of *Lactobacillus* and *Bifidobacterium*, thus resulting in an increase in their numbers and activity in the gut.

Such carbohydrates are known in the scientific and commercial literature with the term "prebiotics" and could be regarded as nutritional tools for the beneficial modulation of the composition and metabolic activity of the host's intestinal microflora.

From the three principal macronutrients of foods (i.e. proteins, carbohydrates and fats) only carbohydrates are currently represented in the prebiotic concept. However, as it has been the case with carbohydrates, the potential of proteins and fats as well as of their respective oligomers to modulate the composition and metabolic activities of the intestinal microflora is under examination. Recently the antimicrobial and bifidogenic properties of major milk proteins (e.g. lactoferrin, κ-casein glycomacropeptide) have been reviewed (Mountzouris et al., 2002a).

Prebiotics are non-viable food/feed ingredients and have been defined as: "non-digestible food ingredients that beneficially affect the host by selectively stimulating the growth and/or activity of one or a limited number of bacteria in the colon, and thus improve host health" (Gibson and Roberfroid, 1995). Therefore they should not get confused with probiotics that have been defined as live microbial feed supplements that are beneficial to health (Salminen et al., 1998).

According to the concept of prebiotics, introduced by Gibson and Roberfroid in 1995, a food ingredient in order to be classified as prebiotic it should: a) be neither hydrolysed or absorbed in the upper part of the gastrointestinal tract, b) be a selective substrate for one or a limited number of beneficial bacteria commensal to the colon, which are stimulated to grow and/or are metabolically activated, c) consequently, be able to alter the colonic flora in favour of a healthier composition and d) induce luminal or systemic effects that are beneficial to host health.

Therefore prebiotics could also act as dietary bioactive components beyond their role as selective substrates for the beneficial (e.g. probiotic) bacteria in the colon.

4.1. Origin and manufacturing processes

Prebiotic ingredients exist in nature in a variety of foods such as garlic, onions, leeks, asparagus, chicory, artichokes, bananas, human milk, wheat, maize and soy and are therefore part of a normal diet. Commercially available carbohydrates that fall in the concept of prebiotics are shown in Table 3.

Generally oligosaccharides of non-starch origin dominate the prebiotic arena with the exception of isomaltooligosaccharides that also get partly digested in the

TABLE 3. Names, chemical formulas and manufacture information for common prebiotic food ingredients

Prebiotic Name	Chemical formula	Manufacturing process
Trans-galactooligosaccharides	α-D-Glu-(1→4)[β-D-Gal-(1→6ή 1→4)-]$_{n = 2-5}$	Enzymatic synthesis from lactose
Inulin	β-D-Fru-(1→2)-[β-D-Fru-(1→2)-]$_{n = 1->60}$	Extraction from chicory, leeks and artichokes
	α-D-Glu-(1→2)-[β-D-Fru-(1→2)-]$_{n = 2->60}$	
Lactulose	β-D-Gal-(1→4)-β-D-Fru	Chemical modification of lactose
Lactosucrose	β-D-Gal-(1→4)-α-D-Glu-(1→2)-β-D-Fru	Enzymatic synthesis from lactose and sucrose
Fructooligosaccharides or oligofructose	α-D-Glu-(1→2)-[β-D-Fru-(1→2)-]$_{n = 2-4}$	Enzymatic synthesis from sucrose
	α-D-Glu-(1→2)-[β-D-Fru-(1→2)-]$_{n = 2-9}$	Enzymatic hydrolysis of inulin
	β-D-Fru-(1→2)-[β-D-Fru-(1→2)-]$_{n = 1-9}$	
Isomaltooligosaccharides	[α-D-Glu-(1→6)-]$_{n = 2-5}$	Enzymatic hydrolysis & subsequent transglycosylation of starch
Xylooligosaccharides	[β-D-Xyl-(1→4)-]$_{n = 2-9}$	Enzymatic hydrolysis of xylan
Soybean oligosaccharides	[α-D-Gal-(1→6)-]$_{n= 1-2}$ α-D-Glu-(1→2)-β-D-Fru	Extraction from soybean

Glu: Glucose, Gal: Galactose, Fru: Fructose, Xyl: Xylose

upper gut. Higher molecular weight molecules of the isomaltose series, termed oligodextrans that have been manufactured by controlled dextran hydrolysis (Mountzouris et al., 2002b) are expected to display higher resistance to digestion in the upper gut whilst promoting saccharolytic bacterial metabolism in the distal colon (Olano-Martin et al., 2000).

Prebiotic polysaccharides include inulin (i.e. storage carbohydrate consisting mainly from fructose moieties found in certain plants such as chicory) which is one of the most widely studied prebiotics (Gibson et al., 1995; Kolida et al., 2002) and high-amylose starch (Brown et al., 1997; Wang et al., 1999) that has not been included in Table 3.

Prebiotics can be produced by (Sako et al., 1999):

- extraction from their natural sources (e.g. inulin from chicory and leeks, soybean oligosaccharides from soybean whey and mannanoligosaccharides from yeast cell wall).
- enzymatic processes either as a result of enzymatic synthesis (e.g. oligofructose from sucrose and trans-galactooligosaccharides from lactose) or as a result of enzymatic hydrolysis (e.g. oligofructose from inulin and xylooligosaccharides from xylan).
- chemical modification (e.g. lactulose from the isomerisation of the glucose moiety in lactose to fructose). Lactulose is used therapeutically for the treatment of constipation and hepatic encephalopathy and is therefore not used in food formulation despite its prebiotic potential.

Average prebiotic and in particular fructan consumption through the normal diet has been evaluated to amount to several grams per day (Van Loo et al., 1995). They are legally classified as food or food ingredients in all EU countries and therefore can be used without specific limitations as ingredients in foods and drinks (Franck, 2002).

Whilst all of the prebiotics described in Table 3 are being used as food ingredients by the food industry in Japan, only oligofructose, inulin and trans-galactooligosaccharides are used by the European food industry. This could be due to the fact that for the above three prebiotics there is accumulating scientific proof of their function *in vivo* in animals and humans that could enable food manufacturers make various claims for their products. In addition they are produced in large quantities by companies based in Europe which provide a high level of technical support for food applications involving their products.

4.2. Physicochemical characteristics and properties

Commercially available prebiotics are supplied in the form of powder and/or solutions of variable purity and concentration level.

The physicochemical composition of the prebiotic carbohydrates is important not only for their effects in gastrointestinal physiology and function but also for their application in food science.

For example factors such as the type of sugar moieties that make the building blocks, the type of glycosidic linkages present, the structural arrangement and the molecular weight of the prebiotic molecule, determine their selective or non-selective fermentation by the intestinal microflora. In addition, these factors have an effect on the extent and the rate that prebiotics get metabolised by the bacteria. This in turn results in the development of different profiles of saccharolytic activity along the colon (Olano-Martin et al., 2000).

The differences in the enzymatic machinery found among bacteria of different genera, even among species of the same genus, could explain the large variation seen in the ability of bacteria to use different prebiotics as substrates for their growth. Furthermore, the physicochemical composition and properties of prebiotic carbohydrates also affects the final products produced by the microbial fermentation (Cummings and Macfarlane, 1991).

Some simple sugars and oligosaccharides are known to be potent inhibitors of bacterial adhesion to epithelial cells by acting as receptor analogues to mucosal adhesion molecules (Kunz and Rudloff, 1993; Peterson *et al.*, 1998; Kunz, 1998, Naugton et al., 2001; Lee and Puong, 2002). For example among the human milk oligosaccharides lacto-N-tetraose and lacto-N-neotetraose act as cell surface receptors for *Streptococcus pneumoniae*, fucosylated oligosaccharides are receptors for *E. coli* and sialated oligosaccharides are recognised receptor sites for influenza viruses A,B and C, *Campylobacter pylori* and *Mycoplasma pneumoniae* (Kunz and Rudloff, 1993).

Finally the functionality of the prebiotic molecules in food systems will depend on their physicochemical properties such as viscosity, solubility, solution stability, crystallisation, freezing point depression sweetness, humectant properties, heat and acid resistance. Therefore development of functional foods with prebiotics would have to account for the above in the research and development effort.

4.3. Legislation and safety issues linked with the use of prebiotics in foods

Prebiotics can be naturally found in many foods and are produced commercially by natural ingredients (Table 3). They are classified as food ingredients and their consumption is generally considered as safe with the restriction that an abuse in consumption could cause intestinal discomfort (Hammes and Hertel, 2002).

Food products that contain prebiotics must comply with the general legislation that governs the labelling requirements and claims for foods. Although all aspects of food labelling are to a very great extent harmonised by European legislation, there is currently no consistent legal framework regarding claims and particularly health claims within the European Union (Gibson et al., 2000). As a result, every product should comply in this respect with the national law of each European country that is distributed. This poses a serious drawback for the marketing of functional food products that relies heavily on claims linking consumption of the products with enhanced physiological functions or reduction of disease risk.

However development of such a legal framework is expected as a consequence of a recent proposal from the European Commission 2003/0165 (COD) on nutrition and health claims made on foods.

Japan is the only country in the world that has, since 1991, developed a legal framework that permits the commercialisation of selected functional foods under the term FOSHU "Food for Specified Health Use". The legislation for FOSHU allows the presentation of a health claim for each approved product.

4.4. Physiological effects and health benefits

The potential health benefits linked with prebiotic consumption stem from the direct, indirect and systemic effects that these ingredients have on gastrointestinal physiology and function.

The direct effects could be attributed to the prebiotic function as soluble dietary fibre that promotes better intestinal function and relief of constipation (Roberfroid, 1993; Teuri and Copella, 1998, Van Loo et al., 1999). In addition, prebiotics contribute to less energy production since they are not digested in the upper gut but get fermented in the colon. In healthy humans fructo oligosaccharides were found to have an energy value of 9.5 kJ/g (i.e. about half that of sucrose) (Molis et al., 1996). Due to their non-digestible nature, prebiotic ingredients could contribute to lower glycaemia compared with equal amounts of digestible carbohydrates and therefore their inclusion could be considered in diets for diabetics.

Some prebiotic oligosaccharides could prevent attachment of pathogenic bacteria to the intestinal enterocytes by blocking the bacterial lectins, thus acting as anti-adhesive (Kunz, 1998; Naugton et al., 2001; Lee and Puong, 2002).

The indirect effects that prebiotics have on the host result from the beneficial modulation of the composition and metabolic activities of the intestinal microflora in a way that it contributes to enhancement of the gut barrier function (Figure 1).

Research in humans and animals demonstrates that consumption of prebiotics stimulates and enhances the growth of probiotic bacteria in the intestinal microflora (Van Loo et al., 1999, Bouhnik et al., 1999). In humans the supplementation of diet with 10-15 g fructooligosaccharides or inulin was found adequate for a significant increase in *Bifidobacterium* (Bouhnik et al., 1999; Gibson et al., 1995).

Enhancement of the beneficial microflora results in stimulation of host's immune system (Schley and Field, 2002), while inhibition of pathogens occurs as a result of competition for growth substrates and production of antimicrobial compounds, lactate and SCFA (Figure 1). Therefore, prebiotics represent an opportunity for the fortification of the indigenous microflora with its beneficial members through diet. This is of relevance to food safety since consumption of foods that have prebiotic ingredients could improve resistance to gastrointestinal infection.

There is accumulating evidence for a number of beneficial systemic effects for host's health linked with prebiotic consumption. The most well researched prebiotics so far are the fructans inulin and oligofructose. Results from research work under way, mainly in laboratory animals, indicate that prebiotics could have protective roles against: a) cardiovascular disease through the lowering of synthesis of VLDL cholesterol and blood triglycerides (Williams and Jackson, 2002), b) osteoporosis through the improvement of mineral calcium and magnesium absorption (Scholz-Ahrens et al., 2001) and c) chemically induced cancer (Femia et al., 2002; Cherbut et al., 2003).

4.5. *Future prebiotics with enhanced functionality*

There is currently a need to increase our knowledge on the role of structure – function relationships for known and candidate prebiotics and to gather more scientific evidence to substantiate the potential health benefits of prebiotics. Once there is a better understanding of the above relationship, then "better" prebiotic molecules could be designed (Rastall and Maitin, 2002, Mountzouris et al., 2002b) and optimal combinations with other food ingredients could be considered/developed.

Currently, a substantial amount of evidence clearly indicates that prebiotics improve gastrointestinal conditions (Arai, 2002). The prebiotics, as food ingredients, should combine enhanced health function properties with good sensory and food technology properties.

Examples of desired properties for a prebiotic ingredient include: a) selective fermentation by beneficial bacteria, b) slow fermentability so that it persists and gets totally fermented to the distal colon, c) Inhibition of pathogen attachment to the gut wall, d) effective at low dosage, e) no side effects (e.g. flatulence, discomfort), f) good storage and self-life stability. Other desirable properties will depend on the food application in which the prebiotics will be used; that is, to heat and acid resistance viscosity, solubility, solution stability, crystallisation, freezing point depression sweetness and humectant properties.

5. Functional Foods: Probiotic and Prebiotic Applications in Foods

Generally, the development of products containing probiotics and prebiotics falls under the functional food concept. The roots of this modern concept go centuries back, perhaps to the time of Hippocrates when the link between diet and health was firstly realised. Because it is difficult to define what exactly is a functional food the following definition by the Functional Food Science in Europe (FUFOSE) project has been adopted: "A food can be regarded as functional if it is satisfactorily demonstrated to affect beneficially one or more target functions in the body, beyond adequate nutritional effects, in a way that is relevant to either improved stage of health and well being and/or reduction of risk of disease. A functional food must remain food and it must demonstrate its effects in amounts that can normally be expected to be consumed in the diet: it is not a pill or a capsule, but part of the normal food pattern." (Diplock et al., 1999).

The research and development of functional foods is trying to address topic areas such as: defence against oxidative damage to the body; effects on the cardiovascular system; psychological effects on behaviour and mood; effects on growth and development and effects on the gastrointestinal tract (Ottaway, 1997). As a result antioxidant substances (i.e. beta-carotene, vitamins C and E), minerals (e.g. calcium, magnesium, zinc, selenium), phytochemicals (e.g. flavonoids), probiotics (i.e. bifidobacteria and lactobacilli), fatty acids and lipids (e.g. fish oils), and a range of macromolecules (e.g. dietary fibers and oligosaccharides) are under investigation (Richardson, 1996).

It is clear that a major effort in research and development is needed to obtain scientific evidence that will validate the specific claims made by functional foods. The European consensus on scientific concepts of functional foods by the FUFOSE project has defined two types of claims: a) the type A claim which concerns beneficial improvement of a target function (i.e. physiological or psychological function) in the body and b) the type B claim that concerns the reduction of a disease risk (Roberfroid 2002). In Japan, the Ministry of Health and Welfare uses the term "Foods for Specified Health Use (FOSHU)" instead of the term "functional food" (Kojima, 1996).

In terms of probiotic and prebiotic incorporation into foodstuffs the market is significantly less well advanced in Europe than in Japan (Gibson et al. 2000; Arai 2002). From 252 foods having FOSHU status in Japan 110 foods that claim to improve gastrointestinal conditions have probiotics (i.e. 49 foods) and prebiotics (i.e. 61 foods). Examples of foods containing probiotics include mainly fermented milks and yoghurt, whereas foods containing prebiotics include soft drinks, yoghurts, biscuits, cookies, table sugar, chocolate, tablet candy, vinegar and powdered soup (Arai, 2002).

In Europe, most products containing probiotics and prebiotics belong to the dairy sector (i.e. yoghurts and milk). Other products with prebiotics include cereal bars, breakfast cereals, biscuits, various ready meals, confectionary, powdered drinks, infant formulae feeds and weaning foods (Young, 1998; Gibson et al., 2000). From the prebiotic products, inulin, oligofructose and trans-galactooligosaccharides are very popular. Other food applications for inulin and oligofructose include baked goods and breads, fillings, fruit preparations, salad dressings, meat products, dietetic products and meal replacers, table sugar and chocolate (Franck, 2002).

In some of the foods described above the prebiotics have been used in conjunction with probiotic bacteria, a combination known as synbiotic (Gibson and Roberfroid, 1995).

It is expected that as our knowledge in the role of nutrition in health and disease expands, many new food preparations using probiotics and prebiotics will be developed. The market success of these products will depend on valid scientific evidence substantiating the claims made and on consumer acceptance. Consumers will have firstly to be educated on nutrition basics in order to be able to understand and assess the benefits claimed. Last but not least, for these kinds of products to succeed the credible science and effective communication would have to be combined by uncompromised taste and convenience.

6. References

Adams, M.R. and Marteau, P., 1995, On the safety of lactic acid bacteria. *International Journal of Food Microbiology* **27**:263.

Arai, S., 2002, Global view on functional foods: Asian perspectives. *British Journal of Nutrition* **88**:S139.

Benno, Y., Sawada, K., and Mitsuoka, T., 1984, The intestinal microflora of infants: composition of faecal flora in breast-fed and bottle-fed infants. *Microbiology and Immunology* **28**:975.

Berg, R.D., 1996, The indigenous gastrointestinal microflora. *Trends in Microbiology* **4**:430.

Bouhnik. Y., Vahedi, K., Achour, L., Attar, A., Salfati, J., Pochart, P., Marteau, P., Flourie, B., Bornet, F., and Rambaud, J.C., 1999, Short-chain fructo-oligosaccharide administration dose-dependently increases fecal bifidobacteria in healthy humans. *Journal of Nutrition* **129**:113.

Brown, I., Warhurst, M., Arcot, J., Playne, M., Illman, R.J., and Topping, D.L., 1997, Fecal numbers of bifidobacteria are higher in pigs fed *Bifidobacterium longum* with a high amylose cornstarch than with a low amylose cornstarch. *J. Nutr.* **127**:1822.

Cebra, J.J., 1999, Influences of microbiota on intestinal immune system development. *Am J Clin Nutr.* **69**:1046S.

Cherbut, C., Michel, C., and Lecannu, G., 2003, The prebiotic characteristics of fructooligosaccharides are necessary for reduction of TNBS – induced colitis in rats. *Journal of Nutrition* **133**:21.

Collins, M.D. and Gibson, G.R., 1999, Probiotics, prebiotics, and synbiotics: approaches for modulating the microbial ecology of the gut. *American Journal of Clinical Nutrition* **69**:1052S.

Commission of the European Communities 2003/0165 (COD), Proposal for a regulation of the European parliament and of the Council on nutrition and health claims made on foods. COM (2003) 424 final, Brussels 16.7.2003

Cummings, J.H. and Macfarlane, G.T., 1991, The control and consequences of bacterial fermentation in the human colon. *Journal of Applied Bacteriology* **70**:443.

Cunningham-Rundles, S. and Lin, D.H.L., 1998, Nutrition and the immune system of the gut. *Nutrition* **14**:573.

Diplock, A.T., Aggett, P.J., Ashwell, M., Bornet, F., Fern, E.B., and Roberfroid, M.B., 1999, Scientific concepts of functional foods in Europe: consensus document. *British Journal of Nutrition* **81**:S1.

FAO/WHO 2002, Joint Working Group Report on Drafting Guidelines for the Evaluation of Probiotics in Food. London, Ontario, April 30 and May 1, 2002.

Farr, D.R., 1997, Functional foods. *Cancer Letters* **114**:59.

Femia, A.P., Luceri, C., Dolara, P., Gainnini, A., Biggeri, A., Salvadori, M., Clune, Y., Collins, K.J., Parlierani, M., and Caderni, G., 2002, Antitumorigenic activity of the prebiotic inulin enriched with oligofructose in combination with the probiotics *Lactobacillus rhamnosus* and *Bifidobacterium lactis* on azoxymethane-induced colon carcinogenesis in rats. *Carcinogenesis* **23**:1953.

Franck, A., 2002, Technological functionality of inulin and oligofrtuctose. *British Journal of Nutrition* **87**:S287.

Fuller, R., 1989, Probiotics in man and animals. *Journal of Applied Bacteriology* **66**:365.

Gibson, G., Ottaway, P., and Rastall, R., 2000, Legal Aspects In: Prebiotics New developments in functional foods, Chandos Publishing Limited, Oxford, pp. 77-92.

Gibson, G.R., Beauty, E.R., Wang, X., and Cummings, J.H., 1995, Selective stimulation of bifidobacteria in the human colon by oligofructose and inulin. *Gastroenterology* **108**:975.

Gibson, G.R. and Roberfroid, M.B., 1995, Dietary modulation of the human colonic microbiota: introducing the concept of prebiotics. *Journal of Nutrition* **125**:1401.

Hammes, W.P. and Hertel, C., 2002, Recent approaches for pre- and probiotics: challenges and outlook. *Food Research International* **35**:165.

Jay, J.M., 1992, Microbiological safety of foods. In: Modern Food Microbiology, fourth edition, Chapman & Hall, New York, pp. 434-478.

Kim, W.J., 1993, Bacteriocins of lactic acid bacteria: their potential as food biopreservative. *Food Reviews International* **9**:299.

Kojima, K., 1996, The eastern consumer viewpoint: the experience in Japan. *Nutrition Reviews* **54**:S186.

Kolida, S., Touhy, K., and Gibson, G.R., 2002, Prebiotic effects of inulin and oligofructose. *British Journal of Nutrition* **87**:S193.

Kunz, C. and Rudloff, S., 1993, Biological functions of oligosaccharides in human milk. *Acta Paediatrica* **82**:903.

Kunz, C., 1998, Complex oligosaccharides in infant nutrition. *Monatsschrift Inderheilkunde* **146**:S49.

Lee, Y.K. and Puong, K.Y., 2002, Competition for adhesion between probiotics and human gastrointestinal pathogens in the presence of carbohydrate. *British Journal of Nutrition* **88**:S101.

Lilly, D.M. and Stillwell, R.H., 1965, Probiotics: growth promoting factors produced by microorganisms. *Science* **147**:747.

Macfarlane, G.T. and McBain, A.J., 1999, The human colonic microbiota. *In*: Colonic microbiota, nutrition and health, Kluwer Scientific Publishers, Holland, pp. 3-28.

Mackie, R.I., Sghir, A., and Gaskins, H.R., 1999, Developmental microbial ecology on the neonatal gastrointestinal tract. *Am J Clin Nutr* **69**:1035S.

Madden, J.A.J. and Hunter, J.O., 2002, A review of the role of gut microflora in irritable bowel syndrome and the effects of probiotics. *British Journal of Nutrition* **88**:S67.

Marteau, P., Seksik, P., and Jian, R., 2002, Probiotics and intestinal health: a clinical perspective. *British Journal of Nutrition* **88**:S51.

Metchnikoff, E., 1907, The Prolongation of Life. William Heinemann, London.

Molis, C., Flourie, B., Quarne, F., Gailing, M.F., Lartigue, S., Guibert, A., Bornet, F., and Galmiche, J.P., 1996, Digestion excretion and energy value of fructooligosaccharides in healthy humans. *American Journal of Clinical Nutrition* **64**:324.

Mountzouris, K.C., Gilmour, S.G., and Rastall, R.A., 2002b, Continuous production of oligodextrans via controlled hydrolysis of dextran in an enzyme membrane reactor. *Journal of Food Science* **67**:1767.

Mountzouris, K.C., McCartney, A.L., and Gibson G.R., 2002a, Intestinal microflora of human infants and current trends for its nutritional modulation. *British Journal of Nutrition* **87**:405.

Naughton, P.J., Mikkelsen, L.L., and Jensen, B.B., 2001, Effects of nondigestible oligosaccharides on *Salmonella enterica* serovar typhimurium and non-pathogenic *Escherichia coli* in the pig small intestine in vitro. *Applied and Environmental Microbiology* **67**:3391.

Olano-Martin, E., Mountzouris, K.C., Gibson, G.R., and Rastall, R.A., 2000, *In vitro* fermentability of dextran oligodextran and maltodextrin by human gut bacteria. *British Journal of Nutrition* **83**:247.

Ottaway, P.B., 1997, Functional foods and health claims in Europe. *Nutraceuticals International* **2**:14.

Ouwehand, A.C., Kirjavainen, P.V., Shortt, C., and Salminen, S., 1999, Probiotics: mechanisms and established effects. *International Dairy Journal* **9**:43.

Peterson, J.A., Patton, S., and Hamosh, M., 1998, Glycoproteins of the human milk fat globule in the protection of the breast fed infant against infections. *Biology of the Neonate* **74**:143.

Pletincx, M., Legein, J., and Vandenplas, Y., 1995, Fungemia with *Saccharomyces boulardii* in a 1-year-old girl with protracted diarrhea. *Journal of Pediatic Gastroenterology and Nutrition* **21**:113.

Rastall, R.A. and Maitin, V., 2002, Prebiotics and synbiotics: towards the next generation. *Current Opinion in Biotechnology* **13**:490.

Richardson, D.P., 1996, Functional foods-shades of gray: an industry perspective. *Nutrition Reviews* **54**:S174.

Roberfroid, M.B., 2002, Global view on functional foods: European perspectives. *British Journal of Nutrition* **88**:S133.

Roberfroid, M., 1993, Dietary fiber, inulin, and oligofructose: a review comparing their physiological effects. *Critical Reviews in Food Science and Nutrition* **33**:103.

Rowland, I., 1999, Probiotics and benefits to human health – the evidence in favour. *Environmental Microbiology* **1**:375.

Saarela, M., Matto, J., and Mattila-Sandholm, T., 2002, Safety aspects of Lactobacillus and Bifidobacterium species originating from human oro-gastrointestinal tract or from probiotic products. *Microbial Ecology in Health and Disease* **14**:233.

Sako, T., Matsumoto, K., and Tanaka, R., 1999, Recent progress on research and applications of non-digestible galacto-oligosaccharides. *International Dairy Journal* **9**:69.

Salminen, S., Bouley, C., Boutron-Ruault, M.C., Cummings J.H., Franck, A., Gibson, G.R., Isolauri, E., Moreau, M.C., Roberfroid, M., and Rowland, I.R., 1998, Functional food science and gastrointestinal physiology and function. *British Journal of Nutrition* **80**:S147.

Salyers, A.A. and Whitt, D.D., 1994, Virulence factors that promote colonisation. In: Bacterial pathogenesis a molecular approach, ASM Press, Washington DC, pp. 16-29.

Sanderson, I.R., 1999, The physicochemical environment of the neonatal intestine. *Am J Clin Nutr* **69**:1028S.

Savage, D.C., 1977, Microbial ecology of the gastrointestinal tract. *Annu Rev Microbiol* **31**:107.

Scheinbach, S., 1998, Probiotics: functionality and commercial status. *Biotechnology Advances* **16**:581.

Schley, P.D. and Field, C.J., 2002, The immune enhancing effects of dietary fibres and prebiotics. *British Journal of Nutrition* **87**:S221.

Scholz-Ahrens, K.E., Schaafsma, G., van den Heuvel, E.G.H.M., and Schrezenmeir, J., 2001, Effects of prebiotics in mineral metabolism. *American Journal of Clinical Nutrition* **73**:459S.

Southgate, D.A.T., 2000, Pathogenic agents in foods. In Human nutrition and dietetics, 10th edition. Churchill Livingstone, Edinburgh, pp. 411-416.

Tamura, Z., 1983, Nutriology of Bifidobacteria. *Bifidobacteria Microflora* **2**:3.

Teuri, U. and Korpela, R., 1998, Galacto-oligosaccharides relieve constipation in elderly people. *Annals of Nutrition and Metabolism* **42**:319.

Van Loo, J., Coussement, P., De Leenheer, L., Hoebregs, H., and Smits, G., 1995, On the presence of inulin and oligofructose as natural ingredients in the Western diet. *Critical Reviews in Food Science and Nutrition* **35**:525.

Van Loo, J., Cummings, J., Delzenne, N., Englyst, H., Franck, A., Hopkins, M., Kok, N., Macfarlane, G., Newton, D., Quigley, M., Roberfroid, M., van Vliet, T., and van den Heuvel, 1999, Functional food properties of non-digestible oligosaccharides: a consensus report from the ENDO project (DGXII AIRII-CT94-1095). *British Journal of Nutrition* **81**:121.

Wang, X., Conway, P.L., Brown, I.L., and Evans, A.J., 1999, In vitro utilisation of amylopectin and high amylose maize (amylomaize) starch granules by human colonic bacteria. *Applied and Environmental Microbiology* **65**:4848.

Williams, C.M. and Jackson, K.G., 2002, Inulin and oligofructose: effects on lipid metabolism from human studies. *British Journal of Nutrition* **87**:S261.

Yildirim, Z. and Johnson, M.G., 1998, Characterisation and antimicrobial spectrum of bifidocin B, a bacteriosin produced by *Bifidobacterium bifidum* NCFB 1454. *Journal of Food Protection* **61**:47.

Young, J., 1998, European market developments in prebiotic- and probiotic-containing foodstuffs. *British Journal of Nutrition* **80**:S231.

8
Exploitation of Microorganisms by the Food and Beverage Industry

George Kalantzopoulos[1], Manuela Pintado[2], and Ana Gomes[2]

1. Introduction

The food and beverage industry exploits non-pathogenic microorganisms for the production of fermented foods. These foods are prepared from raw or headed raw materials and acquire their characteristic properties by a process that involves microorganisms. In certain cases the endogenous enzymes of the wild microflora of the raw material may play a decisive role. [3]It is believed that fermented foods originated from the Orient and date back to the prehistoric times. Originally, these were fermented "*spontaneously*" by autochthonous strains found in the raw materials or the environment; this was the start of traditional biotechnology. Historical reports show that long established artisanal and religious practices utilized cultures, which were very similar to the starter cultures applied today. The most important were cheese, yoghurt, wine, vinegar, beer, bread and the traditional fungal fermentation used in Asia and Africa, for the production of food.

In general, all kinds of microorganisms are used for fermentations, but in Europe bacteria and yeasts are more commonly used than moulds. Through the ages, people gained more and more experience to control these processes and fermented foods became an independent class of foodstuffs. As regards taste, aroma, visual appearance, texture, consistency, shelf-life and safety, these different products possess characteristic properties compared to the raw materials or to other similar products.

The use of fermented milks and cheese dates back many centuries, although there is no precise record of the date when they were first made. In the Bible there are various references to fermented foods. When the Patriach Abraham entertained three angels, he put before them soured and sweet milk (Genesis VIII, 8). Also, Moses having considered the food given by Jehovah to his people mentions the soured milk of cows and goats (Deuteronomy, XXXII 14), (Davis, 1952).

---

[1]Agricultural University of Athenas 118.55 Botanilcos Athens Greece
[2]Esola Superior de Biotecnologia. Rua Dr. António Bernardino de Almeida. 4200-072. Porto. Portugal

Homer in Odysseus mentioned the production of cheese from the first cheese-maker of the world Polyfemus. From this period until today, cheese has become one of the most important fermented products of the world.

The art of producing beers and wines has developed over 5000-8000 years. There must have been several independent discoveries on how fermented beverages arose from exposing fruit juice or cereal extracts to the air. The explanations for the fermentations were available during the nineteenth century but such did not promote a steady improvement in manufacturing techniques, but only in volume. During the height of the Egyptian and Babylonian civilizations some 4300 years ago, the details of brewing were well-illustrated. During Greek and later Roman domination, wine became an important item of international commerce. The beverages were attractive, particularly for those individuals who enjoyed few pleasures, in that they produced alcoholic euphoria (Hough et al., 1982).

The production of bread started at the same time with the cultivation of cereals. The Egyptians, and later the Jews, prepared the "galets" with the use of beer yeasts. Later on, the Greeks discovered bread and produced it from different kinds of cereals. Relative history reports that in this period in Athens 72 different kinds of bread were produced (Derieux , 1988).

In the 1st century A.D., Plinius the Elder (Derieux , 1988), described storage of white cabbage in special earthen vessels, which were used only for this purpose. Nowadays, it has been established that the conditions described by Plinius, were those of the fermentation of cabbage to sauerkraut by microorganisms, which were located in the pores of the vessels and which had persisted from a former fermentation.

Plinius also described the production of bread in Gaule and in the Iberian countries via the use of foam taken from the beer production process.

Presently, it is still possible to observe primitive forms of starter application during preparation of several fermented beverages in some parts of Africa. During a religious ceremony preceding the fermentation, the medicine man dips cult objects into the liquid, which again contains microorganisms from a previous batch.

In the Middle Ages brewing was an art or mystery, the details of which were jealously guarded by the master brewers and their guilds. Mystery it certainly was because there was virtually no appreciation of the reasons for the various processing steps, most of which, like fermentation, had been discovered by chance. Thus, malting involved immersing barley in water and permitting it to sprout but the reasons for the barley becoming soft and sweet was not understood. Similarly, the reasons for drying the germinated barley under relatively cool conditions were cloaked in mystery (Hough et al., 1982).

It is important to mention that besides these well-known products many other traditional fermented products were produced all around the world, with an important contribution to local human nutrition.

The systematic use of starter cultures has only been exploited since the middle of the 20th century. At the turn of the century microbiology science advanced rapidly. The role of microorganisms in disease and food production was being discovered and elucidated. Microbial hygiene was understood and pasteurization of perishable foods was introduced to combat infectious diseases. Initially, starter cultures were isolates from earlier fermentations that were maintained and propagated at the site of production. Christian Hansen was the first to isolate and propagate special yeasts to be used in breweries. The first starter culture was born. During the same period microbial enzymes were also developed such as proteases, amylases and invertases (Mongensen et al., 2002).

Fermentation or supplementation with selected intestinal flora started around 1935, by the Shirota Institute for Research on Protective Bacteria, and fermentation at a large scale with these probiotic microorganisms started after1990 (Bosschaert et al., 2001).

2. Why Fermented Products?

Fermented foods derived from plant or animal materials are an acceptable and essential part of raw materials as substrates, using technology, from the most primitive to the most advanced, and achieving an astounding range of sensory and textural qualities in the final products.

The fundamental reasons for the development and acceptability of fermented foods can be variably ascribed to preservation, improved nutritional properties, improved flavour/aroma, upgrading of substrates to added-value products, improved health aspects and in more recent times, in many affluent communities to novelty and/or trendy considerations (see as example Table 1). All these properties are mainly the result of the activity of a population of bacterial strains or of yeasts. These microorganisms present a very important biodiversity, which contributes to the production of different types of products with different qualities.

A very important aspect is the contribution that Biotechnology has given over the past years, which offers the possibility to select and use new sources, to increase the yield of sources already used, to introduce specific functional properties in raw materials or ingredients and to improve the nutritional value, bioavailability of nutrients and flavours (Huis in't Veld et al., 1989).

The fermented foods and their derivatives constitute a staple meal or a meal supplement and many be used as a snack, drink, dessert or condiment to be spread or to be added as an ingredient in cooked dishes.

3. Bacteria

3.1. Advantages for health and safety

The nutritional value of a particular food depends on its digestibility and its content of essential nutrients; both digestibility and nutrient content may improve by fermentation, during which the enzymatic activity of microbial culture may digest the macronutrients.

TABLE 1. Contribution of lactic acid bacteria as natural ingredients, health/ nutrition additives and specialty fermented foods (Lee, 1996a)

Items	Applications
Ingredients	
Preservatives	
(nisin, bacteriocins, Lactoperoxide-thyocyanate)	Preservation of various foods
Enzymes	
Proteinases, exopeptidases, esterases	Cheese ripening, enzymes-modified cheese flavours, new protein hydrolysates, functional and bioactive peptides.
Lactase, Superoxide dismutases	Lactose-hydrolyzed whey syrups Antioxidants for lipids.
Polysaccharides (dextran).	Gums and thickeners, culture viscosity stabilizers.
Flavors (diacetyl, acetoin)	Butter and yoghurt flavours.
Health/Nutrition	
Probiotic cultures	Lactose digestion, control of intestinal pathogen systems, hypercholesterolemia reduction tumour inhibition, reduction of protein allergenicity, reduction of osteoporosis, increase in vitamins K, B complex, (yoghurt, Bifidus and kefir).
Lac. acidophilus, Lac. casei, Bif. bifidum	
Special foods	
Reduction of toxic compounds	Removal of pesticides in wine/ penicillin in milk
Foods for elderly and babies	Immunostimulating easily digestible,less constipating pasted foods, cereal-carrying cultures
Pet foods	Bioprotection ingredient against pathogens,upgrade by-products dried foods containing live lactic acid bacteria for disaster scene and unsanitary conditions.
Emergency foods	
Ice cream	*Bif.* spp. and *Lac. acidophilus* added.

The beneficial effect of pre-digestion of proteins is relatively small for healthy people, with normal gastro-intestinal function. Nevertheless, it is known that the clots of curd that form in the stomach after digestion of yoghurt are notably smaller than after the ingestion of milk and this will increase proteolysis in the stomach.

Predigestion of carbohydrates may also greatly improve the tolerance for certain foods or the possibility to produce alcoholic beverages. For example, it is a well known fact that due to the low lactose content of yogurt in combination with the lactase activity of the microbial culture, yogurt is better tolerated than milk by lactose deficient individuals. The lactase activity of *Lactobacillus bulgaricus* and *Streptococcus thermophilus* resists passage through the stomach and assists in the digestion of lactose in the intestine. Several lactobacilli, such as strains of *Lectobacillus buchneri, L, brevis, L. cellobiosis, L. fermentum and L. salivarius,* posses alpha-galactosidase activity and can be used to eliminate the gas-producing carbohydrate fraction in legumes and Soya milks. The acid formed during fermentation of carbohydrates may help lowering the pH of the gastric contents and this is certainly beneficial for elderly people, who often have a reduced gastric acid secretion.

In what concerns the fats many bacteria, yeasts and moulds used in fermentation display lipolytic activity, the nutritional significance of this is probably small.

Fermentation may reduce the content of non-digestible material in plant foods, like cellulose, hemicellulose and polygalacturonic and glucuronic acids, Breakdown of these compounds may lead to an improved bioavailability of mineral and trace elements.

Fermented milks are claimed to contain a number of biologically active compounds, which contribute to human health. These compounds include bacteria used for fermentation, their metabolic products and components derived from milk. Certain strains of lactic acid bacteria are probiotic *in vivo*, nevertheless fermented milks peptides derived from hydrolyzed milk proteins may also contribute to the probiotic growth and subsequent probiotic properties. In fact, peptides with different activities have been detected in various milk proteins upon proteolysis with digestive enzymes or starter bacteria.

These peptides include opiod, antimicrobial, anti-cancer, anti-hypertensive, immunomodulatory and mineral carrier activity. Both casein fractions and whey proteins are known to act as precursors for abovementioned bioactive peptides.

The potential to repress pathogens and spoilage organisms is another health benefit mainly of Lactic acid bacilli (LAB), which possess a number of antagonistic properties, which can contribute to the increased of the shelf safe life of fermented foods. These properties are:

– Decreasing of the pH by the production of organic acids, lactic acid, acetic acid, etc,
– Consumption of the available nutrients,
– Decreasing the redox potential,
– Production of hydrogen peroxide, under aerobic conditions,
– Production of special inhibitory components like antimicrobials and bacteriocins.

3.2. Starter cultures

From a modern point of view starter cultures are defined as: preparations which contain living microorganisms, which are applied with the intention of making use of their microbial metabolism.

Generally, the preparations used as starters today can be classified within three categories:

– **Undefined cultures,** which in the dairy industry are called *mixed-strain cultures*. These starters are based on the use of fermenting substrate, taken from selected process that resulted in good quality end products.
– ***Single strains cultures*** which contain one defined strain

- ***Multiple strain cultures,*** whi ch contain one or more defined strains, respectively. However, bacteriophage attack, mutations and seasonal variations in composition of the inoculation material (milk, vegetables, grapes etc), may often constrain stability of such multi-strain, wild types cultures. As consequence, development of starters based on single strains or defined multi-strain starter cultures were initiated.
- ***Back slopping cultures,*** These are inoculation processes based on the continuous inoculation of foods from a previous batch culture and is still currently in industrial use. This method is used in Italy for the preparation of *Sieroinnesto.*

In general, all types of microorganisms have been used for food fermentations, but Lactic Acid Bacteria (LAB) and yeasts are more common than moulds (see Table 2). A combination of LAB and yeasts are used for some products and in a few cases lactic fermentation combined with moulds make up the flora.

In order to explore the potential development of cultures of single strains, it is important that sufficient numbers of acceptable strains are available. These are strains that raise no question as to their use in a food context, meaning that they should be safe.

The GRAS status (Generally Recognized As Safe) in granted to species or cultures that have a record of safe use in foods.

Lactic acid bacteria are consumed in a large extent, primary through consumption of fermented foods. According to the latest statistics as published in Bulletin No 355 of the I.D.F. the average annual consumption of fermented milk products

TABLE 2. Fermented foods which are produced by a process involving LAB, Yeasts and moulds (Buckenhuskes ,1993)

Foods	Raw material	LAB +
Sauekraut	White cabbage	Back slopping
Various vegetables	Various vegetable	Back slopping
Olives	Green and Black	Back slopping
Soy products. (Similar to dairy products)	Soy protein	
Soy sauce	Soybeans	Yeast and mould
Sour dough	Wheaten flour or rye flour	Yeasts
Kwass	Malt, bread	Yeasts
Cacao		Yeasts
Coffee		
Wine	Grapes	Yeasts
Beer	Malt	Yeasts
Fermented sausages	Meat from various animals	
Fishes	Fishes	
Cheese	Milk	
Yogurt	Milk	
Cream	Milk	
Butter	Milk	
Kefir	Milk	Yeasts
Wine Malolactic fermentation	Wine	
Yogurt + probiotic	Milk	Probiotic bacteria

is 22 kg per capita in Europe. In total, this amount is about 8.5 billion kg fermented milk per year. With an average microbial content in these fermented products of 10^8 bacteria/g or ml, this represents ca. 8.5×10^{20} of LAB. Assuming one bacteria cell weighs 4×10^{-12} g, this means that 3400 tones of pure lactic acid bacteria cells are consumed every year in Europe, and this is only for fermented milk.

During the last decade we have witnessed a dramatic progress in the development of methodologies for the genetic modification of bacteria. Obviously the modifications one would like to take place are dictated by the applications whereby bacteria are used. Without any further specification of the sort of application only very general features for strain improvement can be given, like:

- Improvement of product yield,
- Combination of properties into a single strain,
- Elimination of undesirable properties,
- Introduction of novel properties,
- Improvement of existing properties (see Table 3).

3.3. Genetic engineering of species

In order to make the ideal culture for any particular food application, it is necessary to understand the function that is required of the culture, and to have tools to improve the function of the culture. Both aspects have been advanced considerably through scientific achievements during the last few years. The search for a starter culture has until recently relied on the screening of a large number of isolates in small-scale food fermentations. The starter culture finally selected would be the one giving a satisfactory performance in the process and also giving an acceptable organoleptic evaluation of the food product. Excellent cultures have been isolated this way, and the method will certainly also in the future be used to expand the pool of microorganisms to be used as starter cultures. The last two decades of research have, however, generated tools allowing us to specifically target the individual genes and metabolic pathways responsible for desired performance parameters of a starter culture. Specific targeting makes screening by high throughput methods possible, and it opens the possibility to use mutant selection and genetic engineering to construct starters that are superior to the ones found in nature (Hansen, 2002).

Recombinant DNA technology, developed in the 1970s, together with traditional mutagenesis and selection have impacted industrial microbiology and have been major influences in constructing bacterial strains with enhanced fermentative qualities. Conjugation involves the transfer of plasmids from one microorganism to another and necessarily implicates that a close relationship exists between the donor and recipient microorganism. Chromosomal genes may also be transferred, and in this case pure DNA, modified or unmodified, is introduced into recipient cells, without the need for closely related species (Dass, 1999)

Genetic engineering may be a powerful tool but it should not at all affect the GRAS status of those microorganisms.

TABLE 3. Genetic transfer systems in Lactic Acid Bacteria (Lee B, 1996b)

Process	Major use	Limitation
	IN VIVO	
Conjugation		
Donor/recipient cell- to cell contact	Transfer of transmissible plamids	Cell surface recognition required The host DNA must be homologous for recombination to occur
Conjugal plasmid transfer at high frequencies in lactococci and leuconostocs	Genetic transfer of chromosomal DNA with homologous DNA Mobiliation of non transmissible plasmids	
Transduction		
Bacteriophage-mediated transfer of DNA	Useful technique for making fine structural changes in genetic material.	Only small fragment of DNA transfer
Transformation		
Uptake of naked DNA by host	General DNA transfer procedure	Plasmid DNA maintained, but chromosomal DNA must be homologous for recombination to occur
These systems (protoplast, whole cell, electroporation)	Electroporation technique is current choice for plasmid DNA transfer into lactic acid bacteria	
	IN VITRO	
Recombinant DNA		
Insertion of foreign DNA into a plasmid or phage cloning vector, transformation into the host; Screening of the target gene.	Transfer of single gene or groups of linked genes from any source without barrier to a desirable host strain.	No theoretical limitation, but must overcome stability and expression of introduced gene; must meet safety requirements

Among the microorganisms that have been genetically altered through protoplast fusion, one may find *Aspergillus oryzae*, *Penicillium caseicolum*, *P. nalgiovense* and *Rhizopus niveus*. *Penicillium roqueforti*, *Mucor circinelloides* and *A. sojae* are starter organisms that have been manipulated by transfection. Hybrids produced form the mesophilic *Saccharomyces cerevisiae* and the cryophilic *S. bayanus* have promising wine-making qualities at 10°C, releasing more flavour compounds than either of the parent species.

The undesirable properties of microorganisms may also be removed by using gene disruption. In yeasts and moulds, homologous DNA fragments tend to integrate at the homologous site on the fungal genome, resulting in activation of the gene. This process, called homologous recombination is being introduced for the development of starter strains without particular undesirable qualities (Dass, 1999).

Important aspects to be taken into account when applying genetic engineering include the need to carefully compare the activity of genetically manipulated microorganisms with that of parent strains, as well as ensure that no changes occur in the food products wherein they are included that may be harmful to consumer's health or to the organoleptic quality of the product.

4. Starter Culture Production

4.1. Introduction

Microorganisms are important to us for many reasons, but one of the principal ones is that they manufacture products that are valuable for us. These may be very large materials such as proteins, nucleic acids, carbohydrate polymers, or even cells, or they can be smaller molecules which we usually separate into metabolites essential for vegetative growth and those inessential (i.e. primary and secondary metabolites, respectively) (Demain, 2000). These production processes normally occur during fermentation processes.

The direct addition of selected starter cultures to raw materials has been a breakthrough in the processing of fermented foods, resulting in a high degree of control over the fermentation process and standardization of the end product. Strains with the proper physiological and metabolic features are isolated from natural habitats (recent trend exists in the isolation of wild-type strains from traditional products) or from successfully fermented products.

Recently, the use of functional starter cultures in the food fermentation industry is being explored. Functional starter cultures are starters that possess at least one inherent functional property. The latter can contribute to food safety and/or offer one or more organoleptic, technological, nutritional, or health advantages, as previously mentioned. The implementation of carefully selected strains as starter cultures or co-cultures in fermentation processes can help to achieve *in situ* expression of the desired property

Many different species and strains of bacteria are used in making fermented dairy products. Thus starters must be properly identified, not only to ensure reproducibility of starter performance but also for maximum protection of manufacturers under "product liability" laws. Traditional species classification on the basis of microscopic analysis and phenotypic characteristics (especially sugar fermentation abilities and other biochemical properties) is very useful and remains the most widely recognized approach, but it is increasingly being augmented by molecular methods of bacterial identification (Pot et al. 1994).

Important properties of a microorganism to be used as a starter culture include GRAS status, non-pathogenicity, lack of production of toxic compounds, absence of production of compounds that may reduce the desirability of product, ease of handling, ease of storage and transport, period required for raw material to be converted to final edible product, genetic stability of the species, a certain degree of natural resistance to inhibitory factors that may be present in the raw material

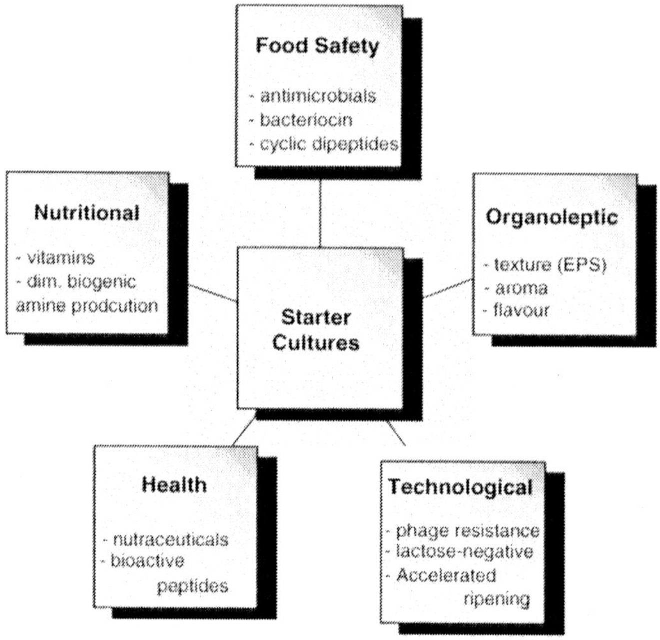

FIGURE 1. Functional properties of starter cultures for food fermentation.

or in the metabolic products as fermentation proceeds, ability to grow in symbiosis with other starter microorganisms, reproducibility in terms of production of different batch cultures and lack of spoilage of the food product throughout storage. The selection of strains is constrained by the product manufactured and its intended use,. Several examples of selection criteria are described below.

4.2. *Starter cultures for wine*

To ensure complete fermentation and to avoid wine aroma defects winemakers encourage the growth and dominance of the yeasts, namely *Saccharomyces cerevisiae*. Selected strains of *Saccharomyces cerevisiae* can be added to the must to initiate fermentation. Active dry wine yeast is available since 1964. The main requirements for yeast starter cultures selection are: (i) complete fermentation (alcohol tolerance) (ii) tolerance to high sugar musts, (iii) tolerance to SO_2, (iv) tolerance to low temperature and high temperature (v) tolerance to high pressure (sparkling wines) (vi) production of small amounts of acetic acid, acetaldehyde, H_2S and mercaptans, diacetyl, SO2, higher alcohols (vii) low foaming, (viii) good flocculation following fermentation to help clarification, and (ix) suitability for drying (Romano *et al.*, 2003).

4.3. Starter cultures for vegetables

Lactic acid bacteria (LAB) are mainly responsible for the fermentation of vegetables.

The LAB selection process includes: (i) evaluation of growth in vegetable juices media, (ii) analysis of acidification profiles in vegetable juice, and (iii) performance of compatibility tests between strains.

The selected culture has to be dried (generally freeze-dried) and finally tested in vegetables in order to confirm its efficiency (Gardner *et al.*, 2001).

4.4. Starter culture for bread

Recent years have witnessed the development from continuous propagation of mixed-cultures in bakeries to commercial, well-defined starter cultures and from simple, manual operations, to continuous processes with advanced control. In the main starter cultures, sourdoughs contain both LAB and yeasts. Strain development has initially been based mainly on their capacity for acid formation (a good starter culture should lower the pH rapidly down to ca. 4), but a good sourdough baking also gives bread a longer shelf-life and improved flavour (the ratio of lactic acid and acetic acid is also important for the flavour of the final bread (Linko *et al*, 1997).

4.5. Starter culture for meat

Starter cultures are of major importance for ensuring the stability and safety of fermented sausages. They consist of pure or mixed strains of lactic bacteria, *Micrococcaceae*, yeasts and moulds which are added for their essential biochemical and, simultaneously, to stabilize fermentation by inhibiting growth of spoilage and pathogenic bacteria.

The starter culture should have the following abilities: nitrate reduction, acid production and decrease of pH value, O_2 consumption, H_2O_2 degradation, proteolytic and lipolytic activity and delayed rancidity (Samelis et al., 1998)

4.6. Starter culture for dairy products

Lactic acid bacteria (LAB) are predominant in microflora of milk and its products, many species are involved in the daily manufacturing of dairy products. Screening tests of LAB for its suitability as a starter culture are based on milk fermentation ability: acidification rate, flavour/texture promotion, proteolytic activity, compatibility with other strains (absence of inhibition and synthesis of bacteriocin), resistance to bacteriophage, salt sensitivity, stability of characteristics and agllutinine resistance (Beresford, et al, 2001) and special attributes (production of exopolysaccharide, probiotic effect, etc) and concentrated culture production: yield in industrial fermentation, maintenance of viability/activity during processing, stability in freeze-dried/frozen format and activity in Direct Vat Inoculation (DVI).

All relevant criteria are taken carefully into account when strains are selected for industry use. Collating all the available data is facilitated by a computerized database. Importantly, strains are evaluated not only on their individual merits but also on the combined properties of all the strains used in a multiple-strain starter. In summary, the main goals are reliable acid production, consistent flavour (especially absence of flavour defects) and minimal inhibition by phage.

4.7. Composition of starter cultures

Starter cultures are used either from a bulk starter vessel or inoculated directly into the mixture. In commercial applications it is possible to find different culture systems described below (Wigley, 1999).

4.7.1. Daily propagated cultures

Daily propagated cultures are prepared from a mother culture, purchased at intervals of two weeks and a year. Mother cultures, liquid or freeze-dried are daily subcultured in order to maintain the activity of the culture, which is also subcultured on through a feeder stage, for inoculation into the bulk starter vessel. Their use is now restricted to small industries.

4.7.1. Deep-frozen cultures

Frozen concentrated cultures are packed in aluminium cans ring-pulls despatched in insulated boxes containing solid CO_2 to maintain the temperature at about –70°C. On receipt, the cans must be stored in a freezer operating at 50°C or lower. The shelf life of the cultures depends on the temperature of storage, but is typically 3 months at –50°C. Bulk starter is prepared by thawing the contents of a can and using them for inoculation.

4.7.3. Lyophilized concentrated cultures

Freeze-dried concentrated cultures have a significant advantage over deep frozen concentrates: they can be stored in a conventional refrigerator and can be sent by post, reducing transport and storage costs. The cultures are packed in foil sachets, in quantities suitable for the direct inoculation of bulk starter medium. Bulk starter is prepared by sprinkling the contents over the surface of the medium.

4.7.4. Deep-frozen direct vat cultures

These cultures must be stored at –50°C or below. They are supplied either in a 360 ml aluminium can with a ring-pull, or as granulates in cardboard box, polythene bottle or aluminium sachet, typically containing 500 g. Some suppliers offer cultures that are packed to reflect customer requirements, with the weight of culture tailored to the size of the batch to be fermented.

4.7.5. Lyophilized direct vat cultures

These cultures are packed in aluminium foil sachets, and can be stored in the domestic refrigerator at 4°C. They are packed in quantities according to units of activity, with a specified number of units being required per 100 g of food to be fermented.

4.8. Manufacture of starter cultures

The need for investment in research into isolating culture strains that perform better in a changing environment has resulted in specialization and rationalization within the industry. As a result many of the old manufacturing companies have been acquired by multinational corporations who work closely with the food industries towards a customized approach to each factory's needs.

During the production of cultures for food manufacture different steps have to be considered: (i) isolation and characterization of microbial strains from of high quality, representative and different type of selected food, and in some cases from different regions; (ii) development of methods for starter production that includes -adequate production of starter cultures in quality supplied with ample nutrients and quantity and assure maintenance of the culture through freezing, drying or freeze drying processes, and (iii) control of activity of starter cultures, checking for contamination by other yeasts and bacteria and check viability and desired fermentation characteristics and production of commercial cultures for bulk starter preparation in factory.

Consistent starter performance in the cheese factory cannot be achieved unless a reliable supply of starter with reproducible characteristics is available. Starter culture companies, supply the inocula used to produce the bulk cultures. These may be purchased as frozen or freeze-dried, concentrated suspensions, containing high numbers ($>10^{10}$) of cells and are used to inoculate bulk starter fermenters in the cheese factory.

Cultures are grown in fermenters under strictly controlled conditions: pH control (using ammonium hydroxide as neutralizer) for 15 hours at constant temperature in proprietary economically feasible media. The resultant cell crop should form a stable pellet after centrifugation or be recoverable in high numbers after a microfiltration process and should be stable under the conservation condition to be used, e.g., freeze drying, spray drying or freezing. The pellet is mixed with a cryoprotectant, and then aseptically packaged into pre-sterilized plastic containers, which are frozen and stored long-term at –80°C or freeze-dried in quantities sufficient to inoculate 300 L, 500 L or 1000 L.

A number of factors influence the viability of lyophilized starter cultures these include, growth medium, freezing rate, drying temperature, composition of freezing medium, together with subsequent storage conditions including temperature, atmosphere, exposure to light and relative humidity. Saccharides are commonly used for protecting lyophilized bacteria during freezing, drying and storage. Maximal survival of organisms in dried starter cultures is obtained by exclusion of air. Oxygen is thought to interact somehow with the membranous system, causing damage to the initiation of DNA synthesis; it has also been shown that lyophilized bacteria produce free radicals, when they are alive prior to lyophilization (Baati *et al.*, 2000). Ascorbic acid, which is commonly found in high concentrations in growth and freezing media, has been described as a two edged sword. In one respect, it acts as a nutritional agent and an antioxidant, but when lyophilized with proteins it is shown to produce free radicals, probably in metal catalyzed reactions.

Modifications in the cell environment (e.g. temperature) can phenotypically alter the composition of cells, and it has been shown that cell membrane composition of microorganisms can be modified in number of ways which can play a role in resistance to freezing. For example cells became more resistant to freezing in water and saline buffer at both rapid and slow cooling rates when they were grown at low temperatures due to their high unsaturated fatty acids (Carvalho et al., 2004).

Spray drying has already been investigated as a method for the production of lactic acid bacteria cultures. Studies have been undertaken on the spray drying of yogurt cultures, cheese cultures, and bacteriocin-producing lactic acid bacteria. The driving force for these studies was mainly to demonstrate the capability of spray-dried cultures in replacing the usual liquid or frozen-bulk starter or freeze-dried cultures in the production of fermented products. In comparison to the latter techniques of culture production, spray drying is claimed to be more cost effective and less time consuming.

However, it is obvious that the exposure to high air temperatures, which are required to facilitate water evaporation during the passage of the bacteria in the spray drying chamber, exerts a negative impact on their viability and hence their activity in the spray-dried product. Furthermore, since water contributes to the stability of biological molecules, the removal of water may cause irreversible changes in the structural and functional integrity of bacterial membranes and proteins. Preservation of these essential functions and structure is crucial for the survival of bacteria and the retention of their functionality. However, there is a body of evidence, which suggests that some strains of lactic acid bacteria (namely probiotic strains) can be spray dried without a drastic loss of viability and activity, or at least show survival rates during spray drying which are comparable to that on freeze-drying (Ananta *et al.*, 2005).

4.8.1. Bulk starter propagation

All starter cultures, other than DVI, must be subcultured in a bulk starter vessel to produce enough cells for addition to the food being fermented. Methods of bulk starter manufacture have evolved with technology. In the 1960s it was common for cheese makers to prepare bulk starter using whole milk in milk churns; however, reconstituted skimmed milk powder became the standard medium in the 1970s upon recognition of its consistency as a substrate and enclosed bulk starter vessels began to be used. Nowadays, the use of a phosphated medium is common, since this type of culture medium chelates calcium, thereby making it unavailable to phages, which require it as a catalyst (see Table 4) (Wigley, 1999).

A factory bulk starter system is normally based on use of a stainless-steel bulk starter vessel (typically 5 000 to 20 000 litres) with an inbuilt cleaning and sterilization system (cleaning-in-place, C.I.P.), using chemical sterilization with hydrogen peroxide/peracetic acid at room temperature. The medium used for growth of the bulk starter is typically UHT (sterilized by brief high-heat treatment) milk. Sterile air from a compressor is fed into the fermenter via a HEPA (High Efficiency Particulate Air) filter (maintaining positive pressure within the vessel) and is allowed to escape around the stirrer shaft and inoculation port. Frozen starter cultures from the starter companies are added through the inoculation port.

A pH control system may be adopted in which case constant pH is maintained through the use of a pH electrode connected via control circuitry to an alkali dosing pump. The stirrer is kept on throughout the whole incubation period to ensure thorough mixing of the intermittent alkali additions and stirrer speed is deliberately kept low to prevent excessive air incorporation into the starter medium. At the end of the fermentation, chilled water is used to cool the bulk starter to below 4°C. At this temperature, satisfactory starter activity is maintained for up to 48 hours. Prior to the introduction of this system, it was common practice to heat milk at 85°C to 95°C for 30 to 60 min within the bulk starter vessel. Such treatments subjected the steel vessel to severe stresses of heating and cooling without always achieving sufficient inactivation of phages and spores. The new system is expected to have lower long-term costs, thanks to an expected operational life of at least twenty years due to the use of non-corrosive chemical sterilization of the vessel and external heat treatment of the medium (ASCRC starter strategy, 1999).

4.8.2. Bulk starter system vs. direct vat inoculation

Considerable debate has centred on the question of which starter preparation system is the most cost-effective for the cheese manufacturer: direct-to-the-vat culture concentrates (frozen or freeze-dried off-site by commercial starter suppliers for direct inoculation of the cheese vat) or bulk starter cultures grown in large fermenters on-site in the dairy factory.

TABLE 4. Bulk starter media (adapted from Wigley, 1999)

Substrate	Observations
Whole milk	Not widely used
Skimmed milk powder (SMP) at 12% solids	Still used today
Phosphated media plus SMP at 12% total solids	First generation media
Phosphated media used at 12% total solids	Isolated areas of use
Low solids phosphated media – 5-6% total solids	Most popular media in use
Internal pH control media	Isolated areas of use
External pH control media	Popular in USA

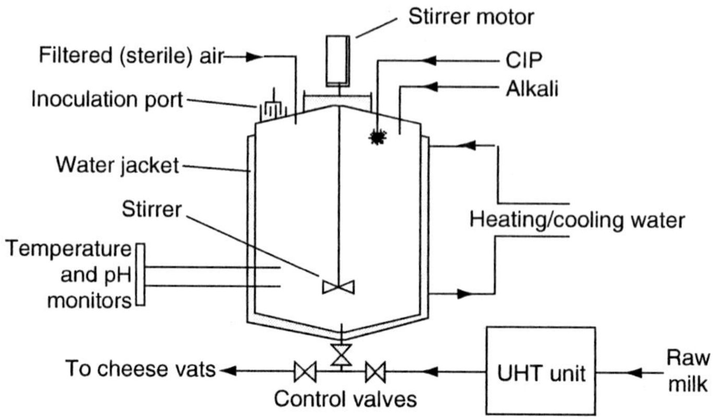

FIGURE 2. Example of a pH-controlled bulk starter fermenter unit (adapted from ASCRC starter strategy).

Statistics state that the proportional cost of in-factory bulk starter preparation decreases with increasing factory size. A moderate-sized cheese factory (10 000 tons of cheese per year) may consider bulk starter system advantageous. On the other hand, direct-to-vat starters are the most cost-effective option in smaller factories, where the cost of installing and operating a bulk starter facility would be too great; although relatively expensive DVC do away with the need to propagate and check cultures for activity in plants. Even in large factories, direct-to-vat starters have a place in making specialty products, adding adjunct cultures, or as a back-up for the bulk starter system. The only handicap is that a small but perceptible lag in acid production is observed.

4.8.3. Starter cultures and associated spoilage problems

When the starter culture, in use, is impure or its activity has deteriorated over time there may be a risk of spoilage of the food product. There is, therefore, need to regularly monitor the quality of the starter culture. The attributes that need to be monitored are activity and purity. To do this, there are various molecular methods available such as PCR and DNA hybridisation, which are said to have great potential for use within the food industry.

Furthermore, culture systems used in cheesemaking, for example, must be monitored regularly, for bacteriophage infection, so that any strain being attacked by phages may be replaced. Besides selection for strains that are resistant to bacteriophage infection industrial producers may also apply culture rotation. Technologists from the culture suppliers help minimize the risk of phage attack by selecting a rotation of unrelated cultures; nevertheless, strict hygiene procedures within the dairy are also required.

5. Yeasts

Yeasts are one of the most important types of microorganisms, for science, technology, medicine and make a significant contribution to the economies of many countries. Some species play beneficial roles in the production of foods, beverages, alcohol and pharmaceuticals products, while others play detrimental roles as spoilage organisms and agent of human diseases. Table 5 shows the most important alcoholic beverages produced by yeast fermentation.

Traditionally, all the strains used in the yeast fermentation belong to the *Saccharomyces, Candida and Kluyveromyces* genera and the choice is made by studying associated properties (Walker, 1998):

- Genetic characteristics, the choice of yeast strain employed;
- Cell physiology, e.g. the stress tolerance of yeast cell, the viability and vitality of the cells and the inoculum cell density.
- Nutritional availability; the concentration and category of absorbable nitrogen and
- Physical environment, temperature, pH and dissolved oxygen.

For each product, the strain with the most suitable properties has to be identified. The diversity of yeast metabolic processes is thus reflected in the diversity of yeast technologies. Today it is possible to develop new strain of *S. cerevisiae* in order to improve fermentation process by biotechnological methods, namely: Hybridization – Mutation and selection – Rate mating – Spheroplast fusion – Single chromosome transfer – transformation – Recombinant DNA technology

The industrial processes commonly carried out by the specific types of yeast can be divided into four major classes (Lee, 1996c) :

TABLE 5. Fermented alcoholic beverages produced by processes involving yeasts (Walker, 1998)

Beverage	Raw material	Carbohydrates	Fermented Yeast	Products
Beer	Barley adjuncts Rice, wheat, maize	Hydrolysed starch	*S. cerevisiae(Ale)* *S. carlsbergensis Lager*	
Whisky	Barley, Barley wheat etc.	Hydrolysed starch	*S. cerevisiae*	
Wine	Grapes	Grapes juice sugars	*S. cerevisiae, wild strains*	
Spirits	Barley, maize, molasses, grapes, sugar can potatoes, cereals etc.	Sucrose Hydrolysed starch various sugars Hydrolysed Inulin	*S. cerevisiae*	Rum, Vodka Raki Tequila
	Whey	Lactose	*K. marxianus*	Neutral spirits, cream liqueurs

1. The production of yeasts as a source of baker's yeast or single-cell proteins. (SCP)
2. The production of nutritional, flavour and bulking aids provided by inactive yeast,
3. The production of alcohol beverages by brewer's and wine yeasts and
4. The production of bread or baked goods by baker's yeast.

The principal characteristics of a **brewing yeast strain are** (Walker, 1998a):

- Rapid fermentation rate without excessive yeast growth.
- Efficient utilization of maltose and maltotriose with good conversion to ethanol.
- Ability to withstand the stresses imposed by the alcohol concentrations and osmotic pressures encountered in brewing.
- Reproducible production of correct levels of flavor and aroma compounds.
- Ideal flocculation properties for the process employed.
- Good handling characteristics. (E.g. retention of viability during storage, genetic stability).

Principal characteristics of a **wine yeast strain**:

- Alcohol tolerant
- Homothallic diploids.
- Correct volatile acidity in relation of ethanol produced.
- Fermentation vigour,
- Aromatic character, esters, terpenes, succinic acid, glycerol etc,
- SO2 tolerance,
- Correct balance of sulphur production.
- Low acetaldehyde,
- Killer character
- Low urea excretion (in view of ethyl carbonate production).
- Sedimentation characteristics that allow easier separation of yeast from wine.

Principal characteristics of a **bread yeast strain**:

- High glycolytic activity, especially with regards to CO2 evolution rate in dough fermentations.
- Rapid utilization of maltose under glucose-repressing conditions.
- Better freeze-tolerance is required in yeasts incorporated into frozen dough's for instant baking.

Chemical tolerance, resistance to bread preservatives and the presence of sodium chloride.Production Of BioethanoIn countries with large agricultural area, such as Brazil, South Africa, Canada and USA, intensive studies are being conducted on the production of ethanol from carbohydrates such as sucrose, starch and cellulose-derived, which until now is not competitive with petroleum, but biotechnology and bioprocess engineering may enhance the future competitiveness. More precisely, we can classified the raw material types : **a**. saccharine materials such as sugar cane, sugar beets, molasses and fruit juice, **b.** starch materials like cereals, potatoes, Jerusalem artichokes and manioca and **c.** cellulose materials such as wood and sulphite liquor.The efficiency of energy conversion by ethanol fermentation varies considerably in efficiency of energy yield, ratio of energy demand to energy produced, is as follows: sugar beet 86%, potatoes 59%, corn 25%, cassava 50% and sugar cane 66% (Data 1989).

Olsson and Hahn-Hagerdal (1996), have evaluated the performance of several pentose-fermenting yeasts, including recombinant strains, in converting lignocellulose hydrolysates to ethanol. *Candida shehatae and Pichia stipitis* were the most successful yeasts studied, although the presence of inhibitory chemicals in the hydrolysate curtailed their growth and metabolic activities.

5.1. Desired properties of fuel ethanol – producing yeasts

Growth: High rate of yeast growth but lowered final growth yield. High cell viability and vitality. Tolerance to high sugar, toxic chemicals/inhibitors and temperature fluctuations. Resistance to bacteria contamination. Genetic stability. Easy to propagate. Minimal heat generation during fermentation. Capacity of appropriate flocculation characteristics depending on process requirements. Capacity of killer character.

Fermentation: Fast fermentation rates and high and reproducible ethanol yields in very high-gravity media, to produce > 18% v/v ethanol. High ethanol tolerance. Low pH and high temperature optima for fermentation . Efficient utilization of varied substrates. Reduced levels of minor fermentation metabolites, organic acids, glycerol, higher alcohols, esters, aldehydes. Derepression for a variety of saccharides in presence of glucose. Amylolytic and cellulolytic activities.

Environmental factors: which play important roles in dictating yeast fermentation performance are: assimilable nitrogen, Magnesium ions, oxygen, lipid compounds, temperature.Theoretically, from one gram of glucose, 0.511 grams of ethanol can be obtained. When pure substrates are fermented, the yield is 95% and reduces to 91% when industrial-grade starting materials are used. One hundred grams of pure glucose will yield 48,4 grams of ethanol 46,6grams of CO, 3,3 grams of glycerol and 1,2 grams of biomass (Crueger W. and Crueger A, 1989). The flow chart in Figure 3 shows a typical series of events. ....

6. Inventory of Microorganism with a Documented History of use in Food

The IDF in collaboration with EFFCA, European Food and Fed Cultures Association, prepared an inventory of microorganisms with documented history of use in food (Bulletin of the IDF 377 p. 10-14). The history of use in foods of species in the inventory is documented by scientific literature references and statements in good faith from companies.

Glucose 100 gram.

⇓ Glycolysis

Pyruvate

Pyruvate decarboxylase

⇓ Mg^{2+} Thiamine pyrophosphate

Acetaldehyde + CO2 (46,6 grams)

⇓ Alcohol dehydrogenase, NADH2

Ethanol 48.4 grams

9.3 gram glycerol.

FIGURE 3. The biosynthesis of ethanol.

The inventory consists mainly of lactic acid bacteria and some other bacterial species belonging to *Enterococcus, Streptococcus as well as yeasts and moulds.* These strains used by the food industry have long history of application in food without any adverse effects (Mogensen et al., 2002).

6.1. Value added products from the microorganisms

After the World War II a large scale production start with the *Penicillin* species, in order to produce this natural antibiotic more economically. Through intense research within the pharmaceutical industry around 20 antibiotics were subsequently put into commercial production.

In the period from 1960 to 1975, new microbial processes for the production of amino acids and 5′nucleosides as flavour enhancers were developed in Japan. During the same period, successful techniques for the immobilization of enzymes and cells were developed. At the same time start the use of continuous fermentation for the production of single –cell protein from yeast and bacteria for use as human and animal food.

Since 1975 biotechnology has entered some important new phases. First was the development of the hybridoma technique for the production of monoclonal antibodies, which was of interest primarily in the medical diagnosis field. The production of human proteins using genetically engineered *Escherichia* coli was also developed around this time. The first product, human insulin, was introduced, followed by Factor III, human growth hormone, interferons and urokinase.

At present, a vast array of human proteins and other products were developed, based in five distinct approaches:

- **Screening** for the production of new metabolites from new isolates and /or new test methods.
- **Chemical modification** of known microbial substances.
- **Biotransformation,** which result in change in a chemical molecule by means of a microbial or enzymatic reaction.
- **Interspecific protoplast fusion**, which means of recombining genetic information from rather closely related producer strains.
- **Gene cloning,** in which genes may be transferred between unrelated strains, which are producers of known substance.

6.2. Some fermentation products of high value added from yeasts

Besides ethanol, other industrially useful alcohols can be produced by **yeast**:

- Higher alcohols,
- Polyhydric alcohols such as glycerol, xylitol, sorbitol, arabinitol, erythritol and mannitol.

Other chemicals produced by **yeasts**:

- **Organic acids:** Citric, Itaconic, Malic, L(+) isotric, a-Ketoglutaric, Brassylic, Sebacic, Fumaric.
- **Fatty acids:** Stearic, Long-chain dicarboxylic.
- **Amino acids:** Lysine, Tryptophan, Phenilalanine, Glutamic acid, Methionine.
- **Vitamins:** Riboflavin, Pyridoxine, D-Erythro-ascorbic acid.
- **Sterol:** Ergosterol, Steroid precursors.
- **Polysaccharides:** Pullulan, Phosphomannan gums,Glycolipids.
- **Single Cell Protein.**
- **Whole** – cell **biomass have a novel application, such as:** Livestock growth factor, Biotherapeutic agent, Chemical reagent, Food pigment, Biocontrol agent, Biosorbent /bioremediation, Bioensor, Bioelectrical fuel cells(Walker, 1998b).

In addition to the abovementioned uses of yeasts, the yeast cells are also extremely valuable as experiments models in biomedical research, such as: Oncology, pharmacology and toxicology, virology genetics and neurodegenerative diseases (Walker, 1998c)

6.3. Some fermentation products of high value added from bacteria

Polysaccharides are used commercially to produce gels and thicken and stabilize foods, medicines and industrial products. At list 20 different microbial polysaccharides with market potential have been described, but the largest part of the market is held by xanthan.

TABLE 6. Microbial polysaccharides use by the Food Industry (Lee, 1996d). (Crueger *et al.*, 1989a)

Polysaccharides	Microorganisms	Composition
Xanthan mannose	*Xanthomonas*	D-Glucose, D-mannose
	Campestris	D-glucuronate
Alginate	*Pseudomonas aeruginosa*	D-Mannuronic acid
		L-Gucuronate
	Azotobacter	
	Vinelandii	
Dextran	*Lactobacillus sp.*	D-Glucose
	Leuconostoc mesenteroides	
	Leuconostoc dextranicum	
	Streptococcus mutans	
Curdlan	*Alcaligenes*	D-Glucose
	Agrobacterium	
Pullulan	*Aurobasidium pullulans*	D-Glucose

The microbial production of polysaccharides from microbial sources offers controllable polymer synthesis from materials in constant supply, yielding products that possess unique physical and chemical properties, improved functional characteristics and low biological oxygen demand.

According to Lee (1996d), the microbial polysaccharides consist of three main types:

- Intracellular polysaccharides, which may provide mechanisms for storing carbon or energy for cell,
- Structural polysaccharides, which are components of cell structures, e.g. lipopolysaccharides and
- Extracellular or exopolysaccharides.

Table 6 presents the most important polysaccharides use by the food industries.

From all this data we can see that the selection of specific starter cultures for substrates such as meat, vegetables, cereals, indigenous crops or combinations of dairy products with cereals substrates and not only, provide good possibilities for the future development of healthy foods with high nutritional and sensorial properties for people in many parts of the world. In the same time for the production of added value products, it seems evident that the microorganisms have become a very important economical asset.

7. References

Ananta, E., Volkert, M., and Knorr, D., 2005, Cellular injuries and storage stability of spray-dried *Lactobacillus rhamnosus* GG. *International Dairy Journal* (in press).

Baati, L., Fabre-Gea., Auriol, D., and Blanc, P.J., 2000, The cryotolerance of *Lactobacillus acidophilus*: effect freezing conditions on the viability and cellular Levels. *International Journal of Food Microbiology* **59**:241-247.

Beresford, T.P., Fitzsimons, N.A., Brennan, N.L., and Cogan, T., 2001, Recent advances in cheese microbiology. *International Dairy Journal* **11**:259-274.

Bosschaert, M.A.R. and Pot, B., 2001, Yakult Europe BV, Almer, the Nederlands

Buckenhuskes H., 1993, Selection criteria for lactic acid bacteria to be used as starter cultures for various food commodities. *FEMS, Microbiology Reviews* **12**:253-272.

Carvalho, A.S., Silva, Joana, Ho, P., Teixeira, P., Malcata, F.X., and Gibbs, P., 2004, Relevant factors for the preparation of freeze-dried lactic acid bacteria. *International Dairy Journal* **14**:835-847.

Crueger, W. and Crueger A., 1989, Organic feedstocks produced by fermentation.)in: *Biotechnology. Book.* Thomas D. Brock ed, Sinuaer Associates Inc. Sunderland. USA. p. 331.

Crueger W. and Crueger A., 1989a, Organic feedstocks produced by fermentation.)in: *Biotechnology. Book.* Thomas D. Brock ed, Sinuaer Associates Inc. Sunderland. USA. pp. 124-129.

Dass, C.R., 1999, Starter cultures: importance of selected genera in: *Encyclopaedia of Food Microbiology,* Academic Press.

Davis, J.G., 1952, *Food* **21**:249(July) 284 (August).

Demain, A.L., 2000, Small bugs, big business: the economic power of the microbe. Biotechnology Advances **18**:499-514.

Derieux, J., 1988, Histoire de la panification et de levure dans *Levure et panification ..* Fould ed., Springer. Tecno-Nathant. pp.14-15

Gardner, N.J., Savard, T., Obermeier, T., Caldwell, G., and Champagne, C.P., 2001, Selection and characterization of mixed starter cultures for lactic acid fermentation of carrot, cabbage, beet and onion vegetable mixtures. *International Journal of Food Microbiology* **64**:261-275.

Hansen, E.B., 2002, Commercial bacterial starter cultures for fermented foods of the future. *International Journal of Food Microbiology* **78**:119-131.

Hough, J.S., Briggs, D.E., Stevens, R., and Young, T.W., 1982, *Malting and Brewing Science* 2nd.ed., Chapman & Hall, London.

Huis in't Veld, J., Hose, H., Schaafsma, G., Silia. H., and Smith, J., 1989, Unpublished personal data.

Lee, B., 1996a, Bacteria – Based processes and Products. p. 221. Fundamentals of Food Biotechnology. Book ed. VCH, USA.

Lee, B., 1996b, *Bacteria – Based processes and Products.. Fundamentals of Food Biotechnology*. VCH, USA p. 233.

Lee, B., 1996c, *Bacteria – Based processes and Products.. Fundamentals of Food Biotechnology* VCH, USA. pp. 183-184.

Lee, B., 1996d, *Bacteria – Based processes and Products.. Fundamentals of Food Biotechnology* VCH, USA. pp. 276-77.

Linko, Y-T., Javanainem, P., and Linko, S., 1997, Biotechnology of bread baking. *Trends in Food Science* **8**:339-344.

Mogensen, G., Salminen, S., O'Brien, J., Ouwehand A., Holzapfel, W., Shortt, C., Fonden, R., Miller G.D., Olsson, L., and Hahn-Hagerdal, B., 1996, Fermentation of lignocellulic hydrolysates for ethanol production. *Enzyme and Microbial Technology*.**18**:312-331.

Pot, B., Ludwig, W., Kersters, K., and Schleifer, K.-H. 1994., Taxonomy of lactic acid bacteria,. In: *Bacteriocins of Lactic Acid Bacteria: microbiology, genetics and applications,* L. De Vuyst and E.J. Vandamme ,eds., Chapman and Hall, London pp. 13-90.

Romano, P.v Fiore, P., Paraggio, M., Caruso, M., and Capece, A., 2003, Function of yeast species and strains in wine flavour. *International Journal of Food Microbiology,* **86**:169-180.

Samelis, J., Metaxopoulos, John., Vlassi, M., and Pappa, M., 1998, Stability and Safety of traditional Greek salami — a microbiological. *International Journal of Food Microbiology* **44**:69-82.

Wainwright, M., 1992, An Introduction in Fungal Biotechnology. John Wiley and Sons, Chichester,U.K.

Walker, G., 1998, Development in Yeast Technologies in: *Yeast Physiology and Biotechnology*. J.Wiley and Sons, England. pp. 283-309.

Walker, G., 1998a, Development in Yeast Technologies in: *Yeast Physiology and Biotechnology*. J.Wiley and Sons, England. pp. 284-288.

Walker, G., 1998b, Development in Yeast Technologies in: *Yeast Physiology and Biotechnology*. J.Wiley and Sons, England. p. 304.

Walker, G., 1998c, Development in Yeast Technologies in: *Yeast Physiology and Biotechnology*. J.Wiley and Sons, England. pp. 309-311.

Wigley, R.C., 1999, Starter cultures use in the food industry. *Encyclopaedia of Food Microbiology*, Academic Press.

9 Pathogenic, Commensal and Beneficial Microorganisms in Foods

Ana M.P. Gomes[1], Manuela E. Pintado[1], and F. Xavier Malcata[1]

1. Introduction

Moulds, yeasts, bacteria, viruses and minute parasites are microorganisms that virtually occur everywhere within the environment. Viruses are the smallest known living organisms (see Figure 1). They do not have a cell wall, a membrane or a nucleus and are defined as obligate intracellular parasites. When they reproduce, they take over the life processes of host cells, which continue to live while producing viral copies. Most cells in food products are dead following processing and therefore simply function as carriers of viral material. Some viruses can be spread by people who handle food and do not follow careful personal hygiene habits. A person may indeed excrete viruses in faeces, urine or even through sneezing; so, if hands are not washed well after using toilet facilities as well as sneezing; any food handled after the event will be contaminated. Foods that are not usually processed thermally after handling – such as bakery products, uncooked oysters or clams, sandwiches, salads and desserts, may therefore carry and hence transmit viral illnesses. Bacteria are single-celled organisms, which multiply and increase in number through cell division given appropriate environmental conditions. Many pathogenic bacteria are facultative anaerobes, so they can grow in either aerobic or anaerobic conditions.

Yeasts are also single-celled organisms, which can convert nutrients into alcohol and carbon dioxide via fermentation. Wild yeast spores are permanently floating in the atmosphere and may land on uncovered liquids and foods – hence resulting in contamination. In general, yeast contamination in food generates slime on the surface, bubbles in the bulk and an alcoholic smell or taste. Yeasts can be destroyed by heating to 121°C for 15 min; nevertheless, in industrial food processing, carefully cultured yeasts are used in the production of beer, wine and bread.

[1]Escola Superior de Biotecnologia, Universidade Católica Portuguesa P-4200-072 Porto, Portugal

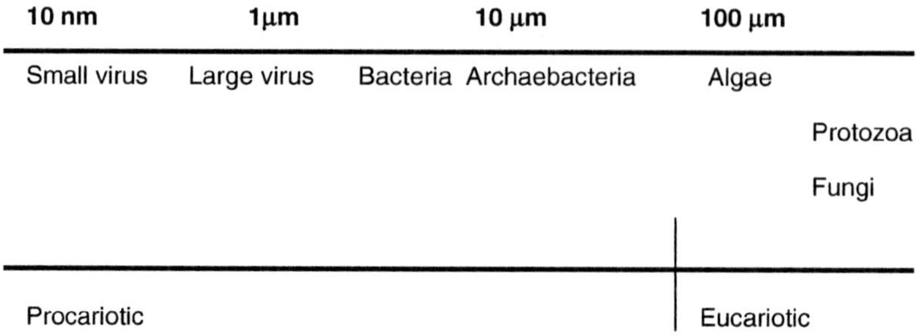

FIGURE 1. Relative size and cell-type of organisms involved in micro-ecology of food.

Moulds are multi-cellular forms of fungi, which can grow on almost any item used as or for food, given suitable conditions. Mould spore casings are present in the environment; when they break, thousands of microscopic mould spores are released, each one capable of germinating and originating a new mould. This is a process that typically occurs in damp, dark environments. As spores on the surface of food ripen, the food develops unpleasant musty odours, which destroy the (normally sought) fresh flavours. Certain moulds may produce poisonous toxins – called mycotoxins. Aflatoxin is one such mycotoxin that is secreted on nuts, corn, wheat and other grains. Aflatoxins may also be found in products made from dry fruits and cereals, such as breads and peanut butter. Ingestion of aflatoxin usually causes low grade fever in humans but can also produce cancer in trout, rats and ducks. Other illnesses thought to worsen via the presence of aflatoxins include Reyes syndrome, cirrhosis and kwashiorkor (Jones, 1992; Jay, 1986).

Parasites are organisms that live or feed off other organisms. In general, they are found in raw animal products or seafood. Parasites, which include *Trichinella spiralis* (a round worm found in wild game or pork) and *Anisakis spiralis* (commonly referred to as "cod fish worm" or "seal worm" and found in fish), are destroyed by thorough cooking.

Most foods harbour a mixture of the aforementioned microorganisms, which play a role in several biological interrelationships ranging from (competitive) amensalism to antagonism and encompass specifically mutualism, commensalism and parasitism (or predation).

Mutualism may be understood as a mutual dependency between two microorganisms, in which each microorganism attains some benefit – trivial or vital, from the other. In a commensalism pattern, only one microorganism is able to obtain a benefit from the association, whereas the other is unaffected by it. Finally, in parasitism, pathogenic microorganisms (considered to be disease-causing agents) obtain support from the host at its expense. As previously mentioned, these harmful microorganisms can invade any food and may survive despite aggressive measures at the processing level and storage – or the food may become contaminated during preparation, cooking or serving. When present in the food at or above their infective dose threshold, they will cause illness – sometimes severe and even life-threatening, especially in young children, older adults and persons with compromised immune systems. In pregnant women, foodborne illness may also endanger their unborn babies. The Centers for Disease Control and Prevention (CDC) in the USA, estimate that 76 million people suffer foodborne illnesses each year in that country, hence accounting for 325,000 hospitalizations and more than 5,000 deaths.

There are more than 250 known foodborne diseases. Bacteria cause most cases, followed by viruses and parasites. Some diseases are caused by toxins (poisons) from disease-causing organisms, others by host reactions to the organism itself. The most common symptoms of foodborne illness are diarrhoea, abdominal cramps, vomiting, head- or muscle-aches and fever. Symptoms usually appear 12 to 72 hr after eating contaminated food but may occur as early as 30 min or as late as 4 weeks afterwards.

2. Microbial Relationships in Food Ecosystems

Most microorganisms are free-living, and do not necessarily form specific associations between themselves or with others. Microorganisms are nevertheless components of a complex ecosystem, in which a continuous interaction with their environment is maintained. With the exception of highly processed products, most foods harbour a mixture of microorganisms – which often includes various species of bacteria, yeasts and filamentous fungi, as well as several strains within each species. In addition, bacteriophages and yeast killers also constitute part of the microflora. In a sustained bid for survival, growth and eventual dominance, interactions occur spontaneously between those strains and species – the outcome of which determines the population levels of any particular microorganism, at any given time during the production and retailing timeframes (Fleet, 1999). It should be noted that the dynamics of survival, growth and biochemical activity of microorganisms in food are the result of stress reactions triggered in response to changing physical and chemical conditions of the surrounding medium, the ability to colonise the food matrix (and to grow into spatial heterogeneity) and the *in situ* cell-to-cell ecological interactions (which often take place within a solid phase).

In food, ecological approaches to the evolution of microbial flora (see Figure 2) would be useful to better understand the microbiological events involved in food processing, to improve microbiological safety by monitoring past and present viability of pathogenic bacteria and to evaluate the effective composition of the microbial populations in stake. These ecological principles are the fundamentals of modern quality assurance, predictive modelling and risk analysis strategies, aimed at preventing outbreaks of food spoilage and foodborne diseases. They also form the basis for use of microorganisms in production of fermented foods and beverages and for use thereof as probiotic, starter and bio-control agents (Giraffa, 2004).

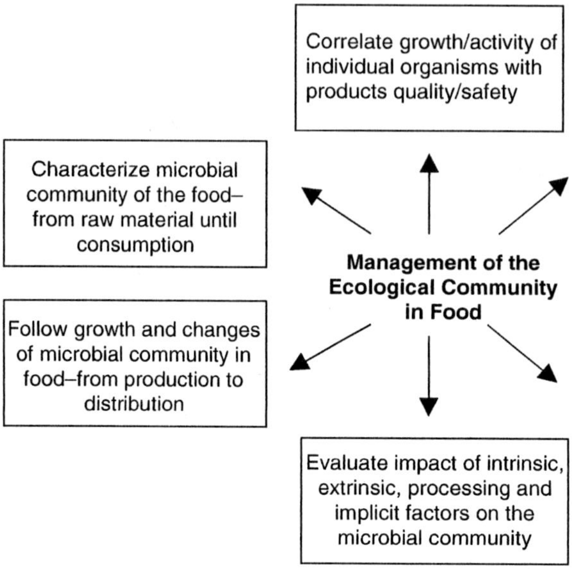

FIGURE 2. Ecological approaches to evolution of the microbial flora in food (adapted from Fleet, 1999).

3. Microbial Interactions – General Considerations

Mutualism, commensalism and parasitism, describe interactions in the microbial world, which are dynamic in nature and dependent on prevailing environmental conditions. There are many examples of these types of interactive associations in every day life; a few case-studies pertaining to these associations are presented below, with a focus on food issues.

Before treating each of these concepts at length, it is important to fully understand the meaning of an old term, *symbiosis*. Originally, symbiosis meant any stable, physical association between different organisms (the symbionts) – regardless of the nature of their relationship. Later, the meaning of the term was restricted to cover only relationships of mutual benefit (mutualism) (Singleton, 1999).

3.1. Mutualism

Mutualism is generally known as a relationship between two dissimilar organisms in which both parties benefit. In the context of microbial pathogenesis, Casadevall and Pirofski (2000) also suggested revision of the definition of mutual – and stated that is a state of infection, whereby both the host and the microbe benefit as a consequence of infection.

One good example is digestion of cellulose (a major component of plant material), by bacteria and protozoa, which are present in the stomachs of domestic ruminants: microorganisms, which on the one hand, have a sheltered and controlled environment and receive a constant supply of nutrients whilst on the other, the ruminant host is, in turn, able to extract nutrients from the eaten plant material ingested.

In leguminous plants (e.g. peas, beans and clover), the roots have small swelling organelles (nodules), which contain bacteria of the genus *Rhizobium*; in this rearrangement, the plant provides nutrients and protection, while *Rhizobium* supplies "fixed" nitrogen – obtained from the atmosphere and made available thereby to the plant (Singleton, 1999).

3.2. Commensalism

The word commensal (from the Latin *com* – meaning with, *mensa* – meaning table and *al* – meaning pertaining to) means literally "eating at the same table". From the different definitions found in the literature (see Table 1), it is clear that in a commensal association one microorganism species benefits from another but the latter derives neither benefit nor harm from the association. In the context of microbial pathogenesis, Casadevall and Pirofski (2000) suggested revision of the definition of commensal – in the spirit of the original meaning of the word and claimed commensalism to be a host-microbial interaction that does not result in perceptible, ongoing and/or persistent host damage.

The commensal microbial flora of *Homo sapiens* is one of the best known examples that illustrate this concept. The associated interactions are so highly specialized that only certain bacterial species are found at particular locations in the human body; such interactions are normally harmless, although they may benefit the host, e.g. prevention of infectious bacteria from colonizing mucosal surfaces (Henderson et al., 1999).

Commensal microbial interactions frequently do occur in food environments. Examples include degradation of complex products: proteins and carbohydrates by some species, to produce simpler substrates for growth of other species; and organic acids taken up by yeasts and moulds, to favour autolytic release of

TABLE 1. Historical definitions of commensalism (adapted from Casadevall and Pirofski, 2000)

The mutual but almost inconsequential association between bacteria and higher organisms.
The presence of microorganisms on skin and mucous membranes.
A symbiotic association between host and microorganism, in which the microorganism is benefited but the host is neither helped nor harmed.
An organism that lives in close association with another of a different species, without either harming or benefiting it.
A form of parasitism, in which no injury is dealt to either participant by the other.
The ability (of a microorganism) to live on the external or internal surfaces of the body, without causing disease.
A symbiotic relationship, in which one species derives benefit and the other, is unharmed.
A state in which one species uses another as its physical environment, normally existing within the larger species.

nutrients by dead cells, as well as production of vitamins, specific amino acids, carbon dioxide and other micronutrients produced by some species that permit growth of other species (Fleet, 1999).

3.3. Antagonism

Antagonism occurs when a microorganism can actively discourage (at least some of) its competitors by producing substances which are toxic to them. This is probably the best known microbial interaction in food ecosystems, because it can be applied as a natural strategy to enhance food quality and safety – as a matter of fact, it can control spoilage and pathogenic bacteria via secretion of bacteriocins. Despite being a classical example of "bacteria-bacteria" interaction, there are circumstances in which bacteriocins will also inhibit yeasts. The production of killer toxins (i.e. extracellular proteins, or glycoproteins that disrupt cell membrane function in susceptible yeasts) by yeasts also occurs and is somewhat analogous to bacteriocin production by bacteria. These antagonistic interactions were originally thought to be species-specific, however, accumulated experimental evidence has made it clear that they occur across species in different yeast genera and that they can kill various filamentous fungi or even other yeasts.

A less recognised form of release in ecosystems that probably accounts for antagonism is the production of cell wall lytic enzymes. Examples include the production of β-(1-3)-glucanases by bacteria and yeasts – that destroy β-(1-3)-glucans in the cell walls of such fruit spoilage fungi as *Penicillium expansum* and *Botrytis cinerea*. Another (less familiar) form of microbial interaction – that may be significant in food systems, is the ability of yeasts and bacterial cells to agglutinate and aggregate. Most species of *Enterobacteriaceae* and lactic acid bacteria will agglutinate *Saccharomyces cerevisiae,* via reaction with mannoproteins of the yeast surface.

Phage activity causes bacterial cell lysis and nutrient release in ecosystems and probably accounts for the variability of data pertaining to the population that is often obtained from food samples. Deeper consideration ought to be given to the role of bacteriophages in food environments; most studies do indeed concern their ability to destroy starter cultures made of lactic acid bacteria, which are used in milk. Foods that are likely to harbour an enormous diversity of bacteriophages could have a significant impact on the *in-situ* bacterial ecology (Fleet, 1999).

3.4. Microbial interactions – selected case studies

3.4.1. Microecology in humans

The largest and most important interface between a superior organism and its environment is accounted for by surfaces covered by epithelial cells. At birth the foetus is delivered from the essentially sterile uterine environment and interactions of the neonate with microorganisms occur from this point on. The main portals of entry of microorganisms are the skin, as well as mucosal surfaces of the gastrointestinal, respiratory and urogenital tracts. In physiological terms, interaction with bacteria eventually leads to colonisation of such epithelial surfaces – this co-existence is usually harmonious and beneficial to the host (which is an example of commensalism). A complex, open ecosystem – formed by resident bacteria and other microorganisms that interact temporarily with the macroorganism - is thus established. However, the interaction with "endogenous" microorganisms can, under specific conditions, be harmful to the host (i.e. parasitism), so opportunistic infections may occur.

Interactions between the microflora and their host are characterised by active participation of both partners and the strategy of both of them seems similar: evolutionary co-existence has provided both the microorganisms and the immune system of the host with similar mechanisms of diversification and selection. Human individuals are thus complex ecosystems, formed by a normal microflora – that comprises mainly bacteria, as well as viruses, fungi and protozoa to a lesser extent.

Commensal bacteria exhibit an enormous diversity; not less than 1000 species are apparently involved. The commensal microflora is thus an integral part of the complex, natural mechanisms acting on mucosal surfaces and skin that safeguard resistance of the organism against pathogenic microorganisms. When the qualitative and quantitative profiles are at an optimum, attachment and multiplication of pathogenic microorganisms on these surfaces, subsequent invasion of epithelial cells and the circulatory system are prevented. This process is termed "colonisation resistance",. Intestinal microflora play an important role in anti-infectious resistance, both by direct interaction with pathogenic bacteria and by influence upon the immune system – during the early postnatal period, the intestinal microflora stimulates development of both local and systemic immunity; afterwards, these components evoke regulatory (inhibitory) mechanisms to keep both mucosal and systemic immunity in balance (Tlaskalov'a-Hogenov'a et al., 2004).

3.4.2. Microecology in dairy products

Although fermented milk products are regarded predominantly as a result of lactic acid fermentations, the frequent co-occurrence of yeasts and lactic acid bacteria has led to the suggestion of interactions that can influence product characteristics (and quality thereof). Presence of yeasts is obviously necessary for the desirable carbon dioxide and ethanol production in eastern European and Asian products, such as *kefir*, *koumiss* and *airag*. The mechanisms of these interactions may depend on stimulation or else inhibition of growth of one (or both) of those co-cultured species. Those organisms may in fact compete for nutrients that cause growth, or they may produce metabolites that inhibit each other's growth; e.g. yeasts may produce vitamins that enhance growth of lactic acid bacteria. Furthermore, mutual influence of microorganisms on each other's metabolism may lead to different profiles of organoleptically important compounds in the final fermented milk (Narvhus and Gadaga, 2003).

The commensalistic interaction between *L. acidophilus* and a lactose fermenting yeast called *Kluyveromyces fragilis* in acidophilus-yeast milk relies on coexistence of both organisms for the formation of a product with good final quality. The co-culture of *L. acidophilus* with *K. fragilis* reduces the time of coagulation of milk due to acid production by the latter, whereas it raises the number of viable lactic acid bacteria, while inhibiting growth of *Escherichia coli* and *Bacillus cereus*.

Mutualistic synergism occurs between yeasts and lactic acid bacteria during fermentation of kefir. Yeasts provide such growth factors as free amino acids and vitamins for bacteria, which consequently entertain elevated acid production; bacterial end-products are in turn used by yeasts as an energy source. This phenomenon creates a balanced stability of the final product. However, a decrease in alcohol production by yeasts may occur due to excessive lactic and acetic acid production by osmophilic lactic acid bacteria, coupled with competition for the carbon source – or even lysis of yeast cell walls by bacterial enzymes (Viljoen, 2001). A positive effect upon growth and kefiran production by *Lactobacillus kefiranofaciens* was observed (Cheirsilp *et al.*, 2003) in a mixed culture with *S. cerevisiae*; physical contact with the latter is enhanced by capsular kefiran produced by *L. kefiranofaciens*.

3.4.3. Microecology in yoghurt

Yoghurt is produced via fermentation of pasteurised (full or skimmed) milk. The major agents in this process are *Streptococcus salivarius* subsp. *thermophilus* and *Lactobacillus delbrueckii* subsp. *bulgaricus*. During incubation, in which the starter grows as mixed culture, a positive interaction between the two microorganisms is generally observed. Typically, *S. thermophilus* and *L. bulgaricus* are inoculated at a 1:1 ratio and then remain present throughout yoghurt production. When both bacteria grow in association, the times required for milk coagulation are shorter than if either of them is grown separately. This process occurs because during growth, *S. thermophilus* produces formic acid which, in turn, stimulates the growth of *L. bulgaricus*. The activity of the latter on casein releases amino acids, which, in turn, stimulate growth of the former. However, Ginovart et al. (2002) have reported that interaction between such bacterial cells is not only due to growth stimulation related to either formic acid or free amino acids but also to acidity of the medium.

3.4.4. Microecology in cheese

Microorganisms are an essential component of natural cheese varieties and play important roles during both cheese manufacture and ripening. They can be divided into two main groups; starter and secondary microflora. The former, typically composed of *Lactococcus lactis*, *S. thermophilus*, *Lactobacillus helveticus* and *L. delbrueckii* – used either individually, or in various combinations depending on variety in stake, are responsible for acid development during cheese making. Starters may be either blends of defined strains, or (as happens in the case of many cheeses manufactured by traditional methods) composed of more or less undefined mixtures of strains – which are adventitious and are present in cheese milk. During cheese ripening, the starter culture coupled with the secondary microflora promotes a complex series of biochemical reactions that are vital for proper development of both flavour and texture. The secondary flora is normally composed of complex mixtures of bacteria, yeasts and moulds – which are dependent on the particular cheese variety, as they contribute significantly to its specific characteristics. The secondary microflora may also be added in the form of defined cultures; however, in many situations it is composed of adventitious microorganisms that gain access to cheese either from its ingredients or from the environment. During cheese manufacture and ripening, a number of interactions occur between individual constituents of the cheese microflora (Beresford et al., 2001).

Yeasts, which often originate in contamination during cheese making, contribute to ripening by metabolizing lactic acid, producing lipases and proteases, fermenting residual lactose and excreting growth factors, either as viable entities or following autolysis (Viljoen, 2001). All of these characteristics contribute to the sensory quality of the final cheese. The increase in pH arising from lactic acid utilization encourages growth of bacteria – which may not only affect flavour and textural quality but also pose a risk to public health. Studies on the interaction between yeasts and starter cultures in Cheddar and Gouda cheeses indicated that the former also play a significant role during ripening via supporting growth of starter cultures. The large number of viable yeasts present during the later stages of ripening is indicative of a possible mutualistic interaction within the microflora. During ripening, yeasts increase at a faster rate than starter cultures but no inhibition of either population is typically observed; therefore, said mutualistic interaction may contribute to the final product.

3.4.5. Microecology in probiotic foods

Additive and synergistic health-promoting effects, brought about by individual strains in multistrain probiotic foods, may be explained on the basis of possible relationships between strains in those mixed systems. Interrelationships may enhance certain probiotic characteristics, *such as* growth and metabolic activity.

Growth of probiotic microorganisms following inoculation is necessary to maintain sustainable numbers in the gastrointestinal tract. This growth can be stimulated by the presence of other species – as happens with certain starter cultures involved in manufacture of fermented dairy products. For such probiotic bacteria as *L. acidophilus* and *Bifidobacterium* spp., it is known that they grow slowly in milk because they lack proteolytic activity. Addition of typical yoghurt-bacteria – particularly *L. delbrueckii* subsp. *bulgaricus*, will enhance growth of such probiotic strains. This positive interaction is referred to as protocooperation – and is explained by the exchange of certain growth factors, *such as* amino acids, free peptides, formate and CO_2 (Timmerman *et al.*, 2004). A progressive increment of *B. animalis* growth was affected by presence of *L. acidophilus* – which hydrolyzes milk caseins using extracellular proteinases, thus yielding amino acids and peptides that stimulate growth of *B. animalis*. On the other hand, growth of *L. acidophilus* is also enhanced by presence of *B. animalis*, possibly due to production of acetate (Gomes et al. 1998; Timmerman et al., 2004).

3.4.6. Microecology in sourdough

Sourdough is an intermediate food product, whose microflora is composed of stable associations of lactobacilli and yeasts, based on metabolic interactions. As shown for certain industrial sourdough processes, such microbial associations may endure for years – even though the fermentation process is run under non-aseptic conditions. The importance of antagonistic and synergistic interactions between lactobacilli and yeasts is based on metabolism of carbohydrates and amino acids, coupled with production of carbon dioxide. Typical mutual associations involve *Lactobacillus sanfranciscensis* and either *Saccharomyces exiguus* or *Candida humilis*. Maltose is the preferred energy source *for L. sanfranciscensis but* is not utilized at all by either of those latter species (*such as* maltose-negative yeasts, which use sucrose, glucose and fructose). Maltose is continuously released by flour amylases; when there is an excess of maltose coupled with environmental stress, several strains of *L. sanfranciscensis* hydrolyse maltose and accumulate glucose in the medium. This glucose affects the ecological system, as it may be metabolised by its producers, by other lactic acid bacilli (LAB) and by yeasts. It may however, initiate glucose repression in competitors for maltose and glucose may then be utilised by maltose-negative yeasts. Due to the faster consumption of maltose and especially glucose by *S. cerevisiae*, a decrease in metabolism of *L. sanfranciscensis* is expected when the latter is associated with maltose-positive yeasts. However, disappearance of *S. cerevisiae* from the microbial population of sourdough during consecutive fermentations is related to repression of genes involved in maltose fermentation – so that maltose cannot be utilized - and to rapid depletion of sucrose. Sourdough yeasts do not affect cell yield of *L. sanfranciscensis,* because pH is the limiting factor for growth of lactobacilli – note that *L. sanfranciscensis* does not grow below pH 3.8 (de Vuyst and Neysens, 2004).

3.4.7. Microecology in wine

Alcoholic fermentation is dominated by growth of yeasts, because of their ability to develop at low pH (3.0–3.5) – as prevailing in grape juice; they produce ethanol, which inhibits growth of filamentous fungi and bacteria. The first 2-4 days of fermentation are characterized by growth of various species of *Kloeckera/Hanseniaspora*, *Candida*, *Metschnikowia*, *Pichia* and *Kluyveromyces* – which achieve populations of the order of 10^8 cfu /ml, before progressively dying off according to their tolerance to increasing concentrations of ethanol. By this time, they have utilized sufficient amounts of sugars and amino acids in the juice and have generated sufficient end-products to provide the fingerprint of wine. *Saccharomyces cerevisiae* also grows during these early stages; however, its unique tolerance of ethanol permits its continuing growth, until it eventually predominates during the mid-to-final phases of fermentation. Further studies on wine ecology have revealed that each yeast species is normally present as several strains and that a strain succession programme takes place (Fleet, 2001).

3.4.8. Microecology and food safety

The most common approaches to microbial food safety and quality consider acceptable levels of microorganisms and metabolic products thereof in foods as indicators of quality and safety. Malakar et al. (2003) have shown that microbial interactions in broth systems only become important at population densities above 10^8 cfu/ml. Spoilage is apparent at these levels, except for deliberately fermented foods.

Colonial growth occurs as a response to food structure. As a consequence of microbial proliferation, diffusion limitation within colonies occurs that acts as a constraint upon growth and becomes important when colonies reach a specific threshold (typically above10^5 cells), which is known to be dependant upon the initial inoculation density. Intra- and inter-colony interactions can be important factors to ensure safe foods; one of the best applications is inter-colony interactions with an antagonistic microorganism, aimed at preventing a pathogenic microorganism from proliferating in food. If the distance between colonies of antagonistic microorganisms is above a certain level, safety of these foods will be compromised because inter-colony interaction will be delayed; in fact, a pathogenic colony may be allowed to grow unhindered and thus cause problems later. The problem may arise because mixing is incomplete in some foods; instead of a homogeneous spatial distribution of colonies, in actual situations non-homogeneous conditions prevail, so there may be regions of space with sparse colonies. Therefore, there is a finite probability that pathogenic colonies located in these sparse regions will be able to grow and become dangerous. To guarantee adequate safety, the probability of existence of such sparse regions should be kept very low (Malakar et al., 2003).

3.4.9. Microecology and bacteriocins

Biological preservation refers to use of antagonistic microorganisms, or their metabolic products, to inhibit or destroy undesired microorganisms in foods. Lactic acid bacteria can exert a 'bio preservative' or inhibitory effect against other microorganisms, as a result of competition for nutrients and/or of production of bacteriocins and other antagonistic compounds. Bacteriocins are extracellularly-released peptides or protein molecules, that are bactericidal (*i.e.* are destructive) to bacteria that are closely related to the producer microorganism. Hence, bacteriocins produced by lactic acid bacteria may be considered as natural preservatives (or bio preservatives). Biological preservation can also involve use of antagonistic or biocompetitive microorganisms to inhibit mycotoxinogenic fungi – which often cause problems in foods of plant origin and to prevent formation of mycotoxins in foods and agricultural products. Certain antimicrobial enzymes are also considered as potential compounds for use in biological preservation (Schillinger et al., 2002).

3.5. Pathogenic microorganisms

The most generally used definition of a pathogen states that it is a microbe provided with virulence factors (structural, biochemical or genetic traits), which render it pathogenic (or virulent) so as to cause disease in a host (see Table 2); nevertheless, such concepts of virulence and pathogenicity are inadequate, because they do not account for the full complexity of microbial pathogenesis in hosts with and without impaired immunity. A number of alternative definitions made available in the literature for this (and related terms) are depicted in Table 2.

More recently, an integrated view of microbial pathogenesis has been suggested. This definition accounts for the contributions of host and pathogen as well as a classification system for microbial pathogens (Casadevall and Pirofski, 1999). These authors propose that host-pathogen interactions can be analyzed using host damage as the common denominator for characterization of microbial pathogenicity and that they can provide a conceptual framework for incorporating the importance of the host response into the outcome of the host-microbe interaction. In this classification, pathogens are grouped according to their ability to inflict damage as a function of host response, irrespective of their phylogenetic derivation, biological kingdom or previous classification. By combining the concept that host response contributes to pathogen-mediated damage, with the classical view that pathogens have distinct characteristics which define their virulence, the damage-response classification permits a new approach to host-pathogen interactions, that is not constrained by pathogen-and/or host-centred

TABLE 2. Definitions of virulence and pathogenicity (adapted from Casadevall and Pirofski, 1999)

Term	Definition
Pathogen	A microbe capable of causing disease.
	A microorganism that can increase in living tissue and produce disease.
	Any microorganism, the survival of which is dependent upon its capacity to replicate and persist on or within another species – by actively breaching or destroying a cellular or humoral host barrier, that ordinarily restricts or inhibits other microorganisms.
	A parasite capable of causing or producing some disturbance in the host.
Pathogenicity	The capacity of a microbe to produce disease.
Pathogenicity	Degree of pathogenicity.
	Inverse of resistance.
Virulence	Strength of pathogenic activity.
	Relative capacity to overcome available defences.
	Disease severity as assessed by reductions in host fitness following infection.
	Percent of death per infection.
	Synonym of pathogenicity.
	Property of invasive power.
	Measure of the capacity of a microorganism to infect or damage a host.
Virulence factor	Relative capacity to enter and multiply in a given host.
Virulence factor	A component of a pathogen that, when deleted specifically, impairs virulence but not viability.

views of microbial pathogenesis. The proposed classification system categorizes all pathogenic microorganisms into six different classes, according to the damage-response curves depicted in Figure 3.

The various classes of pathogenicity are shown in Table 3. The *y*-axis denotes the amount of damage to the host resulting from the host-pathogen interaction. The *x*-axis denotes the magnitude of the host immune response; "Variable" means that the amount of damage can vary, depending on the individual host (adapted form Casadevall and Pirofski, 1999).

The two main features of pathogenic microorganisms that elicit their disease causing mechanisms are: (i) ability to invade tissues; and (ii) ability to produce toxins. In the former case, invasiveness encompasses mechanisms for colonization (*such as* adherence and initial multiplication), for bypassing of (or even overcoming) host defences and for production of extracellular compounds which facilitate invasion. In the latter case, toxigenesis encompasses two types of toxins, called exotoxins and endotoxins.

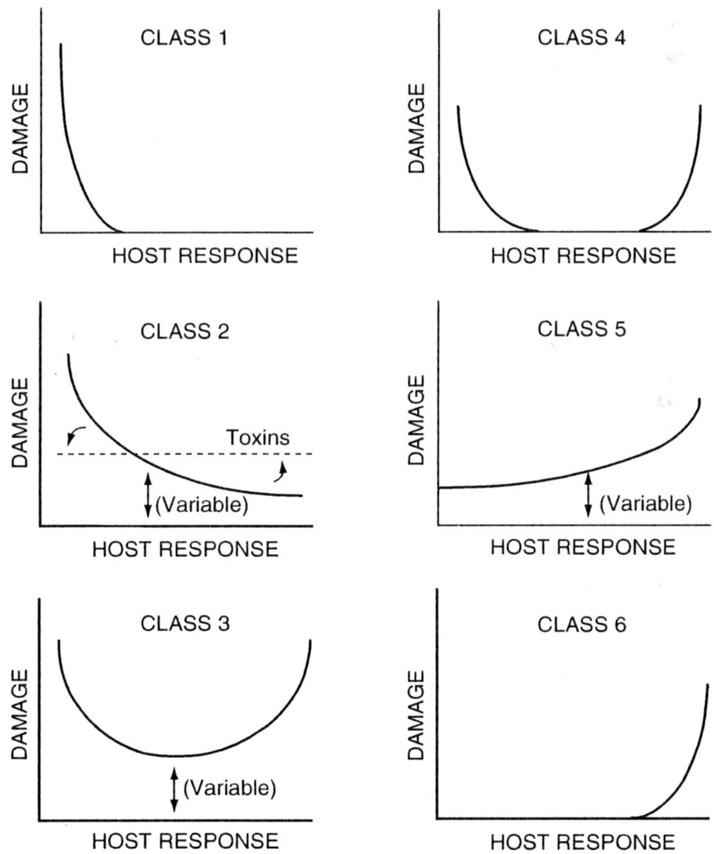

FIGURE 3. Damage-response curves, representing various classes of microbial pathogens.

TABLE 3. Features of various classes of microbial pathogens (adapted form Casadevall and Pirofski, 1999)

Class	Features
1: pathogens that cause damage only in situations of weak immune responses	• usually considered opportunistic or commensal • associated with disease only in individuals with impaired immune function and almost never causing symptomatic or clinically apparent infections in individuals with regular immunity • low virulence e.g. *Pneumocystis carinii*
2: pathogens that cause damage, either in hosts with weak immune responses or in the setting of normal immune responses	• host damage by both host- and pathogen-mediated mechanisms • capacity to cause serious infections in normal hosts but frequently associated with more severe infections in hosts with impaired immune function • viewed as opportunistic because of higher prevalence in groups with impaired immune function • episodic infections caused in normal hosts e.g. *Cryptococcus neoformans, Staphylococcus aureus*
3: pathogens that cause damage in the setting of appropriate immune responses and produce damage at both ends of the continuum of immune responses	• causing disease by both host- and pathogen-mediated mechanisms • vehicles of disease in normal hosts • ability to cause significant damage in setting of both weak or strong immune responses e.g. *Histoplasma capsulatum*
4: pathogens that cause damage, primarily at the extremes of both weak and strong immune responses	➢symptomatic infections only in patients with impaired immunity or protracted immune responses to the pathogen ➢no detectable damage to host e.g. *Aspergillus fumigatus*
5: pathogens that cause damage across the spectrum of immune responses, which can be enhanced by strong immune responses	➢infections resulting in pathogen-mediated damage but associated with protracted or chronic damage resulting from excessive or inappropriate immune response ➢severe gastrointestinal infections in individuals with impaired immunity but most cases solved without permanent damage to gastrointestinal tract or other tissues e.g. *Shigella* and *Campylobacter* spp.
6: microorganisms that cause damage only in conditions of strong immune responses	• theoretical category, not defining any known pathogen • encompassing growing list of diseases, shown to be the result of infectious microorganisms e.g. Crohn's and Whipple's diseases

Exotoxins are released by bacterial cells and may act at tissue sites removed from the site of bacterial growth. Endotoxins are cell-associated substances, which are structural components of cell walls of Gram-negative bacteria. However, endotoxins may be released by growing bacterial cells, or by lysing bacterial cells as a result of effective host defence (mediated by e.g. lysozyme) or of action of certain antibiotics (such as penicillins and cephalosporins). Hence, bacterial toxins – both soluble and cell-associated, may be transported by blood and lymph and thus cause cytotoxic effects at tissue sites remote from the original point of invasion or growth. Some bacterial toxins may also act at the site of colonization and play a role in invasion.

Upon establishment of the pathogen at the appropriate portal of entry – i.e. the digestive tract in the case of foodborne pathogens -colonization of the host, by adhering to specific tissues, is required. In its simplest form, bacterial adherence or attachment to a eukaryotic cell or tissue requires concurrence of two factors: a receptor and an adhesin. Receptors described so far are specific carbohydrate, or peptide, residues lying on the eukaryotic cell surface. Adhesins are typically macromolecular components of the bacterial cell surface, which interact with the host cell receptor. Those two factors usually interact in specific but complementary fashion. Adhesins associated with those pathogens more often implicated in foodborne illness are tabulated in Table 4.

Invasion of a host cell by a pathogenic microorganism may be made easier via production of bacterial extracellular substances – termed invasins, which act against the host by breaking down primary or secondary defences thereof. Invasion of epithelial cells is an active process, induced by pathogenic bacteria. Two main mechanisms of invasion are known: macropinocytosis, as observed in *Salmonella* spp. and *Shigella* spp. and phagocytosis, as happens with *Listeria* spp. and *Yersinia* spp. Evasins are substances (or bacterial structures) that enable bacteria to evade phagocytosis, the complement system and antibodies (e.g. $_{S}$IgA protease). The activities of many bacterial proteins that are known for their contribution to foodborne pathogen bacterial invasion of tissues are listed in Table 5.

There are multiple virulence factors that promote lesion in an infected organism: i) exotoxins; ii) endotoxins; iii) super antigens; iv) hydrolytic enzymes; and v) antigens that induce auto-immune illness. With food microbial interactions, the first three factors are most relevant and will be discussed further.

Exotoxins can be divided into three groups, according to their target site: i) toxins that act on the cytoplasmic membrane and can interfere with cell signalling mechanisms; ii) toxins that alter cellular membrane permeability, or pore-forming toxins; and iii) toxins that act inside the cell via an enzymatic mechanism, thus modifying cytosolic targets. Examples of exotoxins and associated mechanisms of action, are given in Table 6.

TABLE 4. Specific attachment processes of selected bacteria to host cell or tissue surfaces

Bacterium	Adhesin	Receptor	Attachment site	Disease
Staphylococcus aureus	Cell-bound protein	Amino terminus of fibronectin	Mucosal epithelium	Various
Enterotoxigenic *Escherichia coli*	Type-1 fimbriae	Species-specific carbohydrate(s)	Intestinal epithelium	Diarrhoea
Vibrio cholerae	N-methylpheny-lalanine pili	Fucose and mannose carbohydrate	Intestinal epithelium	Cholera
Campylobacter jejuni		Fucose carbohydrate	Intestinal epithelium	Diarrhoea

TABLE 5. Specific invasion processes of selected bacteria

Invasin	Bacterium	Activity
Hyaluronidase	Streptococci Staphylococci Clostridia	Degrades hyaluronic acid of connective tissue
Collagenase	*Clostridium* spp.	Dissolves collagen framework of muscles
Neuraminidase	*Vibrio cholerae* *Shigella dysenteriae*	Degrades neuraminic acid of intestinal mucosa
Coagulase	*Staphylococcus aureus*	Converts fibrinogen to fibrin, which causes clotting
Kinases	Staphylococci Streptococci	Converts plasminogen to plasmin, which digests fibrin
Leukocidin	*Staphylococcus aureus*	Disrupts neutrophil membranes and causes discharge of lysosomal granules
Hemolysins	Streptococci Staphylococci Clostridia	Destroys red blood (and other) cells by lysis
Lecithinases	*Clostridium perfringens*	Destroys lecithin in cell membranes
Phospholipases	*Clostridium perfringens*	Destroys phospholipids in cell membranes
Anthrax EF	*Bacillus anthracis*	Causes increased levels of intracellular cyclic AMP

TABLE 6. Sources and activities of bacterial exotoxins

Toxin	Bacterium	Activity
Cholera enterotoxin	*Vibrio cholerae*	Promotes ADP ribosylation of G proteins and consequent stimulation of adenlyate cyclase and increase of cAMP in cells of GI tract, which causes secretion of water and electrolytes
E. coli LT toxin	*Escherichia coli*	Similar to cholera enterotoxin
C. jejuni enterotoxin	*Campylobacter jejuni*	Presents immunological relationship similar to cholera enterotoxin and *E. coli* LT toxin
Shigella toxin	*Shigella dysenteriae*	Promotes enzymatic cleavage of rRNA, which results in inhibition of protein synthesis in susceptible cells
Botulinum toxin	*Clostridium botulinum*	Zn^{++}-dependent protease, which inhibits neurotransmission at neuromuscular synapses, thus causing flaccid paralysis
Tetanus toxin	*Clostridium tetani*	Zn^{++}-dependent protease, which inhibits neurotransmission at inhibitory synapses, thus resulting in spastic paralysis
Diphtheria toxin	*Corynebacterium diphtheriae*	Promotes ADP ribosylation of elongation factor 2, which causes inhibition of protein synthesis in target cells
Staphylococcus enterotoxins	*Staphylococcus aureus*	Promotes massive activation of immune system, including lymphocytes and macrophages, which causes emesis (vomiting)

Endotoxins are invariably associated with Gram-negative bacteria and are constituents of the outer membrane of their cell wall. They can be lipopolysaccharides (LPS) – as in the case of *E. coli*, as well as the genera *Salmonella*, *Shigella*, *Pseudomonas*, *Neisseria* and *Haemophilus*. Such LPS participate in a number of outer membrane functions that are essential for bacterial growth and survival, especially within the context of a host-parasite interaction. Toxicity is associated with their lipid component (Lipid A), whereas immunogenicity (or antigenicity) is associated with its polysaccharide counterpart. LPSs activate the complement system via the alternative (properdin) pathway and may indeed be a step of the pathology process of most Gram-negative bacterial infections. Most endotoxins remain associated with the cell wall until disintegration of the bacterium; *in vivo*, this happens via autolysis, external lysis, or phagocytic digestion of bacterial cells. It is known, however, that small amounts of endotoxins may be released in soluble form, especially by young cultures.

Compared with the classic exotoxins of bacteria, endotoxins are less potent and less specific in their action – since they do not act via an enzymatic pathway. Endotoxins are heat-stable (boiling for 30 min does not lead to significant destabilization) but such powerful oxidizing agents as superoxide, peroxide and hypochlorite are able to degrade them. Although strongly antigenic, endotoxins cannot be converted to toxoids. These toxins are responsible for clinical manifestations, *such as* fever, inflammation and shock. The cellular receptors of endotoxins on the surface of macrophages are CD14 molecules.

Super antigens are proteins produced mainly by *Streptococcus* spp. and *Staphylococcus* spp. They are not processed and are presented to macrophages in association with histocompatibility complex class II molecules (MHC). Super antigens can interact with this type of molecule and with the lymphocyte receptors that recognize them.

3.5.1. Foodborne pathogens

The Centers for Disease Control and prevention (CDC) have listed four classical sources of foodborne illness: disease-causing bacteria, viruses, parasites and toxins. At least 30 pathogens are commonly associated with foodborne illness; a few of these are very common in nature and account for the majority of the cases of illness reported. Nevertheless, pathogens are under permanent evolution, because the bacteria and the host population, as well as the ecological conditions that provide their mutual interplay, undergo changes constantly. Such an evolution gives way to new and emerging foodborne pathogens, that were not described two decades ago but which are of great concern nowadays – *examples include Campylobacter jejuni*, *Cryptosporidium parvum*, *Cyclospora cayetanensis*, *E. coli* O157:H7 and related strains (e.g. O111:NM and O104:H21), *Listeria monocytogenes*, *Nitzchia pungens*, *Salmonella enteritidis*, *S typhimurium* DT 104, *Vibrio cholerae* O1, *V. vulnificus*, V. *parahaemolyticus* and *Yersinia enterocolitica*, as well as Noroviruses (Norwalk-like viruses) and prions.

The most commonly recognized foodborne infections are those caused by the bacteria *C. jejuni, S. enteritidis* and *E. coli* O157:H7 and by Noroviruses – with substantial variations throughout geographical area and season. Other bacterial pathogens, such as *V. vulnificus, Y. enterocolitica, Clostridium perfringens, Shigella* spp. and *S. aureus* are also of importance. Emerging infectious diseases are most likely caused by microorganisms that are opportunistic, or true pathogens that have acquired additional DNA elements that encode a 'true virulence determinant' – e.g. toxin-converting bacteriophage encoding cholera toxin, or *Shigella* toxins (Wassenaar and Gaastra, 2001).

Bacteria in food can cause infections when the microorganism is eaten and establishes in the body – usually multiplying in the intestinal tract and irritating the intestinal lining. Two well-known bacterium genera involved in these types of infections are *Salmonella* and *Campylobacter*. The former is widespread in the intestines of birds and reptiles and of such mammals as pigs and cattle. It can contaminate humans via a number of different foods of animal origin; illnesses caused thereby include salmonellosis – which typically includes fever, diarrhoea and abdominal cramps. In people with poor underlying health or weakened immune system, it can invade the bloodstream and cause life-threatening infections. *Campylobacter* spp. are bacterial pathogens that also cause fever, diarrhoea and abdominal cramps; in fact, they are the most commonly identified bacterial cause of diarrhoeal illness in the world. These bacteria live in the intestines of healthy birds and can be found in most raw poultry meat. Ingestion of undercooked chicken, or other food that has been contaminated with juices dripping from raw chicken is the most frequent cause of infection thereby. *Escherichia coli* O157:H7 is a bacterial pathogen, for which a reservoir exists in cattle and other similar animals. Human illness typically follows consumption of food (e.g. ground beef) or water previously contaminated even with trace amounts of bovine faeces. Such an illness often causes severe and bloody diarrhoea, as well as painful abdominal cramps but is not accompanied by fever. In 3-5% of the cases, a complication termed haemolytic uraemic syndrome may occur several weeks after the initial symptoms; this severe status includes temporary anaemia, profuse bleeding and kidney failure.

In addition to direct infection, a few foodborne diseases are caused by presence of toxins in the food which were produced by a microorganism that lived therein but is no longer active. One such example is *S. aureus*, which can grow in food and produce a toxin that causes intense vomiting. The rare but deadly disease termed botulism occurs when *Clostridium botulinum* grows and produces a powerful paralytic toxin in food, typically in neutral/alkaline pH under anaerobiosis – as happens in poorly sterilized canned meat or fish.

Viral pathogens are often transmitted by infected food handlers, or via contact with sewage. However, only a few viruses, such as Hepatitis A and Noroviruses, have been proven to cause foodborne illnesses. Calicivirus, or Norovirus, is a rather common cause of foodborne illness – although it is rarely diagnosed, because suitable laboratory tests are not widely available. It causes acute gastrointestinal illness, usually with more vomiting than diarrhoea, which normally resolves within two days. Unlike many foodborne pathogens that have animal reservoirs, it is believed that Noroviruses spread primarily from one infected person to another – typically at the production and preparation levels. Infected kitchen workers can contaminate a salad or sandwich as they prepare it, if the virus is present on their hands. Furthermore, infected fishermen have contaminated oysters during harvest.

Parasites, such as *Trichinella spiralis* – which causes trichinosis, can occur in microscopic forms (eggs and larvae) in pork meat. A few common diseases are occasionally foodborne, even though they are usually transmitted by other routes; these include infections caused by *Shigella* spp. and by the parasites *Giardia lamblia* and *Cryptosporidia*, as well as hepatitis A virus. Strep throats have been transmitted occasionally through food. The most important foodborne diseases are depicted in Table 7.

4. Beneficial Microorganisms

The uniqueness of several microorganisms and their often unpredictable nature and biosynthetic capabilities, given a specific set of environmental conditions, have made them candidates in attempts to solve difficult problems in life sciences and other fields. Microorganisms have been used in various ways over the past 50 years, to advance medical technology, human and animal health, food processing, food safety and quality, genetic engineering, environmental protection and agricultural biotechnology. The use of beneficial microorganisms in the food sector has a long tradition, namely lactic acid bacteria and yeasts in fermentation processes; the former are widely used in the manufacture of fermented food and are among the best studied microorganisms. Detailed knowledge of a number of physiological traits has opened novel potential applications for these organisms in the food industry, while other traits might be specifically beneficial for human health.

4.1. Functional cultures with technological advantages

The use of functional starter cultures in the food industry has been on the rise. These starters possess at least one inherent functional property, which can contribute to food safety and/or offer one or more organoleptic, technological, nutritional or health advantage(s). Implementation of carefully selected strains as starter cultures or co-cultures in fermentation processes can help achieve *in situ* expression of desired properties, while maintaining a natural and healthy product. The main examples encompass the lactic acid bacteria (see Table 8), which are able to synthesize antimicrobial substances, sugar polymers, sweeteners, aromatic compounds and useful enzymes (Leroy and de Vuyst, 2004). Such activity allows the replacement of chemical additives by natural compounds, while providing the consumer with new, attractive food products.

TABLE 7. Main foodborne pathogens, source of illness and associated symptoms (adapted from Anonymous, 2004)

Disease Responsible microorganism	Vehicle	Symptoms
Bacteria		
Botulism	Wide-spread spores	**Onset:** Generally 4-36 hr upon ingestion
Botulinum toxin (produced by *Clostridium botulinum* bacteria)	Production of toxin only in anaerobic environment of low acidity Found in considerable variety of canned goods, *such as* corn, green beans, soups, beets, asparagus, mushrooms, tuna and liver paté Found in processed food, *such as* luncheon meats, ham, sausage, stuffed eggplant, lobster and smoked and salted fish	**Symptoms:** Neurotoxic symptoms, including double vision, inability to swallow, speech difficulty and progressive paralysis of respiratory system **Advice:** urgent medical help, as botulism is frequently fatal
Campylobacteriosis *Campylobacter jejuni*	Found in meat and milk of contaminated poultry, cattle and sheep Found in raw poultry, meat and unpasteurized milk	**Onset:** Generally 2-5 d upon ingestion **Symptoms:** Diarrhoea, abdominal cramping, fever and sometimes bloody stools, for 7-10 d
Listeriosis *Listeria monocytogenes*	Found in soft cheese, unpasteurized milk, imported seafood products, frozen cooked crab meat, cooked shrimp and cooked surimi (imitation shellfish) Resistant to heat, salt, nitrite and acidity Survival and growth at low temperatures	**Onset:** From 7-30 d upon ingestion but most symptoms reported by 48-72 hr **Symptoms:** Fever, headache, nausea and vomiting, with strongest effects on pregnant women and their foetuses, newborns, elderly, people with cancer and people with impaired immune systems. **Advice:** frequently fatal to foetus and infants.
Perfringens food poisoning *Clostridium perfringens*	Caused by failure to keep food hot (*such as* gravies and stuffing below 60°C) Organisms often present after cooking, with multiplication to toxic levels during cooling down and storage of prepared foods Found in meats and meat products	**Onset:** Generally 8-12 hr upon ingestion **Symptoms:** Abdominal pain and diarrhoea and sometimes nausea and vomiting, for 1 d or less and usually mild **Advice:** Serious in older or debilitated people
Salmonellosis *Salmonella* spp.	Found in raw meats, poultry, milk and other dairy products, shrimp, frog legs, coconut, pasta, chocolate, tomatoes and alfalfa sprouts	**Onset:** Generally 8-12 hr upon ingestion **Symptoms:** Mild abdominal pain and diarrhoea and sometimes nausea and vomiting, for 1 d

TABLE 7.—*Continued*

Disease Responsible microorganism	Vehicle	Symptoms
		Advice: Serious in older or debilitated people
Shigellosis (bacillary dysentery) *Shigella* spp.	Found in milk and dairy products, poultry and potato salad Contamination after poor hygiene of human handlers and lack of thorough cooking afterwards Fast multiplication in food left at room temperature	**Onset:** 1-7 d upon ingestion **Symptoms:** Abdominal cramps, diarrhoea, fever, sometimes vomiting and blood, pus or mucus in stools
Staphylococcal food poisoning Staphylococcal enterotoxin (produced by *S. aureus*)	Caused by leaving food too long at room temperature Found in meats, poultry, egg products, tuna, potato and macaroni salads and cream-filled pastries	**Onset:** Generally 30 min-8 hr upon ingestion **Symptoms:** Diarrhoea, vomiting, nausea, abdominal pain, cramps and prostration, for 24-48 hr
Vibrio infection *Vibrio vulnificus*	Transmitted through open wounds exposed to coastal waters or consumption of contaminated seafood **Advice:** High risk to people with liver condition, low gastric (stomach) acid and weakened immune system Multiplication at warm conditions	**Onset:** Abrupt **Symptoms:** Chills, fever and/or prostration
Protozoa		
Amoebiasis *Entamoeba histolytica*	Found in intestinal tract of humans, expelled in faeces and spread via polluted water and vegetables grown in polluted soil	**Onset:** 3-10 d upon ingestion **Symptoms:** Severe cramp pain, tenderness over colon or liver, loose morning stools, recurrent diarrhoea, loss of weight, fatigue and (sometimes) anaemia
Giardiasis *Giardia lamblia*	Found in contaminated water Transmitted through uncooked foods that were contaminated while growing or after cooking, especially in cool, moist conditions	**Onset:** 1-3 d upon ingestion **Symptoms:** Sudden onset of explosive watery stools, abdominal cramps, anorexia, nausea and vomiting, especially in hikers, children, travellers and institutionalized patients
Virus		
Hepatitis A virus	Found in molluscs (*such as* oysters, clams, mussels, scallops and cockles) harvested in polluted or untreated sewage, when eaten raw or after light cooking	**Onset:** 1-2 d upon ingestion **Symptoms:** Malaise, appetite loss, nausea, vomiting and fever; jaundice with darkened urine after 3-10 d **Advice:** Potential liver damage and death

(*Continued*)

TABLE 7. Main foodborne pathogens, source of illness and associated symptoms (adapted from Anonymous, 2004)—*Continued*

Disease Responsible microorganism	Vehicle	Symptoms
"Stomach flu," or gastroenteritis Norovirus	Found in any type of contaminated food or drinking liquid; transmitted through touching contaminated surfaces or objects followed by oral contact, including direct contact with infected person and via sharing foods or eating utensils	**Onset:** 12 h-2 d upon ingestion **Symptoms:** usually nausea, vomiting (more frequent in children than adults), diarrhoea and some stomach cramping; less frequently low-grade fever, chills, headache, muscle aches and general sense of tiredness; for 1-2 d

One good example of *functional starters* with technological advantage relates to bacteriophages – a serious problem in the dairy industry. Resistance to intracellular phage development may be brought about by natural mechanisms (e.g. restriction and modification enzymes), phage adsorption and abortive phage infection or by intracellular defence strategies. Strains that have acquired natural mechanisms of phage resistance, e.g. through *in vivo* recombination (conjugation) or *in vitro* self-cloning, are currently applied on a large scale in the dairy industry (Moineau, 1999).

Another example is undesirable post-acidification – ascribed to *L. delbrueckii* subsp. *bulgaricus* during yoghurt storage that leads to acid and bitter tastes. Lactose-negative mutants enable production of mild yoghurts, since such cells can, given their protocooperation, grow only in the presence of actively lactose-fermenting *S. thermophilus* cells (Leroy and de Vuyst, 2004).

4.2. Functional cultures with health advantages

A probiotic is a live microbial feed supplement that beneficially affects the host beyond correcting for traditional nutrient deficiencies, via improvement of its intestinal balance.

Human gut microbiota can be influenced by diet and it is the improved resistance to pathogens that offers most promise for efficacious development of probiotic cultures. Such beneficial effects include such factors as antagonistic effects, competition, immune effects and attenuation. The importance and increased application of probiotics have supported the intense search in recent years for novel strains that have beneficial effects ascribed thereto (see Table 9).

TABLE 8. Selected functional starter cultures or co-cultures in the food industry and advantages thereof (adapted from Leroy and de Vuyst, 2004)

Features	Functionality	Microorganism
Preservation	Bacteriocin production	*Lactococcus lacti* subsp. *Lactis.*
	Dairy products	*Enterococcus* spp.
	Fermented meats	*Lactobacillus curvatus, Lb. Sakei, P. acidilactici*
	Fermented olives	*E. faecium Lb. plantarum*
	Fermented vegetables	*L. lactis*
Organoleptic	Exopolysaccharide production	Several lactobacilli and streptococci
	Amylase production	Several lactobacilli
	Aroma generation	Several strains
	Sweetness enhancement	
	Homoalanine-fermenting starters	*Lactococcus lactis*
	Galactose-positive/glucose-negative starters	*Lactobacillus delbrueckii* subsp. *bulgaricus, Streptococcus salivarius* subsp.*thermophilus*
	Malolactic fermentation	*Oenococcus oeni*
Technological	Bacteriophage resistance	Several strains
	Overacidification prevention in lactose-free yoghurt	Lb. delbrueckii subsp. Bulgaricus
	Starter autolysis	
	Phage-mediated	L. lactis subsp. lactis
	Bacteriocin-induced	L. lactis
Nutritional	-Nutraceutical production	*Lb. plantarum*
	Low-calorie sugars (e.g. sorbitol and mannitol)	*L. lactis*
	Oligosaccharide production	*L. lactis* *Lb. delbrueckii* subsp. *bulgaricus,*
	B-group vitamin production e.g. folic acid)	*S. Streptococcus salivarius* subsp. *thermophilus*
	-Bioactive peptide release	Several strains
	-Toxic and anti-nutritional compound reduction	
	(+)-lactic acid production	lactic acid-producing isomer strains
	Lactose and galactose removal	Several strains
	Soy raffinose removal	*S. sali varius* subsp *thermophilus*
	Phytic acid content reduction, amylase inhibitor	*Lb. Plantarum, Lb. acidophilus*
	Decreased biogenic amines production	*Enterococcus faecalis*

TABLE 9. Selected probiotic strains and advantages thereof for human health (adapted from Mattila-Sandholm *et al.*, 1999)

Microorganism	Functionality
Lactobacillus GG ATCC 53103	Adherence to human intestinal cells
	Lowering of faecal enzyme activities
	Prevention of antibiotic-associated diarrhoea
	Treatment and prevention of rotavirus diarrhoea
	Prevention of acute diarrhoea
	Immune response modulation
Lactobacillus johnsonii LJ-1 (LA-1)	Prevention of traveller's diarrhoea
	Modulation of intestinal flora
	Alleviation of lactose intolerance symptoms
	Improvement of constipation
	Immune enhancement
	Adjuvant in *Helicobacter pylori* treatment
Bifidobacterium lactis (*bifidum*) Bb-12	Prevention of traveller's diarrhoea
	Treatment of viral diarrhoea, including rotavirus diarrhoea
	Modulation of intestinal flora
	Improvement of constipation
	Modulation of immune response
Lactobacillus reuteri ATCC 55730	Colonisation of intestinal tract
	Shortening of rotavirus diarrhoea
	Treatment of acute diarrhoea
Lactobacillus casei Shirota	Modulation of intestinal flora
	Lowering of faecal enzyme activities
	Positive effects on superficial bladder cancer
Lactobacillus plantarum DSM 9843	Adherence to human intestinal cells
	Modulation of intestinal flora
Saccharomyces boulardii	Prevention of antibiotic-associated diarrhoea
	Treatment of *Clostridium difficile* colitis

5. References

Anonymous, 2004, *Diagnosis and Management of Foodborne Illnesses: A Primer for Physicians and Other Health Care Professionals.* American Medical Association.

Beresford, T.P., Fitzsimons, N.A., Brennan, N.L., and Cogan, T.M., 2001, Recent advances in cheese microbiology. *International Dairy Journal* **11**:259-274.

Casadevall, A. and Pirofski, L-A., 1999, Minireview: host-pathogen interactions: redefining the basic concepts of virulence and pathogenicity. *Infection and Immunity* **67**: 3703-3713.

Casadevall, A. and Pirofski, L-A., 2000, Minireview: host-pathogen interactions: basic concepts of microbial commensalisms, colonization, infection and disease. *Infection and Immunity* **68**:6511-6518.

Cheirsilp, B., Shoji, H., Shimizu, H., and Shioya, S., 2003, Interactions between *Lactobacillus kejiranofaciens* and *Saccharomyces cerevisiae* in mixed culture for kefiran production. *Journal of Bioscience and Bioengineering* **96**:279-284.

de Vuyst, L. and Neysens, P. 2005, The sourdough microflora: biodiversity and metabolic interactions. *Trends in Food Science and Technology* (In press).

Fleet, G.H., 1999, Microorganisms in food ecosystems. *International Journal of Food Microbiology* **50**:101-117.

Ginovart, M., López, V., Valls, J., and Silbert, M., 2002, Simulation modelling of bacterial growth in yoghurt. *International Journal of Food Microbiology* **73**:415-425.

Giraffa, G., 2004, Studying the dynamics of microbial populations during food fermentation. *FEMS Microbiology Reviews* **28**:251-260.

Gomes, A.M.P., Malcata, F.X., and Klaver, F.A.M., 1998, Growth enhancement of *Bifidobacterium lactis* Bo and *Lactobacillus acidophilus* Ki by milk hydrolyzates. *Journal Dairy Science* **81**:2817-2825.

Henderson, B., Wilson, M., McNab, R., and Lax, A.J., 2001, *Cellular Microbiology: bacteria-host interactions in health and disease*. Wiley. Chichester, UK.

Kramer, J. and Gilbert, B., 1995, *Bacillus cereus* and other *Bacillus* species. In *Foodborne Bacterial Pathogens* Doyle, M.P. (Ed.), Marcel-Dekker, New York, pp. 21-70.

Leroy, F. and de Vuyst, L., 2004, Lactic acid bacteria as functional starter cultures for the food fermentation industry. *Trends in Food Science & Technology* **15:**67-78.

Malakar, P.K., Barker, G.C., Zwietering, M.H., and van't Riet, K., 2003, Relevance of microbial interactions to predictive microbiology. *International Journal of Food Microbiology* **84**:263-272.

Mattila-Sandholm, T., Matto, J., and Saarela, M., 1999, Lactic acid bacteria with health claims interactions and interference with gastrointestinal flora. *International Dairy Journal* **9**:25-35.

Moineau, S., 1999, Applications of phage resistance in lactic acid bacteria. *Antonie van Leeuwenhoek International Journal of General and Molecular Microbiology* **76**: 377-382.

Narvhus, J.A. and Gadaga, T.H., 2003, The role of interaction between yeasts and lactic acid bacteria in African fermented milks: a review. *International Journal of Food Microbiology* **86**:1-60.

Singleton, P., 1999, *Bacteria: in Biology, Biotechnology and Medicine*. Wiley Chichester, UK.

Schillinger, U., Geisen, R., and Holzapfel, W.H., 1996, Potential of antagonistic microorganisms and bacteriocins for the biological preservation of foods. *Trends in Food Science & Technology* **71**:58-64.

Timmermana, H.M., Koningb, C.J.M., Mulderc, L., Romboutsd, F.M., and Beynena A.C., 2004, Monostrain, multistrain and multispecies probiotics – a comparison of functionality and efficacy. *International Journal of Food Microbiology* **96**:219-233.

Tlaskalová-Hogenová, H., Stepánková, S., Hudcovic, T., Tucková, L., Cukrowska, B., Lodinová-žádníková, R., Kozáková, H., Rossmann, P., Bártová, J., Sokol, D., Funda, D. P., Borovská, D., Reháková, Z, Šinkora, J., Hofman, J., Drastich, P., and Kokešová, A., 2004, Commensal bacteria (normal microflora), mucosal immunity and chronic inflammatory and autoimmune diseases. *Immunology Letters* **93**:97-108.

Todar, K., 2002, *Todar's Online Textbook of Bacteriology. The Nature of Host-parasite Interactions*. University of Wisconsin-Madison Department of Bacteriology.

Viljoen, B.C., 2001, The interaction between yeasts and bacteria in dairy environments. *International Journal of Food Microbiology* **69**:37-44.

Wassenaar, T.M. and Gaastra, W., 2001, Bacterial virulence: can we draw the line? *FEMS Microbiology Letters* **201**:1-7.

10
Foodborne Viruses: An Emerging Risk to Health

Leen Baert, Mieke Uyttendaele, and Johan Debevere*

1. Prevalence and Importance of Foodborne Viruses

1.1. Introduction

The essence of food producing and processing industries is to bring safe food products according to international and national regulations on the market to prevent any kind of food poisoning to the consumers. The issue of food safety focuses most of the time on microbial micro-organisms, moulds or toxins. Previously, an overlooked and certainly underestimated causative agent of foodborne illnesses in humans concerned viruses present on food items. For the last three decades foodborne and waterborne viral infections have been increasingly recognized as causes of illness in humans. Reasons for this increase include the improved diagnostic methods that have enhanced detection of some virus groups and the increased marketing of fresh and frozen foods (Norrung, 2000).

Viruses are obligate intracellular organisms and are not able to grow on food products. This implies that replication can not occur on the food or in the environment but needs living cells. Viruses interfering with human health and transmitted by food can be shed in the faeces of infected persons. These viruses have been classified to the human enteric viruses. The most commonly implicated are listed in Table 1(Sair et al., 2002).

The food- and waterborne viruses can be classified in categories according to the clinical symptoms they cause: The Noroviruses, Sapoviruses, Astroviruses, and Rotaviruses cause gastroenteritis. The HAV and HEV, on the otherhand cause hepatitis.

The importance of the contribution of viral pathogens to illness was reviewed by Mead and co-workers (1999) who reported that 67% of total foodborne illness was due to viruses. The data from ten surveillance systems, executed in ten different European countries, established that Noroviruses were responsible for >85%

*Leen Baert, Mieke Uyttendaele and Johan Debevere, University of Ghent, Ghent, Belgium B-9000

TABLE 1. Significant human enteric viruses transmitted by food

Noroviruses, previously called Norwalk-like virus (NLV)
Hepatitis A virus (HAV)
Sapoviruses
Astroviruses
Rotaviruses (group A-C)
Hepatitis E virus (HEV)

of all non-bacteriological outbreaks of gastroenteritis reported from 1995 to 2000 (Lopman et al., 2003).

The Centers for Disease Control and Prevention (CDC) reported that between January 1997 and June 1998, 96% of the acute non bacterial gastroenteritis in the US was caused by **Norwalk-Like Virus** (NLV) (Greene et al., 2003; Frankhauser et al., 1998). From 1992 to 1997, noroviruses accounted for one-third of all gastroenteritis outbreaks reported to the PHLS Communicable Disease Surveillance Centre and the number of outbreaks of noroviruses exceeded the number of outbreaks of salmonellosis (Seymour and Appleton, 2001).

Hepatitis A virus (HAV) is also involved in many foodborne outbreaks. In the US, HAV causes an estimated 83,000 illnesses per year (Richards, 2001). From 1980 to 2001, an average of 25,000 cases each year was reported to the CDC, but when corrected for underreporting and asymptomatic infections, an estimated average of 263,000 HAV infections occurred yearly. Relatively few reported cases (2%-3% per year) are identified through routine surveillance as part of common source outbreaks of disease transmitted by food or water. However, some Hepatitis A transmission attributed to personal contact or other risk factors is likely to have been foodborne, occurring when an HAV infected person contaminated food eaten by others. The proportion of sporadic cases that might be from foodborne sources is unknown, but could be considerable; approximately 50% of reported patients with Hepatitis A do not have an identified source of infection (Fiore, 2004).

However, the main problem of HAV infections implies the shifted age at which the infections most probably occur. In most cases infected children < 6 years with HAV don't show any symptoms. Among the majority of young adults, clinical manifestations occur. It is not easy to report the outbreaks and particularly the source and transmission routes of HAV because of the long incubation period (time from exposure to onset of symptoms) of 28 days (range: 15-50 days).

Data about outbreaks can be accessed on the internet in the Eurosurveillance weekly archives and Eurosurveillance monthly archives. The described outbreaks take place at schools, restaurants, nursing homes, cruise ships, youth camps.

Outbreaks due to NLV and HAV spread by water or food have the potential to involve large numbers of people, by widely geographically spread and perhaps, introduce new variants to an area. There are, as mentioned above, numerous reports of foodborne outbreaks of NLV and HAV but the real incidence of these diseases and the contribution of these outbreaks to the disease burden remain unclear. As result, a European database was set up to map outbreaks geographically and elucidate transmission routes (Lopman et al., 2002).

From the data received from surveillance studies, it seems that noroviruses and HAV are clearly the most important cause of illness transmitted by food. Those foodborne viruses will be discussed in detail.

1.2. Noroviruses

1.2.1. Historically – classification

The Norwalk agent was discovered in 1972 when a gastroenteritis outbreak, in a primary school, was reported in Norwalk. Through the use of electron microscopy, researchers found viral micro-organisms in the collected faeces samples received from the infected persons. In the following years other viruses with a similar morphology were identified and named after the place where they are initially recognized (for example Hawaii, Snow Mountain, Sapporo, Tounton) and were grouped together and termed the small round structured viruses (SRSV's). In the 1990's, molecular techniques established the further typing and classification of the SRSVs.

The SRSV group was shown to be member of the family *Caliciviridae*. The viruses are characterised by small spherical viruses, measuring between 28 and 35 nm in size, containing a positive-stranded, polyadenylated RNA genome of 7.7 kilobases protected by a protein capsid but do not contain a lipid envelope. The genome contains coding information for a set of non structural proteins, located at the 5′-end and one major structural protein at the 3′-end.

The *Caliciviridae* family is subdivided into 4 genera: (1) *Vesivirus*; (2) the *Lagovirus*; (3) the "Norwalk-like viruses", abbreviated NLV, having the Norwalk virus as the prototype strain; and (4) the "Sapporo-like viruses" (SLVs), represented by the Sapporo virus. The Norwalk-like viruses and Sapporo-like viruses are now called *Norovirus* and *Sapovirus* respectively. NLV and SLV are the human pathogenic genera and are grouped as the human Caliciviruses (HuCV) while the Vesiviruses and Lagoviruses are considered to be animal Caliciviruses (Sair et al., 2002). Noroviruses are the most important causative agent of human foodborne disease. Within the genus of the noroviruses, at least 5 genetic groups have been identified based on the sequence similarities across highly conserved regions in the genome (the RNA-dependent RNA polymerase and the shell domain of the capsid protein (Hutson et al., 2004). Two genetic groups, genogroup I (GI) and genogroup II GII) comprise the most human noroviruses. Some animal virus strains are closely related to the human virus strains and consequently can be classified in the same division. For example, the bovine Calicivirus cluster has been placed in genogroup III and which is closely related to GI. Genogroup IV (GIV) includes at least two human strains (Alphatron and Ft. Lauderdale), while a fifth genogroup has been proposed for the porcine noroviruses.

Human Noroviruses within the same genogroup share at least 60% amino acid sequence identity in the major capsid protein, whereas most GI share less than 50% amino acid identity with those of GII. GI and GII can be further subdivided into genetic clusters. Those Norovirus strains within the same cluster share at least 80% amino acid sequence identity with the cluster reference strain (Hutson et al., 2004). The undeniable conclusion of the phylogenetic analysis points out the enormous heterogeneity among the circulating human Noroviruses.

1.2.2. Clinical features

Low-grade fever, vomiting, abdominal cramps, diarrhoea and headache are the prominent symptoms after a 1 to 3 days incubation period. Patients can experience vomiting alone, a condition first identified as winter vomiting disease, for which the average attack rate is high (typically 45% or more). The virus is shed in the faeces and could last up to 10 days after the symptoms disappeared. As a consequence, recovered persons recently infected with Norovirus can act as carriers for the time as long as the virus is shed in the faeces. Moreover, NLV infections are highly contagious, resulting in high occurrence of secondary infections. Nevertheless, the illness is self limiting. A possible fatal outcome can occur when severe dehydration appears as a result of the infection but is only possible in susceptible people (e.g. elderly, immunocompromised people ...)

1.2.3. Transmission

Reports of NLV outbreaks have been epidemiologically associated with various items of fresh produce (Seymor and Appleton, 2001), such as washed salads (Lieb, 1985; Lo et al., 1994), imported frozen raspberries (Pönkä et al., 1999; Le Guyader et al., 2004), coleslaw (Currier, 1996), green salads (Griffin et al., 1982, fresh cut fruits (Herwaldt et al., 1994) and potato salad (Patterson et al., 1997). Many outbreaks of foodborne disease attributed to NLV have been associated with the consumption of contaminated bivalve molluscs (e.g. clams, cockles, mussels and oysters) (Doyle et al., 2004; Prato et al., 2004; Atmar et al., 1995; Le Guyader et al., 2000; Goswami et al., 2002). Also commercial ice, sandwiches (Parashar et al., 1998), bakery products and salads (Jaykus, 2000; Sair et al., 2002) are reported to be implicated.

1.3. Hepatitis A

HAV is frequently implicated in cases of acute gastroenteritis around the world. The virus is transmitted among humans via the faecal-oral route and infection by HAV represents the most serious form of viral illness acquired from foods (Jean et al., 2001). HAV is classified as a Picornavirus, measures 27 nm and is non-enveloped. The virus is related to the Poliovirus. The virus contains four capsid proteins encompassing a positive polarity single-stranded RNA genome (Bidawid et al., 2000). There is only one HAV serotype and primates are the only natural host. After ingestion, uptake of HAV particles takes place in the gastrointestinal tract, and subsequent replication in the liver. HAV will be secreted in the bile, and finally high concentrations are found in stool specimens. HAV has contributed to numerous outbreaks often associated with raw or lightly cooked shellfish.

It causes a potentially severe but controllable loss of liver function and general malaise. Proper medical care will generally result in full recovery of liver function and full clearance of virus from the host, with effective and lifelong immunity against reinfection (Fiore, 2004).

1.3.1. Clinical features

The symptoms start with fever, anorexia, nausea, vomiting, diarrhoea, myalgia and malaise. Jaundice, dark coloured urine or light coloured stools might be present at the onset or might develop/occur within a few days. For the most people infected with Hepatitis A virus, illness lasts for several weeks. The mortality rate stands at about 0.3% of reported cases. If people above fifty years are considered than the mortality rate rises to 1.8%.

1.3.2. Transmission

Many outbreaks of HAV have been associated with foods handled by infected restaurant workers. Food products can be contaminated at any time from the pre-harvest to the post-harvest stage by improper irrigation or fertilization practices, by infected pickers or processors, or by contact with contaminated surfaces. Raw or lightly cooked food, such as shellfish (Desenclos et al., 1991; Halliday et al., 1999; Coelho et al., 2003), fruit and vegetables (Hutin et al., 1999; Croci et al., 2002; Hernández et al., 1997), salads or post contamination after cooking, for example frosted bakery products (Feinstone, 1996; Cliver, 1997) are reported sources of gastroenteritis.

1.4. Emerging hepatitis E virus

The Hepatitis E virus (HEV) is found throughout the world and has caused significant epidemics in India and Russia through problems with drinking water. Only a few foodborne cases have been reported to date (Chan, 1995; Stolle and Sperner, 1997). Furthermore, no cases or outbreaks have been unequivocally shown to be due to food consumption. It is however, possible that foods may act as vehicles for the HEV. Transmission is thought to occur through foods washed or processed with HEV-contaminated water, produce irrigated with HEV-contaminated water or shellfish harvested from HEV-contaminated waters. Furthermore, if those foods are eaten raw or inadequately cooked then the foods themselves can be a source of disease. HEV can be introduced more and more in industrialised countries through travel to an endemic area or the increasing globalization of the world market place (Smith, 2001).

HEV is a non-enveloped spherical virus of 27-34 nm diameters and contains a single stranded, positive sense RNA genome of approximately 7.5 kb and is still an unclassified virus. Sequence similarity studies subdivided HEV into 4 genotypes

(Yazaki et al., 2003). The appearance of a specific genotype is correlated with a specific geographical region.

1.4.1. Clinical features

The incubation period varies from 15 to 60 days with a mean of 40 days. Most infected persons show jaundice accompanied with malaise, anorexia, abdominal pain, liver enlargement, and fever. The syndrome is similar to that seen with hepatitis A. Recovery from acute infection is generally complete, and there is no evidence of chronic infection following the acute phase. The case-fatality rate for hepatitis E infection ranges from 0.5-3% but rises to 15-25% in pregnant women. Hepatitis E during pregnancy is a serious health risk. Pregnant women are more likely to be infected, more likely to develop fulminant hepatic failure (FHF), and more likely to die with loss of the foetus (Smith, 2001).

1.4.2. Transmission

The main transmission route in developing countries is the faecal-oral waterborne route caused by the pollution of drinking water with faeces contaminated with HEV. In developed countries, the maintenance of good hygiene of the water supply and sewerage systems makes the likelihood of waterborne infections low (Tei et al., 2004).

Hepatitis E infections have been reported after the consumption of shellfish from faecal contaminated waters (Cacopardo et al., 1997; Mechnik et al., 2001). HEV is also found in raw pig liver (Yazaki et al., 2004) and an outbreak of HEV occurred by eating raw meat of an infected wild deer in Japan (Tei et al., 2003).

Direct person-to-person transmission during epidemics plays a minor role in the spread of HEV, with an attack rate of less than 2.5% within households containing primary cases. Zoonotic spread of HEV has been suggested as human and swine HEV strains are genetically closely related. Experimental cross-species infections of swine HEV to non-human primates and that of human HEV have been demonstrated (Yazaki et al., 2003).

1.5. Other foodborne viruses

1.5.1. Sapoviruses

Sapoviruses, previously called Sapporo-like viruses (SLV), represent one of the 4 genera belonging to the family of the *Caliciviridae* (Green et al., 2000). Together with the noroviruses described as the human caliciviruses. SLV reveal typical calicivirus morphology and are also called "typical" or "classical" caliciviruses. NLV do not reveal typical calicivirus morphology and are called small, round-structured viruses (Jiang et al., 1999). The genome of SLV is slightly different organized, compared to NLV. The genomic organization is more similar to that found in an animal calicivirus, rabbit hemorrhagic disease virus belonging to the genus Lagovirus. The non-structural proteins as well as the capsid protein are coded in the first Open Reading Frame (ORF) while the second ORF encodes a predicted small, highly basic protein of unknown function, similar to ORF3 for NLV (Atmar and Estes, 2001). Sapoviruses cause acute gastroenteritis mainly in children (under the age of 1 year) (Chiba et al., 1979; Cubitt et al., 1979, 1980; Nakata et al.; 1985; Cubitt and McSwiggan, 1987; Nakata et al., 1996) and in the elderly (Cubitt et al., 1981; Gray et al., 1987). Sapoviruses are a minor causative agent of gastroenteritis in comparison to noroviruses. In the Netherlands a study estimated a contribution of noroviruses of 11% while 2% for Sapoviruses and another 5% by rotaviruses group A in the year 1999 when the overall incidence of gastroenteritis was estimated on 283 per 1000 persons annually (de Wit, 2003).

1.5.2. Rotaviruses

Rotaviruses are non-enveloped, wheel like viruses. The genome consists of 11 segments of double stranded RNA which code for six structural and five non-structural proteins.

Group A, B and C Rotaviruses are three serological groups of the 6 known, infecting humans. Group A rotavirus is the most common cause of childhood diarrhoea worldwide, infecting > 90% of children by the age 3 years (Kapikan et al., 1996). About half of the cases of Group A Rotavirus requires hospitalization. Over 3 million cases of Rotavirus gastroenteritis occur annually in the U.S. The disease is characterized by a self-limiting disease accompanied with vomiting, watery diarrhoea, and low-grade fever. Humans of all ages are susceptible to rotavirus infection. Children from 6 months to 2 years of age, premature infants, the elderly, and the immunocompromised are particularly prone to more severe symptoms caused by infection with group A Rotavirus. The infective dose is presumed to be 10-100 infectious viral particles. A person with rotavirus diarrhoea often excretes large numbers of virus (10^8-10^{10} infectious particles/ml of faeces). Consequently, infectious doses can be readily acquired through contaminated hands, objects, or utensils. Asymptomatic Rotavirus excretion has been well documented and may play a role in perpetuating endemic disease (FDA, cfan:a). The number attributable to food contamination has not yet been elucidated but it is estimated that only 1% of the rotavirus infections are foodborne (Mead, 1999).

1.5.3. Astroviruses

Astroviruses are non-enveloped star shaped viruses consisting of a single positive strand of RNA surrounded by a protein capsid (FDA, cfan:b). Astroviruses have been mainly associated with illness in young children, often under 1 year old. Reports of astrovirus infection in older children and adults are infrequent, although outbreaks have been reported in the elderly. Astroviruses have been seen in some adults following the consumption of shellfish or contaminated water, but these incidents appear to be comparatively rare (Seymour and Appleton, 2001).

2. Sources of Viral Foodborne Contamination

Viruses are parasites. They need living cells in order to replicate. This characteristic does not exclude the ability to resist in the environment for a long time. Viruses, particularly non enveloped viruses, are extremely stable outside their host and act as inert particles providing time to transfer from one host to another. Viruses associated with foodborne illness, adapted to survive several environmental stresses and transmittable through the faecal-oral route are the human enteric viruses. Consequently, when carried in food, the amount of virus particles can persist but can not replicate. Therefore the contamination level remains unchanged or decreases. The drawback of such viral contamination is the unawareness of the potential hazard (by the human host) because the contaminated food products will look, smell and taste the same. Contrary to the situation that prevails with microbial contamination in which, bacteria grow to high numbers and change the features of the food which is termed food spoilage.

The major transmission possibilities are (1) shellfish contaminated by faecal polluted water; (2) human sewage pollution of drinking, washing or irrigation water; (3) ready-to-eat (RTE) and prepared foods contaminated as a result of poor personal hygiene of infected food handlers (Jaykus, 2000).

The feasibility of human enteric viruses to survive outside the host depends on the environmental factors and the properties of the intended virus. Conditions such as heat, moisture, and pH will influence the survival of virus particles.

2.1. Shellfish

Bivalve molluscan shellfish such as oysters, mussels and cockles grown in estuarine waters are a major source of foodborne gastroenteritis. These grow and are harvested in waters that may be discharge sites for human sewage as a result of illegal throwing/dumping of waste overboard by ships, the failure of septic systems along the shoreline, treated and untreated municipal waste water and sludge discharges occurring mostly during heavy rainfall (Jaykus et al., 1994; Shieh et al., 2000).

Bivalve molluscan shellfish are filter-feeders. They take up surrounding water and filter out nutritive components. When human pathogens, for example enteric viruses, are present in the coastal growing waters, an accumulation can occur by those animals. The method of filter-feeding is considered as a concentration method of those pathogenic micro-organisms (Mullendore et al., 2001). The common seafood processing procedures of icing and freezing are likely to enhance survival of viruses as these are widely used laboratory preservation techniques for viruses (Lees, 2000). In general, the oysters are consumed raw or might be lightly cooked. Also mussels or other shellfish are prepared by cooking but most of the time they are just heated until the shells opened, which is usually achieved at temperatures under 70°C at 47 ± 5 s (Koff and Sear, 1967). This heating process is not sufficient for shellfish viral decontamination (Croci et al., 1999).

2.2. *Drinking, washing and irrigation waters*

Contaminated drinking water as source of human viral disease is of major concern in developing countries. Lower hygienic standards, little available recourses for purification of water and especially the flooding of certain areas mediate the contact of water with human faecal sewage.

In developed countries, the drinking water is not the main problem. However, the way water is used for agricultural purposes can be a possible source of viral contamination. Fertilization with manure containing contaminated human waste could come directly into contact with the produce (fruits, vegetables) or indirectly with the irrigation water and could bring about the unsafe products.

Processing of fruits and vegetables usually includes a wash step before further handling. Using contaminated washing water enables the transfer of virus particles to the fresh produce. Without a proper heat treatment or other adequate preservation technique, the consumer ingests the viral dose. There are few published studies that report the stability of viruses in sewage composed and used on agricultural lands. It is however, known that adsorption is a major factor in virus removal and persistence in soils, with adsorptive capabilities known to be dependant upon both the virus and soil type. There is little data available to estimate the relative importance of contaminated water and fertilizer in the propagation of foodborne viral disease (Sair et al., 2002).

2.3. *RTE-food and post-contamination of food products: lack of personal hygiene*

Foods that don't need a heat treatment, washing or additional preparation before consumption are termed/called Ready-to-Eat foods (RTE-foods). These foods are very susceptible to levels of good hygienic practices adopted by food handlers. When the food handler is infected with enteric viruses, transmission is possible through improper hygienic actions. After visiting the bathroom, hands should be thoroughly washed with soap. Faecal material can be left on the hands, under nails or on fingers which can come in contact with the food (Jaykus, 2000, Bidawid et al., 2000). Bare hands have been identified as an important source of pathogen transfer in the handling of cooked or RTE-food products (Bryan, 1995). Aerosol particles originating from vomiting can also spread virus particles, so a decent cleaning of the working and producing environment should be taken into account when one of the personnel falls ill with a gastrointestinal complaint. Patterson (1997) described an incident of an employee vomiting in a sink that afterwards was cleaned with chlorine-based disinfectant and even then the potato salad made in that sink contributed to a Norovirus outbreak. The spread of aerosols contaminated with virus particles is described by Marks and co-workers (2000) with an incident in which one of the customers had a bout of vomiting in the restaurant itself. The further away the people were sitting from this unfortunate customer, the lesser ill people could be later observed. While the people sitting nearby, were observed to have a greater tendency to be contaminated and later could be observed to be ill. This outbreak was due to noroviruses and in this case the spread was reciprocally correlated with the distance. Other researchers concluded also the evidence for airborne transmission of viruses causing acute gastroenteritis (Sawyer et al., 1988; Chadwick et al., 1994; Caul, 1994).

Expelling infected food handlers from work during the transmissible period of those foodborne viruses is hardly possible to achieve because food handlers can shed virus after the recovery of the illness (e.g. shedding of Norovirus until at least 10 days after recovery) and moreover, it is possible that food handlers are infected without showing symptoms. Asymptomatic infections are common for all foodborne viruses. For example, carriers of hepatitis A typically shed high quantities of the virus 10-14 days after infection during the incubation period; in the weeks following this period carriers may or may not develop symptoms (Koopmans et al., 2004).

3. Correlating Foodborne Illness to Sources of Contamination

The diagnosis of viral gastroenteritis can be made based on the epidemiological criteria described by Kaplan and co-workers (1982). Characteristic features are: acute onset after a 24-36 h incubation period, vomiting or diarrhoea lasting a few days, a high attack rate (average 45%), and a high number of secondary cases (Koopmans et al., 2002). The clinical samples, mostly stools or vomitus as supplement to the stool specimens or if feasible serum specimens, collected from the ill persons showing the symptoms mentioned above can be examined. If the same viral agents are found in samples from patients correlated geographically or some correlation is seen in time, it might be an outbreak. In that case it is interesting to search for the common source of the outbreak. Food or water samples or other environmental specimens should be examined. If the same viral origin is found in clinical samples as in environmental samples, the cause of the viral gastroenteritis infection is found and the outbreak is cleared up. All the outbreaks should be centralized in a database to set up a good surveillance system. It is very useful to understand the behaviour of the virus intended, the evolution in time and the prevalence. The more the image of the virus is comprehended, the better the anticipation can be with regard to prevention of viral contamination (e.g. formulate strict regulations about hygienic practices) to avoid outbreaks.

The linking of clinical samples to environmental sources needs appropriate detection methods. The matrix and the level of contamination of stools are different than that of food items. This implies the necessity of different extraction methods and detection methods with matching detection limits.

3.1. Detection of infections in clinical samples

Previously, the detection of viruses was based on a specific immune response to the virus or the detection of virus particles or antigen in the stool. Norovirus infection was detected in the stool by the use of electron microscopy (EM). Subsequently antibody rich convalescent-phase sera were used to aggregate Norovirus particles for detection by IEM. Afterwards more sensitive immunoassays, including RIA, biotin-avidin immunoassay, immune adherence haemagglutination assay, enzyme immuno assay and enzyme linked immuno sorbent assay (ELISA) replaced the formerly methods. In contrast to Norovirus detection, where the detection is frequently based on the detection of viral antigen or viral nucleic acid in stool, diagnosis of hepatitis A infection is primarily done by detection of IgM anti-HAV antibodies in serum (Sair et al., 2002).

In general, the molecular technique RT-PCR for the detection of human enteric viruses in clinical specimens is frequently applied.

3.2. Detection of viruses on food

The detection of virus particles on food includes more problems in comparison with the examination of the viral load present in stools. In general infected persons shed at least 10^6 virus particles per ml. To guarantee food safety towards human enteric viruses, ingestion of 10 to 100 virus particles is sufficient to cause disease. Consequently, the food is considered safe if the contamination does not exceed that amount. Considering the low infectious dose of only 10 to 100 virus particles, the detection methods of human enteric viruses in food sources should be sensitive enough. Food safety towards microbial contamination uses techniques which includes enrichment steps and selective plating to enumerate the microbial pathogens until a detectable amount (increasing the needed detection limit of the used method) and to decrease the possibly disturbing accompanying flora. This strategy is not feasible for human enteric viruses because they need mammalian cells in order to replicate instead of culture media, just containing nutritive constituents, regarding to bacteria. The need of a cell culture and of course the intended virus should be able to grow in a certain cell line hampers this approach. For instance human caliciviruses are not succeeded to grow in a cell line demanding other fields of detecting approaches. Gene probes and nucleic acid hybridization methods have been reported for the detection of enteroviruses (Margolin et al., 1989), HAV (Jiang et al., 1987), and rotaviruses (Zhou et al., 1991) in environmental water and shellfish (Sair et al., 2002) but the detection limit exceeded usually 10^3 to 10^4 particles per sample which is too high. Techniques feasible to detect low entities and pre-eminently used is the nucleic acid amplification technique RT-PCR. Therefore, the virus particles should be concentrated into a small volume and the food matrix should be eliminated as much as possible preventing inhibition of the RT-PCR reaction.

Bivalve molluscan shellfish such as oysters, mussels, and cockles are focussed because the way the animals feed themselves results in a natural concentration of virus particles. Nevertheless, actually any kind of food product is exposed to viral contamination if food handlers did not care about the hygienic practices. Consuming the food without any treatment destroying the viruses, (e.g. thoroughly heating) can result in illness.

3.3. Protocols associated with viral identification in foods

In literature, many protocols to separate virus material from the food are described. Two different approaches can be distinguished. The first proceeding includes the isolation of virus particles prior to the extraction of the viral RNA. The alternative procedure extracts the total RNA content of the food followed by the selection of the viral RNA through the use of specific primers in the nucleic acid amplification technique (e.g. RT-PCR).

Limitations of these concentration methods is the loss of virus particles during the manipulations of adding chemical agents in order to exclude most of the food matrix, preventing the presence of inhibitory components, and to concentrate the virus material in a low volume to make it useful for RT-PCR. Besides RT-PCR, some investigators developed a nested RT-PCR protocol increasing the sensitivity. The disadvantage of a nested procedure is the implication of more handling steps, enhancing the risk of cross-contamination.

The critical point about the use of molecular techniques based on the detection of nucleic acids for virus detection is the correlation between the infectivity of the virus with the detected viral RNA. However, nowadays the main issue is the underestimation of the viral safety aspect of food. The presence of viral RNA indicates that a food–virus contact occurred, implying a possible hazard to human health.

4. Prevention and Regulations

4.1. Shellfish and fishery products

Most food related products with acute viral gastroenteritis are shellfish. In the European Union (EU) control on the production of live bivalve molluscan shellfish is covered by Council Directive 91/492/EEC. In the US similar regulations are set in the FDA National Shellfish Sanitation Program Manual of Operations. Requirements for harvesting area classification, bivalve transport, wet storage, depuration, relaying, analytical methods, movement and marketing documentation and provision of suspension of harvesting from polluted classified areas or in case of public health emergency are included. These requirements are applied until the point of processing. Processed bivalve molluscs termed "fishery products" are covered by 91/493/EEC within the EU comprising legislative requirements for the hygienic production and marketing of seafood (other than live bivalve molluscs). The Directive sets bacteriological standards related to food spoilage, food processing and food handler hygiene issues rather than sewage contamination in the harvesting area.

A major feature of these controls is the use of traditional bacterial indicators of faecal contamination, such as faecal coliforms or *E. coli* to determine viral contamination. Faecal indicators are either measured in the shellfish themselves (EU approach) or in the shellfish growing waters (US FDA approach). It has been internationally accepted that harvested shellfish, which meet a microbiological standard of less than 230 *E. coli* or 300 faecal coliforms in 100 g of shellfish flesh, can be placed on the market for human consumption. It should be noted that viral standards are not currently set in either EU or US legislation. In Table 2 regulations about shellfish operative in the EU and the US are summarised (EU, 2002).

Shellfish from EU class B and C needs to be heat processed (cooked) by an approved method prior to sale. HAV could be inactivated by more than 4 log10 infectious units by raising the internal temperature of shellfish meats to 85-90°C for 1 min. Since noroviruses can not be cultivated, heat inactivation studies for these viruses are not available. Instead studies are carried out on a surrogate model, namely feline calicivirus. It showed that norovirus is more heat sensitive than HAV (Slomka and Appleton, 1998). The approved heat treatment is 90°C during 1.5 min.

On the other hand when shellfish harvested from Class B areas are intended for sale live, purification must be applied and is done by depuration. Depuration involves the transfer of shellfish to a tank containing clean seawater to purge out contaminants. Depuration periods may vary from 1 to 7 days, with 2 days being probably the most widely used period. Relaying involves the transfer of harvested animals to cleaner estuaries or inlets for self-purification in the natural environment. Shellfish from class C areas, which are intended for sale live, can only be placed on the market following extended two months relaying. This process can also be used as an alternative to depuration for class B shellfish.

4.2. *Other foods*

Ready-to-eat (RTE) foods or fresh produce are susceptible food products that contribute to viral infectious disease. Food handlers have to establish Good Agricultural Practices (GAP), Good Hygienic Practises (GHP) or Hazard Analysis Critical Control Points (HACCP) based systems especially focussed on water quality or organic fertilisers.

TABLE 2. Legislative standards for bivalve molluscan shellfish in the EU and the US

Shellfish treatment	US FDA classification	Microbial standards per 100 ml water (US FDA)	EU classification	Microbial standard per 100 g shellfish (EU)
Non required	Approved	GM<14 FC[a] and 90%<43 FC	Category A	All samples <230 *E. coli* or all samples <300 FC
Purification or relaying	Restricted	GM<88 FC and 90%<260 FC	Category B	90% <4600 *E. coli* or 90% <6000 FC
Protected relaying (> 2 months)	–	–	Category C	All samples >60,000 FC

[a]FC: Faecal coliforms, GM: geometric mean, 90%: 90%-ile compliance

4.3. *Alternative indicators for viral contamination*

The correlation between the faecal indicators and the presence of enteric viruses is not consistent. Sometimes safe clarified shellfish based on the faecal indicators contributed to disease. Investigators proposed the use of bacteriophages as indicators. Bacteriophages are viruses that infect bacteria. They are present in substantial numbers in human stool samples and, in some respects, are similar to viruses pathogenic to humans. But an important remark is that there are clear observed differences in stability in different human pathogenic viruses which illustrate that extrapolation of data from one virus to another are not always reliable (Slomka and Appleton, 1998; Koopmans et al., 2004).

4.4. *Case study: Norovirus in food caused an outbreak of gastroenteritis*

Outbreak

In the Netherlands during 2004, the events which lead to an outbreak of gastroenteritis were solved by the Food and Consumer Product Safety Authority (VWA). More than forty people had become ill within 24-40 hours after attending a party at a city hall (Boxman et al., 2004). There were complaints about vomiting, diarrhoea and some individuals complained of a headache and a slight fever. The company contracted to cater the event had delivered soup and sandwiches to the venue. The person responsible for the delivery later confirmed that a week before the party he had the same symptoms and in the days between, his wife and children fell also ill. Four faecal samples of persons present at the party were taken and tested; these samples were positive for Norovirus (GGIIb).

Besides one sample of complaint (sandwich with salami), fourteen other food samples were taken from the caterer. The outer side of the salami, ham and several cheeses were cut and the soup powder, cooked eggs, meat salad, egg salad and the sandwich were tested. An appropriate virus extraction and concentration method was applied for the above-mentioned food products.

Results

Norovirus was found on the ham and on the salami samples. The sequence present on the salami (163/163 nt) was 100 % identical to the overlapping region with sequences found in four faecal samples of the visitors. The Norovirus detected belongs to the GGIIb group. A difference was found between the sequence on the ham (162/163 nt) and the sequence on the salami of one nucleotide at the 5' end of the PCR product. These results and recent history of disease of the caterer and his family, makes it very likely that the sandwiches were the source of this outbreak.

This outbreak was very similar to an outbreak that occurred in 1998 in Texas (Daniels et al., 2000). The outbreak was attributed to sandwiches prepared by an infected person.

This case in the Netherlands gives evidence that Norovirus infections are highly contagious and people can stay infectious for a long time. That is why good hygienic practices are very important.

To clear out an outbreak, it is necessary to use the same RT-PCR primer sets for the clinical samples (faeces samples) as the food samples or primer sets of which the amplicons overlap (applied in this described outbreak), only then certainty exists if the Norovirus contaminated sample was the source of the outbreak.

TABLE 3. Solving non bacteriological gastroenteritis outbreaks[a]

	Outbreak ↓	
Norovirus related	Diarrhoea and/or vomiting?	
	Yes	
	Persons got ill at the same time (within a few hours)	No: no samples from the food should be taken for norovirus tests *Inform also if there were sick people in the kitchen before the outbreak*
	Yes	
	1 or more collective mealtimes or water consumption?	No: no samples from the food should be taken for norovirus test, outbreak probably person-to-person
Foodborne	Yes	
	Time between sampled meal and getting ill 12-48 h?	No: no samples from the food taken for norovirus test, contamination not attributed to one food source
Time of incubation	Yes	
Preparation	Consumed products were not sufficiently heated (vegetables, fruit, shellfish) or after preparation manually handled, heated again or presented on a counter?	No: these meals should not be considered as a source
	Yes, these samples are relevant for viral testing!	No: no sufficiently heated products should be tested.

[a]Source: Boxman et al., 2004

Conclusion

This case study used techniques feasible to detect noroviruses in food products. Searching for the source of an outbreak, relevant clinical and food samples should be tested to compare the norovirus sequences. Keeping in mind the fact that not all outbreaks are attributed to a foodborne source and that person-to-person transmission is likely to occur. It is also necessary to take into consideration the time of incubation, the kind of the product and the epidemiological relations to decide if food items and which food items could possibly potential sources of the outbreak (Table 3).

5. References

Atmar, R.L., Neill, F.H., Romalde, J.L., Le Guyader, F., Woodley, C.M., Metcalf, T.G., and Estes, M.K., 1995, Detection of Norwalk virus and hepatitis A virus in shellfish tissues with the PCR, *Appl. Environ. Microbiol.* **61**:3014-3018.

Atmar, R.L. and Estes, M.K., 2001, Diagnosis of noncultivatable gastroenteritis viruses, the human caliciviruses, *Clinical Microbiol. Rev.* **14**(1):15-37.

Bidawid, S., Farber, J.M., and Sattar, S.A., 2000, Contamination of foods by foodhandlers: experiments on hepatitis A virus transfer to food and its interruption, *Appl. Environ. Microbiol.* **66**:2759-2763.

Boxman, I.L.A., te Loeke, N.A.J.M., and Tilburg, J.J.H.C., 2004, *Norovirus in food associated with outbreaks of gastroenteritis*, [online] Available at http://www.vwa.nl/download/rapporten/ Voedselveiligheid/OT04H006_norovirus.PDF (accessed 5 january 2005).

Bryan, F.L., 1995, Hazard analysis: the link between epidemiology and microbiology, *J. Food Prot.* **59**:102-107.

Cacopardo, B., Russo, R., Preiser, W., Benanti, F., Brancati, G., and Nunnari, A., 1997, Acute hepatitis E in Catania (eastern Sicily) 1980-1994. The role hepatitis E virus, *Infect.* **25**:313-316.

Chadwick, P.R., Walker, M., and Rees, A.E., 1994, Airborne transmission of a small round structured virus, *Lancet* **343**:171.

Caul, E.O., 1994, Small round-structured viruses – airborne transmission and hospital control, *Lancet* **343**:1240-1242.

Chan, T.Y., 1995, Shellfish-borne illnesses. A Hongkong perspective, *Trop. Geogr. Med.* **47**:305-307.

Chiba, S., Sakuma, Y., Kogasaka, R., Akihara, M., Horino, K., Nakao, T., and Fukui, S., 1979, An outbreak of gastroenteritis associated with calicivirus in an infant home, *J. Med. Virol.* **4**(4):249-254.

Cliver, D.O., 1997, Virus transmission via food, *Food Technol.* **51**(4):71-78.

Coelho, C., Heinert, A.P., Simões, C.M.O., and Barardi, C.R.M., 2003, Hepatitis A virus detection in oysters (Crassostrea gigas) in Santa Catarina State, Brazil, by reverse transcription-polymerase chain reaction, *J. Food Prot.* **66**(3):507-511.

Cubitt, W.D., McSwiggan, D.A., and Moore, W., 1979, Winter vomiting disease caused by calicivirus, *J. Clin. Pathol.* **32**(8):786-793.

Cubitt, W.D., McSwiggan, D.A., and Arstall, S., 1980, An outbreak of calicivirus infection in a mother and baby unit, *J. Clin. Pathol.* **33**(11):1095-1098.

Cubitt, W.D., Pead, P.J., and Saeed, A.A., 1981, A new serotype of calicivirus associated with an outbreak of gastroenteritis in a residential home for the elderly, *J. Clin. Pathol.* **34**(8): 924-926.

Cubitt, W.D. and McSwiggan, D.A., 1987, Seroepidemiological survey of the prevalence of antibodies to a strain of human calicivirus, *J. Med. Virol.* **21**(4):361-368.

Currier, M., 1996, Foodborne outbreak. *Dairy Food Environ. Sanitation* **16**:32-33.

Croci, L., Ciccozzi, M., De Medici, D., Di Pasquale, S., Fiore, A., Mele, A., and Toti, L., 1999, Inactivation of hepatitis A virus in heat-treated mussels, *J. Appl. Microbiol.* **87**: 884-888.

Croci, L.D., De Medici, D., Scalfaro, C., Fiore, A., and Toti, L., 2002, The survival of hepatitis A virus in fresh produce, *Int. J. Food Microbiol.* **73**:29-34.

Daniels, N.A., Bergmire-Sweat, D.A., Schwab, K.J., Hendricks, K.A., Reddy, S., Rowe, S.H., Fankhauser, R.L., Monroe, S.S., Atmar, R.L., Glass, R.I., and Mead, P., 2000, A foodborne outbreak of gastroenteritis associated with Norwalk-like viruses: First molecular traceback to deli sandwiches contaminated during preparation, *J. Infect. Dis.* **181**:1467-1470.

Desenclos, J-CA, Klontz, K.C., and Wilder, M.H., 1991, A multistate outbreak of hepatitis A caused by consumption of raw oysters, American *J. Public Health* **81**: 1268-1272.

de Wit, M.A.S., Koopmans, M.P.G., and van Duynhoven, Y.T.H.P., 2003, Risk factors for norovirus, Sapporo-like virus, and group A rotavirus gastroenteritis, *Emerg. Inf. Dis.* **9**(12):1563-1570.

Doyle, A., Barataud, D., Gallay, A., Thiolet, J.M., Le Guyader, S., Kohli, E., and Vaillant, V., 2004, Norovirus foodborne outbreaks associated with the consumption of oysters from the Etang De Thau, France, December 2002, *Euro Surveill* **9**(1):24-26.

European commission, 2002, Health & Consumer protection directive-general, *Opinion of the scientific committee on veterinary measures relating to public health on Norwalk-like viruses* [online] Available at http://www.europa.eu. int/comm/food/fs/sc/scv/out49_en.pdf (accessed 5 january 2005).

Feinstone, S.M., 1996, Hepatitis A: epidemiology and prevention, *Eur J. Gastroenterol Hepatol* **8**:300-305.

Fiore, A.E., 2004, Hepatitis A transmitted by food, *Clin. Infect. Dis.* **38**: 705-715.

Frankhauser R.L., Noel J.S., Monroe S.S., Ando T., and Glass R.I., 1998, Molecular epidemiology of "Norwalk-like viruses" in outbreaks of gastroenteritis in the United States, *J. Inf. Dis.* **178**:1571-1578.

Food & Drug Administration, Center for food safety & applied nutrition a, Foodborne pathogenic microorganisms and natural toxins handbook, *Rotavirus* [online] Available at http://www.cfsan.fda. gov/~mow/chap33.html (accessed 5 january 2005).

Food & Drug Administration, Center for food safety & applied nutrition b. Foodborne pathogenic microorganisms and natural toxins handbook, *Other gastroenteritis viruses* [online] Available at http://www.cfsan. fda.gov/~mow/chap35.html (accessed 5 january 2005).

Goswami, B.B., Kulka, M., Ngo, D., Istafanos, P., and Cebula, T.A., 2002, A polymerase chain reaction-based method for the detection of hepatitis A virus in produce and shellfish, *J. Food Prot.* **65**:393-402.

Gray, J.J., Wreghitt, T.G., Cubitt, W.D., and Elliot, P.R., 1987, An outbreak of gastroenteritis in a home for the elderly associated with astrovirus type I and human calicivirus, *J. Med. Virol.* **23**:377-381.

Green, K.Y., Ando, T., Balayan, M.S., Berke, T., Clarke, I.N., Estes, M.K., Matson, D.O., Nataska, S., Neill, J.D., Studdert, J.M., and Thiel, H-J., 2000, Taxonomy of the caliciviruses, *J. Infect. Dis.* **181**(2):S322-330.

Greene, S.R., Moe C.L., Jaykus, L.A., Cronin, M., Grosso L., and Van Aarle P., 2003, Evaluation of the Nuclisens Basic kit assay for detrection of Norwalk virus RNA in stool specimens, *J. Virol. Methods* **108**:123-131.

Griffin, M.R., Surowiec, J.J., McCloskey, D.I., Capuano, B., Pierzynski, B., Quinn, M., Wojnarski, R., Parkin, W.E., Greenberg, H., and Gary, G.W., 1982, Foodborne norwalk virus, *Am. J. Epidemiol.* **115**:178-184.

Halliday, M.L., Kang L-Y., Zhou T-K, Hu, M.D., Pan, Q.C., Fu, T.Y., Huang, Y.S., and Hu, S.L., 1991, An epidemic of hepatitis A attributable to the ingestion of raw clams in Shangai, China, *J. Infect. Dis.* **164**:852-859.

Hernández, F.R., Monge, R., Jiménez, C., and Taylor, L., 1997, Rotavirus and hepatitis A virus in market lettuce (Latuca sativa) in Costa Rica, *Int. J. Food Microbiol.* **37**:221-223.

Herwaldt, B.L., Lew, J.F., Moe, C.L., Lewis, D.C., Humphery, C.D., Monroe, S.S., Pon, E.W., and Glass, R., 1994, Characterization of a variant strain of Norwalk virus from a foodborne outbreak of gastroenteritis on a cruise ship in Hawaii, *J. Clin. Microbiol.* **32**: 861-866.

Hutin, Y.J.F., Pool, V., Cramer, E.H., Nainan, O.V., Weth, J., Williams, I.T., Goldstein, S.T., Gensheimer, K.F., Bell, B.P., Shapiro, C.N., Alter, M.J., and Margolis, H.S., 1999, A multistate, foodborne outbreak of hepatitis A, *N. Engl. J. Med.* **340**:595-602.

Hutson, A.M., Atmar, R.L., and Estes, M.K., 2004, Norovirus disease: changing epidemiology and host susceptibility factors, *Trends in Microbiol.* **12**(6):279-287.

Jaykus, L.A., Hemard, M.T., and Sobsey, M.D., 1994, Human enteric pathogenic viruses, in: *Environmental Indicators and Shellfish Safety* C.R. Hackney, M.D. Pierson, eds., Chapman and Hall, New York:. pp. 92-153

Jaykus, L.A., 2000, Enteric viruses as "emerging" agents of foodborne disease, *Irish J. Agr. Food Res.* **39**:245-255.

Jean, J., Burton, B., Darveau, A., and Fliss, I., 2001, Detection of Hepatitis A virus by the nucleic acid sequence-based amplification technique and comparison with reverse transcription-PCR, *Appl. Environt. Microbiol.* **67**(12):5593-5600.

Jiang, X., Estes, M.K., and Metcalf, T.G., 1987, Detection of hepatitis A by hybridization with single-stranded RNA probes, *Appl. Environ. Microbiol.* **53**:2487-2495.

Jiang, X., Huang, P.W., Zhong, W.M., Farkas, T., Cubitt, D.W., and Matson, D.O., 1999, Design and evaluation of a primer pair that detects both Norwalk- and Sapporo-like caliciviruses by RT-PCR, *J. Virol. Methods* **83**:145-154.

Kapikan, A.Z. and Channock, R.M., 1996, Rotaviruses, in: *Fields Virology* B.N. Fields, D.M. Knipe, P.M. Howley, R. M. Chanock, T.P. Monath, J.L. Melnick, B. Roizman, S.E. Straus, eds. 3rd ed., Lippincott-Raven, Philadelphia, Pennsylvania: pp. 1657-1708.

Kaplan, J.E., Feldman, R., Campbell, D.S., Lookabaugh, C., and Gary, G.W., 1982, The frequency of a Norwalk-like pattern of illness in outbreaks of acute gastro-enteritis, *Am. J. Publ. Health* **72**:1329-1332.

Koff, R.F. and Sair, H.S., 1967, Internal temperature of steamed clams, *Med. Intelligence* **273**:737-739.

Koopmans, M., von Bonsdorff, C-H., Vinjé, J., de Medici, D., and Monroe, S., 2002, Foodborne viruses, *FEMS Microbiol. Rev.* **26**:187-205.

Koopmans, M. and Duizer, E., 2004, Foodborne viruses: an emerging problem, *Int. J. Food Microbiol.* **90**:23-41.

Lees, D., 2000, Viruses and bivalve shellfish, *Int. J. Food Microbiol.* **59**:81-116.

Le Guyader, F., Haugarreau, L., Miossec, L., Dubois, E., and Pommepuy, M., 2000, Three-year study to assess human enteric viruses in shellfish, Appl. *Environ. Microbiol.* **66**(8): 3241-3248.

Le Guyader, F.S., Mittelholzer, C., Haugarreau, L., Hedlund, K-O., Alsterlund, R., Pommepuy, M., and Svensson, L., 2004, Detection of noroviruses in raspberries associated with a gastroenteritis outbreak, *Int. J. Food Microbiol.* **97**:179-186.

Lieb, S., 1985, Norwalk virus gastroenteritis outbreak associated with a cafeteria at a college, *Am. J. Epidemiol.* **121**:259-268.

Lo, S.V., Connely, A.M., Palmer, S.R., Wright, D., Thomas, P.D., and Joynson, D., 1994, The role of the pre-symtomatic food handler in a common source outbreak of foodborne SRSV gastroenteritis in a group of hospitals, *Epidemiol. Infect.* **113**:513-521.

Lopman, B., Van Duynhoven, Y., Hanon, F.X., Reacher, M., Koopmans, M., and Brownl D., 2002, Laboratory capability in Europe for foodborne viruses, *Eurosurveillance* **7**(4):61-65.

Lopman B.A., Reacher M.H., van Duijnhoven Y., Hanon F.X., Brown D., and Koopmans M., 2003, Viral gastroenteritis outbreaks in Europe, 1995-2000. *Emer. Infect. Dis.* **9**(1):90-96.

Margolin, A.B., Richardson, K.J., DeLeon, R., and Gerba, C.P., 1989, Application of gene probes to the detection of enteroviruses in water. In: Larson R.A., editor. Biohazards of drinking water treatment. Chelsea: Lewis Publishers pp. 265-270

Marks, P.J., Vipond, I.B., Carlisle, D., Deakin, D., Fey, R.E., and Caul, E.O., 2000, Evidence for airborne transmission of Norwalk-like virus (NLV) in a hotel restaurant, *Epidemiol. Infect.* **124**:481-487.

Mead, P.S., Slutsker, L., Dietz, V., McCaig, L.F., Bresee, J.S., Shapiro, C., Griffin, P.M., and Tauxe, R.V., 1999, Food-related illness and death in the United States, *Emer. Infect. Dis.* **5**:607-625.

Mechnik, L., Bergman, N., Attali, M., Beergabel, M. Mosenkis, B., Sokolowski, N., and Malnick, S., 2001, Acute hepatitis E virus infection presenting as a prolonged cholestatic jaundice, *J. Clin. Gastroenterol.* **33**:421-422.

Mullendore, L.J.L., Sobsey, M.D., and Shieh, Y-S.C., 2001, Improved method for the recovery of hepatitis A virus from oysters, *J. Virol. Methods* **94**:25-35.

Nakata, S., Chiba, S., Terashima, H., and Nakao, T., 1985, Prevalence of antibody to human calicivirus in Japan and South-east Asia detrermined by radioimmunoassay, *J. Clin. Microbiol.* **22**(4):519-521.

Nakata, S., Kogawa, K., Numata, K., Ukae, S., Adachi, N., Matson, D.O., Estes, M.K., and Chiba, S., 1996, The epidemilogy of human calicivirus/Sapporo/82/Japan, *Arch. Virol. Suppl.* **12**:263-270.

Norrung, B., 2000, Foodborne viruses – introduction, *Int.l J. Food Microbiol.* **59**(1-2): 79-80.

Parashar, U.D., Dow, L., Fankhauser, R.L., Humphrey, C.D., Miller, J., Williams, K.S., Eddy, C.R., Noel, J.S., Ingram, T., Bresee, J.S., Monroe, S.S., and Glass, R.I.,1998, An outbreak of viral gastroenteritis associated with consumption of sandwiches: Implications for the control of transmission by food handlers. *Epidemiol. Infect.* **121**(3): 615-621.

Patterson, W., Haswell, P., Fryers, P.T., and Green, J., 1997, Outbreak of small round structured virus gastro-enteritis arose after kitchen assistant vomited, *Communicable Dis. Rep.* **7**:101-103.

Pönkä, A., Maunula L., von Bonsdorff, C-H., and Lyytikäinen, O., 1999, An outbreak of calicivirus associated with consumption of frozen raspberries, *Epidemiol. Infect.* **123**: 469-474.

Prato, R., Lopalco, P.L., Chironna, M., Barbuti, G., Germinario, C., and Quarto, M., 2004, Norovirus gastroenteritis general outbreak associated with raw shellfish consumption in South Italy, *BMC Infect. Dis.* **4**(37), [online] Available at http://www.biomedcentral.com/1471-2334/4/37 (accessed 5 january 2005).

Richards, G.P., 2001, Foodborne pathogens. Enteric virus contamination of foods through industrial practices: a primer on intervention strategies, *J. Ind. Microbiol. Biotechnol.* **27**:117-125.

Sair, A.I., D'Souza H., and Jaykus L-A., 2002, Human enteric viruses as causes of food-borne disease, *Comp. Rev. in Food Sci. Food Saf.* **1**:73-89.

Sawyer, L.A., Murphy, J.J., Kaplan, J.E., Pinsky, P.F., Chacon, D., Walmsley, S., Schronberger, L.B., Philips, A., Forward, K., Goldman, C., Brunton, J., Fralick, R.A., Carter, A.O., Gary, G.W., Glaa, R.L., and Low, D.E., 1988, 25 to 30 nm virus particle associated with a hospital outbreak of acute gastroenteritis with evidence for airborne transmission, *Am. J. Epidemiol.* **127**:1261-1271.

Seymour, I.J. and Appleton, H., 2001, A review. Foodborne viruses and fresh produce, *J of Appl. Microbiol.* **91**:759-773.

Shieh, Y.C., Monroe, S.S., Frankhauser, R.L., Langlois, G.W., Burkhardt, R.S., and Baric, R.S., 2000, Detection of Norwalk-like virus in shellfish implicated in illness, *J. Infect. Dis.* **181**(S2):S360-366.

Slomka, M.J. and Appleton, H., 1998, Feline calicivirus as a model system for heat inactivation studies of small round structured viruses in shellfish, *Epidemiol. Infect.* **121**: 401-407.

Smith, J.L., 2001, A review of hepatitis E virus, *J. Food Prot.* **64**:572-586.

Stolle, A. and Sperner, B., 1997, Viral infections transmitted by food of animal origin: the present situation in the European Union. *Arch. Virol. Suppl.* **13**:219-228.

Tei, S., Kitajima, N., Takahashi, K., and Mishiro, S., 2003, Zoonotic transmission of hepatitis E virus from deer to human beings, *Lancet* **362**:371-373.

Tei, S., Kitajima, N., Ohara, S., Inoue, Y., Miki, M., Yamatani, T., Yamabe, H., Mishori, S., and Kinoshita, Y., 2004, Consumption of uncooked deer meat as a risk factor for hepatitis E virus infection: an age- and sex-matched case-control study, *J. Med. Virol.* **74**: 67-70.

Yazaki, Y., Mizuo, H., Takahashi, M., Nishizawa, T., Saski, N., Gotanda, Y., and Okamoto, H., 2003, Sporadic acute or fulminent hepatitis E in Hokkaido, Japan, may be food-borne, as suggested by the presence of hepatitis E virus in pig liver as food, *J. Gen Virol.* **84**:2351-2357.

Zhou, Y.J., Estes, M.K., Jiang, X., and Metcalf, T.G., 1991, Concentration and detection of hepatitis A virus and rotavirus from shellfish by hybridization tests, *Appl. Environ. Microbiol.* **57**:2963-2968.

Part III
Process Safety

11
Safety Models: HACCP and Risk Assessment

Gabriela Rotaru[1] and Daniela Borda

As compared with other products, food quality and safety have diverse and profound implications on human health (Rotaru, 2003). Food quality affects consumers' health and may also determine the health of the future generations. It is well known that food can cause consumers to become ill if technological, health and hygiene criteria at the production, manipulation, transport, commercialization and utilization stages are not respected. Foodborne illness may range from transient discomfort to life threatening illness which may at times lead to death.

In the last decades, development of the processing methods has emerged, as a consequence of the development of the technology, modernization of the equipment and diversification of the food range. The food chain has become more and more complex and that is why it now represents an important issue which requires specific measures, meant to assure an acceptable standard for food security. The new emerging technologies are today designed to obtain food, which sometimes transcend the food industry barrier. More often, they enter the therapeutic food domain, or they can be probiotics/prebiotics, nutraceutic, or they even can transcend the genetic engineering zone (i.e. genetically modified food – GMF), imposing new measures for consumers' protection (Rotaru and al., 2001).

Although EU food security regulations seek to promote the highest possible standards illness caused by food toxic infections with *Salmonella* and other pathogen agents do still occur. Moreover, recent food scares such as avian influenza, dioxin and BSE, etc. have polarized specialists', public opinions' and authorities' attention. As result, they imposed severe measures concerning the preparation, distribution and commercialization of food. The EU has proposed a new concept in administration of the aspects of food safety. The White paper on food safety (Brussels, 2000) is a publication which sets as goals the concept of a proactive food policy, with the view to achieving the rekindle of public trust in food, food science, food legislation and ultimately the food control system. The

[1]Gabriela Rotaru, University Dunărea de Jos, Faculty of Food Science and Engineering, 111 Domneasca Str, 800201, Galati, Romania, e-mail:gabriela.rotaru@ugal.ro

general legislative framework which will enforce preparation and distribution of food appropriate for consumption must be strengthened by the organizations through specific regulations which require:

- adequate standards concerning food security;
- clear labelling that is easy to read and interpret;
- good manufacturing practice;
- good hygiene practice;
- management system concerning food quality and safety.

1. HACCP System

HACCP is the international wide known safety system, which has proved its efficiency in all the food industry's sectors and on the entire food chain.

HACCP is a systematic interactive scientific method of identification, evaluation and control of the risks associated with the food. HACCP strategy is based on the introduction of a new system of prevention, elimination or minimizing the risks at acceptable levels by taking into account all the hazards, which may affect the consumers' security (Belton, 1999).

1.1. HACCP – past and present

The HACCP system was developed on the basis of the research carried out in the 1960s, by NASA, and the Pillsbury Company in association with Natick laboratories of the US Army for the production of the food characterized by 'zero deficiencies'. The purpose was to obtain safe food for use in space missions.

That research has led to the development of a valuable tool which has changed the approach to the quality and security strategy in the food domain. It led manufacturers to change from post–process control to the application of the forewarning methods (Rotaru and Moraru, 1997). HACCP as it is known today is an efficient solution for the safety assurance of the food which is promoted by the vast majority of international (FAO, WHO, Codex Alimentarius), regional (EU board) and national organisations.

In USA this method has been used since 1973 for preserved fish and meat produced on a large scale. In 1995 the Service of Inspection and Food Safety required the utilization of HACCP principles in the fish and marine food domain, while in 1996 it was used for poultry and other meat products. In 1998 the system was also applied to fruits, vegetables and eggs.

The EU backed the introduction of safety regulations since 1983 and supported the HACCP system implementation in its legislation as far back as 1993 by the EU Council Directives (i.e.: 93/43EEC; 2001/471/EC; 853/2004/EC) which directed/advised the producers, wholesalers and retailers of food regarding the utilization of HACCP. In 1995 the system became an obligatory requirement for fish food at the entrance in EU, while in 2007 HACCP will become mandatory for all the food companies in EU.

1.2. The importance of HACCP

Hazard Analysis of Critical Control Points or HACCP, as it is often referred to, is an important system that both the commercial and consumer sectors of society should understand and appreciate. It is a system that sets goals and boundaries for the sake of health and quality standards and is essentially:

- an operating system of food security management during production, distribution and preparation;
- an efficient control tool of operations, with multiple inspection points designed to confirm the plan and its efficiency;
- a tool to eliminate the traditional limits concerning the quality control;
- a system which is flexible enough to handle the changes/progress in the equipments fabrication and the modernization of the technological processes;
- a way of diminution of outbreaks and of problems which concern food safety;
- a commitment to an efficient utilization of human, material and financial resources and the simplification of the costs system of the organization;
- a key for identification of the foreseen novel risks which makes it very useful for the new products or technologies;
- a solid proof to demonstrate the clients and inspectors that all the potential risks are under control;
- a way of achieving the trust of the consumers in product and producer's practices;
- a global approach for international promotion of the goods and of the business by increasing the food security.

Food safety should be given the highest priority by the companies in order to keep the risks under control in an effective way (Dillon and Griffith, 1996).

1.3. HACCP/ISO 9001:2000 or ISO 22000:2005

The food industry is currently working with two management systems, namely, HACCP and ISO 9001:2000. In September 2005 a new ISO standard has been published ISO 22000:2005, which for the first time provides a framework for the harmonized approach of quality and safety issues applicable in the entire food chain.

Ultimately, it has been understood and backed by a large consensus from specialized international organizations that the quality issues in the food industry can not be addressed separately from the safety issues.

As a result an increased number of organizations will directly apply the standard ISO 22000:2005 to manage both safety and quality problems.

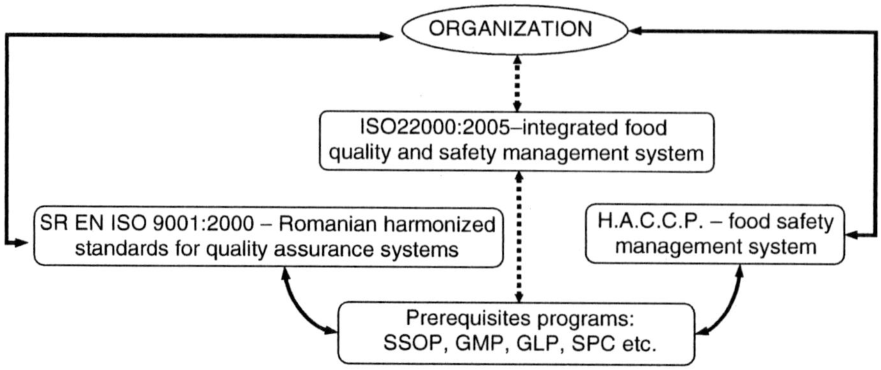

FIGURE 1. Integrated food safety and quality management systems in an organization throughout ISO 22000:2005 (---) or HACCP+ ISO 9001:2000(—) implementation.

In Romania as in many other countries there will be still a large number of organizations in which a general quality management system like ISO 9001:2000 is at work and valid. In this case separate management systems for safety and quality can be used (figure1).

ISO 9001:2000 is a management system of quality, through which errors and faults can be identified and dealt with. If the system is not supported by a safety management system it might allow the fabrication of dangerous products. This can happen only when the safety criteria are not taken into account by the fabrication process.

HACCP on the other hand remains an extremely flexible approach to safety management issues concerning products intended to be consumed as food. HACCP rely on a series of preliminary measures programs (hygiene programs, GMPs, statistic process control – SPCs, etc.) called prerequisites (Wallace and Williams, 2001).

Thus, the managerial strategy can be oriented in three directions, depending on the specific and the resources of the organizations in food industry:

- the implementation of the ISO 22000:2005 standard;
- the implementation of HACCP where ISO 9001:2000 already exists;
- organisation of an ISO 9001:2000 where HACCP has been implemented.

In each of these situations if maximum efficiency is to be reached, the coexisting particularities of the two systems must be taken into account and integrated. It is extremely important to understand that HACCP does not replace the hygiene programs or other prerequisites programs, but if correctly run these programs represent its support network. Once the differences are identified it is easier to see the role and function of the HACCP system in the quality management general plan. The prerequisites represent a chain of support for the HACCP system (figure 2).

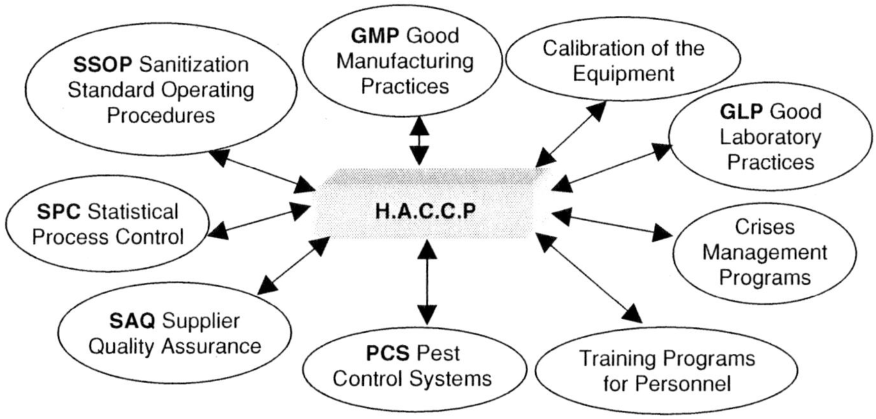

FIGURE 2. The support network of HACCP (Mortimore, 2001).

The prerequisites programs may contain:

1) Hygiene programs which includes cleaning and maintenance of the hygiene practices and may also contain control measures to avoid the insects and the pests presence;
2) GMP (Good Manufacturing Practices) which represents a combination of technological instructions and quality assurance procedures (monitoring of the working practices and the final products, verification of certain steps of the technological process);
3) SQA (Supplier Quality Assurance) programs aiming to ensure the quality of the raw materials and other incoming materials used in the fabrication of the product. It can include audit and inspection schedules of the supplier's practices.

1.4. Hygiene practices

It is necessary that each producer develop specific procedures for good hygiene practices. When these procedures are included in a quality management system they can focus also on certain aspects of the manufacturing process.

Hygiene policy is a written procedure that is adopted in an industrial facility. It shows the level of hygiene that has to be assured in the industry and the pledge to maintain it.

It is necessary to classify the products and the ingredients regarding the magnitude and the frequency of the risk like:

- high risk products;
- medium risk products;
- low risk products.

Because contamination from the staff represents one of the most frequent problems in the food industry, it is necessary to ensure an adequate level of hygiene and to follow the hygiene standard requirements for the staff.

Mandatory measures involve:

⇨ periodical medical checking of the personnel health;
⇨ training programs in good hygiene practices;
⇨ protection equipment;
⇨ good manufacturing practices which will avoid contamination.

Thorough cleaning and washing is efficient when it forms part of hygienisation programs, which involve periodical cleaning of the spaces and equipments.

Rodents, insects and birds represent a major source of biological hazards so their presence must be avoided and limited at all costs (Heggum, 2001).

Measures against pests contamination involve the installation of:

⇨ protection racks at the windows;
⇨ protection screens at all doors and windows against insects/rodents. Also it includes the examination of the received materials (to check if they are contaminated, or not).

Contamination might be a consequence of the inadequate design or maintenance of the equipment used. Equipment which should be periodically tested includes:

⇨ ventilation equipment;
⇨ installations for compressed air;
⇨ water pipes;
⇨ steam pipes;
⇨ residues: maintenance and evacuation;
⇨ electrical lightening;
⇨ washing pools;
⇨ pipes seals, bends.

These hygiene and maintenance protocols are designed to ensure that manufactured products will not put the health of the consumers at risk. The hygiene practices are a solid base line for the accomplishment of the HACCP system (Jouve, 1999). To function well it is necessary to have an organisation and employee complement that can cope and implement the HACCP plan; hence foresight exercises may prove to be beneficial.

SSOPs can be developed together with GMPs which covers the operational procedures and the instructions for the equipments' maintenance.

In USA, CGMP (Current Good Manufacturing Practices) describes the conditions and the measures which the organizations must follow. Its purpose is to avoid the presence of any hazard in the final product.

CGMPs include explanations concerning:

A) General requirements:

- definitions;
- current manufacturing practices;
- personnel;
- exceptions.

B) Buildings and annexes:

- the factory and the terrain;
- hygiene operations;
- hygiene control.

C) Equipment:

- equipment and tools.

D) The production and processes control:

- the process and the process' inspections;
- storage and distribution;

E) Flaws:

- natural or unavoidable flaws of food.

Auditing periodically the practices applied by the system will help identify the weak points. Once the weak points are identified they can serve as focus points where the efforts should be concentrating for avoiding the failure of the hygiene measures (table 1). Supplementary information can be given when analysing the likelihood or the probability of the risk occurrence on a statistic basis together with its severity which can indicate the significance of a risk. Ranking risks by significance will give consistence to the decisions and action taken. When analysing the likelihood of a risk occurrence if there are no statistical data available some arguments from the recent history of the organization can be used to quickly settle the facts. These are:

⇨ the high frequency of some emergency situations;
⇨ frequent inspections;
⇨ a high number of reclamations and big costs for litigations;
⇨ high costs for scraps;
⇨ delay in launching new projected;
⇨ difficulties in the GMP application; negative audit reports.

If the hygiene programmes are in place and effective attention should be given to other prerequisites and their effectiveness, such as:

⇨ good manufacturing practices;
⇨ handling practices;
⇨ complaints registration;
⇨ the appearance of quarantine situation;

TABLE 1. Lists for the verification of hygiene

Conditions[a]	Frequency	Date	Initials	Observations
Is the water coming from a certified source?				
Are there any cross-contamination points of the clean and the dirty flow?				
Are the equipment and tools easy to be hygienised?				
Are the sinks and the toilets in the appropriate places?				
Revised				
Date				

Procedure/ Conditions[b]	Date	Duration	The sanitization substances concentration	Equipment/ correction activities	Signature
Is the surface of contact with the products appropriate?					
Are the toilettes clean enough?					
Are there any washing and disinfection products?					
Are the food and the packages exposed to alteration/ contamination?					
Revised					
Date					

[a]Periodic or annual list of hygiene report
[b]Daily report

- ⇨ the existence of quality management system;
- ⇨ improving the in-line registration methods at the critical control points;
- ⇨ revision and maintenance schedules for equipment;
- ⇨ laboratory tests procedures to certify the findings and the trust level they offer;
- ⇨ periodical revision of the results obtained after the inspections
- ⇨ crises management programmes
- ⇨ traceability of the registrations.

The expected results after the implementation of the prerequisites and HACCP programs:

- o These systems will not completely eliminate the identified risks, because zero risk food still eludes the food industry, so risk is said to be reduced to an acceptable level;
- o Some positive aspects which result after applying hygiene practices and HACCP systems includes:

- identification and validation of the quality of raw material prior to the production process;
- the diminution of problems, so that if these occurs can be dealt with more efficiency;
- the avoidance of the cross-contamination of certain areas,
- the diminution of quarantine situations for the produced products;
- reducing the role and frequency of the final inspection while focussing and according more importance to the preventive measures and in line control measures for the technological process.

2. Foodborne Illnesses

Food can cause illness to consumers if the technological criteria and hygiene at the production, manipulation, transport and utilization stages are not followed carefully. These illnesses can vary from light diseases to very severe forms and sometimes can have even a lethal effect.

Diseases caused by food represent a widespread international problem. The World Health Organization estimates that at least 10^9 cases of gastrointestinal diseases occur each year in third world countries, which lead to mortalities exceeding 5×10^6 especially at an early age (Encyclopaedia of Food Science, 1993). The economic costs of these illnesses are tremendous and they provoke both suffering and financial loss. It is estimated that in USA the annual cost of the foodborne illnesses reach between $6.5 billion and $34.9 billion (FDA Report, 2004) (if productivity loss, the cost of medical care and of epidemiological investigations and also the costs that affect the producers are taken into consideration). These figures represent the equivalent of the contaminated products withdrawn from market, the destruction of the stocks of contaminated products and other associated costs.

In the last two decades there has been an increase in consumers, institutions and related organizations preoccupation to determine the origin of diseases caused by the consumption of certain foods. Good news is that recent statistics (FDA, 2005) demonstrates important declines in foodborne infections due to common bacterial pathogens in 2004 in the US. Basically from 1996-2004, the incidence of *E. coli* O157:H7 infections decreased 42 percent, *Campylobacter* infections decreased 31 percent and *Yersinia* decreased 45 percent while *Salmonella* infections dropped 8 percent. Table 2 show some epidemiological data registered between 1981 and 2004 which demonstrated the general increase in the reported cases number.

The increased trend at the end of 90`s should be considered with caution since it is influenced by the number of reported cases versus the total case number. Nowadays it is well known that the true incidence of the foodborne diseases is still underreported in the third world countries for example.

TABLE 2. Annual food diseases registered from 1981 to 2004*

Year	Country	The agent which provoked the disease	Number of illnesses
1981	Holland	*Salmonella indiana*	700
1984	Canada	*Salmonella typhimurium*	2700
1984	Kuwait	*Brucella melitensis*	720
1985	France	*Trichomonas spiralis*	1200
1985	USA	*Salmonella typhimurium*	16000
1988	China	*Hepatitis A virus*	292000
1991-1992	South America	*Vibrio cholerae*	>600000
1999-2000	USA	*Bacillus cereus*	27,360
2000-2004	USA	*Campylobacter* spp.	1.963.141
2003	USA	*Salmonella* spp.	1.341.873

*FDA report, 2004

To eliminate these disastrous effects, hazards must be identified and their causes must be eliminated. All the possible contamination routes should be analysed and kept under control to avoid presence of the risks in the end-products.

The negative potential of the foodborne agents must be evaluated in relation with their ability to cause illnesses and the likelihood of appearance in the product.

Firstly, the pathogenic or diseases causing microorganisms can be analysed considering the kinds of diseases they can produce: infections or food poisoning.

Infections (i.e. tuberculosis, brucellosis, and typhoid fever) can appear after eating food that contains a low number of viable pathogenic agents. Their reproduction on the product is not necessary.

Food intoxications or poisoning are caused by the ingestion of the food containing preformed toxin, such as the toxins produced by *Staphylococcus aureus* and *Clostridium* spp.

For the apparition of the food intoxications is essential that the microorganisms which provoke it grow on the product. The growth takes place in certain conditions and is influenced by a large number of intrinsic (i.e.: the chemical composition of the product, pH, a_w,) extrinsic (such as: temperature, humidity, ventilation, oxygen presence etc.) and technological factors (packaging, the NaCl addition, the period and the intensity of the thermal treatments applied, and others). Food intoxication can be produced by a large variety of agents of which the most important are listed in the table 3.

Foodborne infections result when the microorganisms are present in a large number in food and the foodborne toxic infections are produced when they are ingested and subsequently produce toxin in the host organism. In Europe and North America the food viruses and bacteria lead to toxic infection very often. The main microorganisms which can provoke food infections and the details regarding these diseases are presented table 4.

TABLE 3. Food poisoning agents*

Agent	Example
Bacteria	*Salmonella, Staphylococcus aureus, Clostridium perfringens, Cl. botulinum, Bacillus cereus*
Viruses	Hepatitis A virus, parvovirus, Norwalk virus
Moulds	*Claviceps purpurans, Aspergillus flavus* (foodborne aflatoxins)
Marine protozoa	*Gonyaulax tamarensis* (paralytic crustacean poisoning), *Gambierdiscus toxicus* (poisoning with ciguatera), *Prorocentrum lima* and some species of *Dinophys* (diarrheic shellfish poisoning)
Parasites (protozoa, amoebas)	*Trichinella spiralis, Taenia solium, Giardia lamblia, Cryptosporidium, Entamoeba histolytica*
Chemical substances	Heavy metals (lead, zinc, copper, cadmium, mercury, arsenic, antimony), pesticides, herbicides, fungicides, substances used for cleaning and disinfections.
Toxic plants	Germinated potatoes, apricot core, red bean (*Phaseolus vulgaris*)
Toxic animals	Scombrotoxin from fish insufficient cooked – herring-(histamine)

*(Macrae et al., 1993)

An important step in identifying the hazards is analysing the emerging pathogens representing those agents which were not historically recognised to produce diseases. This analyse is a task for the specialists which are aware of the continuous emerging of the foodborne microbial pathogens.

In table 5 the main food sources for contamination and the pathogens which cause illnesses are presented.

The incidence of the other agents' presence (for instance parasites, protozoa, natural toxic substances or chemical substances) in food can differ depending on geographical region were they can be found. However a known wide spread high toxic potential is recognised for mycotoxines like aflatoxins, patulin, vomitoxin from which most are produced by fungi. Parasites and protozoa like *Trichinella spiralis* and *Giardia intestinalis* are also wide spread hazards. It also depends on the hygiene measures at the preparation of the food and on the technological level of the production.

An important contamination source with pathogen agents is given by the personnel who had contact with the food during production and handling. The contamination pathways of food by the personnel are presented in table 6.

To keep under control the biological hazards it is safer and cheaper to apply preventative methods so that the biological hazards will be under control rather than relying exclusively on tests and final inspections, which can only show us the finding of some certain problems. A sound analyse of the risks (biological, chemical and physical) together with the preventative measures, effective monitoring systems and the related corrective actions settled will result in a successful HACCP plan implementation in any food organization.

TABLE 4. The main microbial agents which provoke food infection*

Microorganism	Incubation period (h)	Disease length	Symptoms
0	1	2	3
Salmonella (infections)	6÷48(usually 12÷36)	1÷7 days	diarrhoea, abdominal pains, vomiting, fever
Staphylococcus aureus (release toxins in the product)	2÷6	6÷24 h	nausea, vomiting, diarrhoea, swoon, dehydration in the most serious cases
Clostridium perfringens (release toxins in the bowel)	8÷22	24÷48 h	diarrhoea, abdominal pains, nausea
Clostridium botulinum (release toxins in the product)	12÷96(usually 18÷36)	Death in 24h÷8 days or convalescence 6÷8months	tiredness, dizziness, diarrhoea followed by constipation, implication of the central nervous system, perturbation of speaking and sight
Bacillus cereus (release toxins in the product)	Vomit syndrome: 1÷6	6÷24 h	nausea, vomiting, watery diarrhoea
	Diarrheic syndrome: 8÷16	18÷24 h	abdominal pains, nausea sometimes
Bacillus subtilis (release toxins in product)	<1÷14(an average of 2÷5)	1,5÷8 h	vomiting, diarrhoea, abdominal pains, nausea
Bacillus licheniformis (release toxins in product)	2÷14(an average of 8)	6÷24 h	diarrhoea, vomiting, abdominal pains
Escherichia coli (infection + toxin)	12÷72	1÷7 days	diarrhoea with blood and mucus
Vibrio parahaemolyticus (infection)	2÷48(usually 12÷18)	2÷5 days	abundant diarrhoea, abdominal pains, vomit, fever, some times dehydration
Yersinia enterecolitica (infection)	24÷36(3÷5 days)	3÷5 days	from moderate diarrhoea to chronic enteritis
Streptococcus (toxins in product + infection)	3÷22	24÷48	abdominal pains, vomit, diarrhoea
Campylobacter jejuni (infection)	3÷5 days	Days or weeks	fever, head aches, abdominal pains, abundant diarrhoea with blood
*Listeria monocytogenes***	1÷70 days	from few days to years, with a high rate of mortality (approximately 30%)	

*(Macrae et al., 1993)

**Listeria provokes a disease that can be considered more infectious rather than toxic

TABLE 5. Main pathogenic microorganisms which produce food poisoning*

Agent	Source
0	1
Salmonella	raw meat (pork, beef and poultry), uncooked milk, eggs
Staphylococcus aureus	meat products and sweets (which are manipulated a lot during preparation), dairy products (especially the ones obtained from unpasteurized milk)
Clostridium perfringens	raw meat (poultry also), dry foods, vegetables, spices, greens
Clostridium botulinum	inappropriate kept meat, fish and vegetables
Bacillus cereus and other *Bacillus* spp.	cereals, dry foods, diary products, meat and meat products, greens, spices, vegetables
Escherichia coli	agricultural raw materials, especially animal origin
Vibrio parahaemolyticus	raw and cooked fish, crustaceans, sea fruits
Yersinia enterocolitica	raw meat (poultry also), meat products, milk and diary products, vegetables
Streptococcus	uncooked milk, raw meat (pork, beef and poultry), products prepared by infected people
Campylobacter jejuni	raw poultry meat, uncooked milk or unpasteurized, untreated water
Listeria monocytogenes	meat (pork, beef and poultry), soft cheese and other diary products, vegetables, crustaceans
Viruses	crustaceans, cold snacks prepared by infected workers

*(Macrae et al., 1993)

TABLE 6. Pathogenic agents sent by men

Pathway	Disease	Agent	Source
skin, nasal, oral	Staphylococcal foodborne illness	*Staphylococcus aureus*	Nasal ways and skin infections
Faeces, oral	Salmonellosis	*Salmonella* species	Intestinal tract
Unknown (faeco-oral)	Enterocolitis	Escherichia coli	Unknown
Faeces and oral	Bacillary dysentery	*Shigella* species	Intestinal mucosa
Faeces and oral	Hepatitis A	Hepatitis A virus	Unknown

3. References

Belton, P.S., 1999, Safety aspects of food biotechnology, *Food Control* **12**:285-287.

Dillon, M. and Griffith, C., 1996, *How to HACCP*, 2nd edition, M.D. Associates, Grimsby, pp. 3-52.

Heggum, C., 2001, Trends in hygiene management - the dairy sector example, *Food Control* **12**:241-246.

Jouve, F., 1999, Establishment of food safety objectives, *Food Control* **10**:303-305.

Macrae, R., Robinson, R.K., and Sadler, M.J., 1993, Encyclopaedia of Food Science, Food Technology and Nutrition, Academic Press, London, pp. 1980-1985.

Mortimore, S., 2001, How to make HACCP really work in practice, *Food Control* **12**: 209-215.

Rotaru, G., 2003, Systems of quality assurance in cheese making, in *Cheesemaking Science and Engineering* G.M. Costin, ed., Academica, Galati, pp. 564-613.

Rotaru, G., Borda, D. and Sava, N., 2001, *Implementing Quality Management Programmes*, Ed. Academica, Galati, pp. 82-90.
Rotaru, G. and Moraru, C., 1997, *HACCP. Hazard Analysis. Critical Control Points*, Ed. Academica, Galati, pp.11-74.
Wallace, C., and Williams, C., 2001, Prerequisites: a help or a hindrance to HACCP, Food Control **12**:235-241.
***2000 SR EN ISO 9001:2000.
***2004 FDA *Report on the Occurrence of Foodborne Illness Risk Factors in Selected Institutional Foodservice, Restaurants, and Retail Food Store Facility Types.*
***2004 IFT *Scientific status summary, Bacteria associated with foodborne disease.*
***2005 *ISO22000:2005.*

12 Application of HACCP in Small Food Businesses

Vassilis Georgakopoulos[1]

1. Introduction

The **Hazard Analysis Critical Control Point (HACCP)** system is a food safety management system developed in the 1960s to ensure the safety of foods for the United States space programme. Since the 1960s HACCP has evolved into a recognized means to ensure the safety of foods throughout the food industry around the world. Developments in HACCP during the 1990s have been major and some governments perceive its implementation as a remedy for all of their country's food safety issues (Mortimore and Wallace, 1994).

The application of the HACCP system is progressing rapidly; in particular in large and medium scale food industries, and more people than ever are being trained in food safety. At the same time, however, the incidence of foodborne diseases is increasing worldwide. Does this increase in foodborne diseases represent a failure of the way the system is being applied or even the system itself?

The HACCP system can look at the food-processing operation systematically, from the selection of the ingredients right through to the serving of the food to the customer. By carefully analyzing each step of the food-processing operation, anything that might affect the safety of the food is identified. You can then decide when and how to control that hazard (a hazard is anything that may cause harm to a consumer. It may be biological, physical and/or chemical). This technique helps **prevent** safety problems by careful planning in easy steps.

The concern, however, emerging today, is not the use of HACCP but the manner in which it is being implemented (Peters, 1995). HACCP is an effective system, which is frequently misunderstood and badly implemented. The only thing in a food business that is worse than no HACCP system is a HACCP system that does not work.

[1]Vassilis Georgakopoulos, BSc, MBA, PGCE, PhD, MSOFHT, MIFST, Food Safety & Quality Consultant – Trainer, 10 Miniati street, Athens, 11636, Greece. e-mail: vgeorgakopoulos@hotmail.com

On one hand, the frequent occurrence of foodborne disease outbreaks and the continued existence of bacterial food poisoning as a major public health problem, indicate that the food safety system is failing, either because of faults in its design or because of faults in its operation (Griffiths and Wallace, 1998). On the other hand, HACCP training is being seen as an isolated domain which sole purpose is to produce certificated personnel (Ehiri *et al.*, 1997).

In practice HACCP is a tool but in principle it is **a way of thinking**. Therefore, it should come as no surprise that there could be different opinions on how it should be applied. Carrying out a hazard analysis and then implementing a system to control risks is impossible without relevant knowledge and expertise. Food businesses are required to ensure that their staff is trained on the generalities and specifics of HACCP (Council Directive 93/43/EEC). As a result, **training** becomes the most important element in setting up a HACCP system.

The reason for this is quite simple: A manager can only run the lean and flexible organization demanded by today's competitive business environment with an educated workforce. Modern training methods are designed to take workers to peak performance quickly and the company that invests both resources and commitment in training gains a competitive advantage.

2. Present Status of HACCP Implementation

Good **communication** lies at the heart of any successful system. Communication in this case is not just about dialogue between the "HACCP expert" and the person from the food business, it also involves many other people – the legislator, the enforcer, the food worker, the various people in the food chain from producer to retailer, the consumer and the media. There are some basic questions that can be asked before we implement a HACCP programme:

- Are all these people talking to each other and talking the same language?
- Are their priorities the same?
- Are they all aware of their individual responsibilities and the responsibilities of others?
- Is there a clear understanding of the expectations and limitations of the others?
- Have they had any, even basic, instruction in food hygiene?

Sadly the answer to many of these questions is usually 'no' and the simple day-to-day occurrence of communication is, unfortunately, so bad that it is responsible for the majority of any problems encountered with HACCP.

The importance of training cannot be overstated. If all the employees in a food business are not properly trained to understand food hygiene and safety, then all of the time, the work and the money spent in implementing the system have been wasted. In most cases training needs only to cover the basics of food hygiene and safety with the plan to 'keep it simple'. This is because large numbers of people working in the food industry do not have a college education, many do not speak the language of the country fluently and others carry out only relatively menial tasks with little interest in their workplace other than their own immediate responsibilities. Training should always be to a recognized standard and should never be regarded as a 'one off' event. Refresher training is, unfortunately, only a feature in the most progressive businesses with many companies slipping into complacency once they have given all their staff some kind of induction or basic training. Even the best food safety management systems will not work if food handlers ignore, or are ignorant of, the need to wash their hands, never mind the more complicated aspects of food hygiene.

Furthermore, reports have been published (Morrison, 1999) expressing concerns about the ability of food businesses to comply with the Hazard Analysis and Risk Assessment requirements. A recent survey on the status of HACCP implementation in the UK (Panisselo *et al.*, 1999) shows that the number of companies without HACCP increases as the number of employees decreases.

Earlier research (Jouve, 1994; Kirby, 1994) shows that smaller companies lack:

- Knowledge and expertise in HACCP
- Appropriate resources (e.g. capital, time, manpower, management commitment) to obtain this knowledge

On one hand, the requirements of HACCP are far too academic for ordinary food businesses to interpret and adopt unless they are able to see specific examples directly applicable to them (Morrison, 1999).

Larger companies (companies with more than 250 employees), on the other hand, can invest resources and staff time in training for successful implementation of the HACCP principles (Henson *et al.*, 1999). Improved safety has provided a new avenue for them to allow products to capture increased market share. While food safety is a general issue, companies that address it creatively are able to benefit from improved consumer confidence.

Given this background, it is hardly surprising that a legal requirement to impose HACCP principles on small food businesses (businesses with 10-250 employees), which form the majority of food businesses, has been less than successful.

Experience over time will show whether or not the regulations are effective. Emphasis, however, should be placed on the fact that the identification of steps critical to food safety in a food business is a task, **which should be based on education rather than the fear of legal enforcement**.

There is also a danger that HACCP may be perceived as the panacea for food manufacturers because it can provide a system which meets the requirements of enforcement agencies, customers and consumers, and provides the manufacturer with a due diligence defence. Problems are encountered if the benefits of HACCP are oversold, or the system is incomplete or ineffective. Consequently, this leads to a false sense of security, lack of confidence in a manufacturer and a loss of credibility of the HACCP system itself (Easter *et al.*, 1994).

Many of the publications on HACCP examine the implementation of the system in the food industry and give detailed information on the generation of HACCP plans. However, as Khandke and Mayes (1998) observe, there seems to be very little published on the **practical implementation of the HACCP plan** on the 'factory floor'. From the literature review and the case studies, it is evident that in most of the cases HACCP plans fail because they are not fully implemented in practice. Most of the attempts on HACCP stop soon after the generation of the HACCP document, which will be required for inspection by the enforcement authorities.

Ehiri and Morris (1995) suggest that HACCP should be viewed as a process to be initiated and not as a programme to be implemented. HACCP is as much a thought process as it is a food safety management system. Attempts to standardize the system have resulted in defining it in terms of flow diagrams and record-keeping rather than encouraging behaviours that control factors, which contribute to foodborne diseases.

In addition, Ehiri and Morris (1995) state that there is an unfortunate tendency to regard HACCP as a programme to be contracted out or as a one-off paper exercise. There may even be a temptation for food operators to engage the services of outside 'consultants' to develop for them 'paper-based HACCP systems' without themselves understanding the concept.

HACCP's implementation requires diligence and training. HACCP is a discipline and as such, it creates **a culture**, which revolves around good management with a systems orientation (Adams, 1998).

3. Food Safety Law

The law relating to food hygiene has in the past been based on providing detailed rules to be observed but in June 1993 the European Community adopted a Directive on the Hygiene of Foodstuffs (No 93/43/EEC) to protect public health and also try to ensure common standards of hygiene were applied throughout the EC. The Directive has a number of important features:

- 'Hygiene' is widely defined to include not only measures to ensure the safety but also measures to ensure the wholesomeness of foods. This ensures that matters such as damaged or contaminated food are also controlled
- The Directive differs from the approach taken by previous Regulations, which set down in detail how hygienic operations were to be achieved. Now there is a general obligation to ensure food is prepared, handled and sold in a hygienic way
- The Directive requires that businesses must review their activities, identify the critical steps they need to take to ensure food safety and ensure these are properly addressed
- **A new obligation for businesses to carry out supervision and instruction and/or training for food handlers is included**
- For the first time in this area, provision is made for industry guides to be produced to assist compliance and for their recognition by Governments with due consideration being given to them by Food Authorities

It is a criminal offence to fail to meet the requirements of the Law, or to have food for sale which is unfit, contaminated or may injure a person's health. If prosecuted the food handler/retailer may be fined and in serious cases could be imprisoned and/or have their business closed down and be personally banned from running a food business in the future. Just as importantly though the retailer may cause their customers to become seriously ill and in this case they may also be able to claim damages from the retailers or your insurers.

The requirements of the Food Hygiene Directive 93/43/EEC in Europe created a widespread demand for HACCP training. In 1997, however, the Royal Institute of Public Health & Hygiene in the UK criticized the poor knowledge and inconsistent standards of some of the food safety-HACCP training that was on offer (Blanchfield, 1997). A year later an open letter to a scientific journal (Lachlan, 1998) asked the following question:

"... if more food handlers than ever are trained in food safety why has the number of officially notified cases of food poisoning gone up at the same rate? [...] Perhaps it is time for the basic level food safety examination bodies to revise their papers; on most of them a candidate can fail all the temperature control questions and still receive a license to kill – sorry, a certificate ..."

Following several animal disease outbreaks and food contaminant scandals in Europe in recent years, the European Commission adopted the White Paper on Food Safety in 2000. This White Paper contains a number of recommendations aimed at increasing food safety, improving the traceability of food products and regaining consumer confidence in the food industry. To this effect, a package of proposals for new European legislation on food and feed is being prepared with the following characteristics:

- Responsibility of food safety lies with the establishment operator
- The competent authority of the Member State verifies correct implementation of the new rules
- Production should be based on good hygienic practice and HACCP principles
- Products are subject to microbiological criteria and temperature limits

The proposals deal with a large variety of food types and cover the entire food chain ("from farm to fork") and once adopted by the European Parliament and the Council, will replace the Directive on the hygiene of foodstuffs (93/43) as well as 16 product specific Council Directives.

4. Food Safety Training

Reece and Walker (1997) have described how the role of the teacher/trainer has changed over the years and have provided 'food for thought' about **how to create positive attitudes to learning**. They argue that traditionally, the role of the teacher/trainer has been as a purveyor of information: the trainer was the fount of all knowledge and the trainees were listening passively and/or were taking their own notes. This is not true anymore. The modern trainer is a **facilitator**, a person who assists trainees to learn for themselves. In addition, the trainees might well be at different stages in their learning and, in consequence, the learning is individualised to suit individual requirements and abilities. This change has been the result of the following factors:

- It is recognised that adults have a wealth of experience and are able to plan their learning quite effectively
- Not all individuals learn in the same manner
- Everyone learns at his/her own pace and not, of necessity, at the pace set by the teacher/trainer. Hence, the individualising of learning has definite advantages

Food safety-HACCP is a form of 'change management' and this is also how learning is described. **Learning is about change**, the change brought about by developing a new skill, understanding a scientific law, changing an attitude. Learning is a relatively permanent change, brought about intentionally or by experience. HACCP is about changing attitudes with regard to food safety and Davies (in Reece and Walker, 1997) believes that attitudes cannot be changed before the appropriate change of behaviour. Attitudinal change often takes a long time but if participation is encouraged, then this slow process tends to be long term and to all intents and purposes permanent (Reece and Walker, 1997).

Research into the ways people learn has identified that learning is generally more effective if it is based on experiences; either direct or experiences that have been read about (Rogers, 1989). However, concepts that can be practiced or seen are more likely to be learned. Knowles (1984), furthermore, has identified six **assumptions** made about adult learning, which are related to:

- ***The need to know***. Adults need to know why they need to learn something before starting to learn it
- ***Self-concept***. The self-concept moves from teacher dependence to self-direction in the learning process. Adults have a self-concept of being responsible for their own lives. Once they have arrived at this self-concept a need is developed to be seen by others, and treated by others, as being capable of self-direction
- ***Experience***. Adults have a reservoir of experience upon which to draw for their learning
- ***Readiness to learn***. Adults are motivated to learn those things they need to know and be able to do in order to cope effectively with their real-life situations
- ***Orientation to learning***. Adults are motivated to learn when it will help them to perform tasks or deal with problems that they meet at work or in real-life situations
- ***Motivation***. While adults are responsive to some external motivators (like promotion) the best motivators are internal pressures (like increased job satisfaction and self-esteem)

5. Key Elements of a Successful Approach

Food businesses, like any other type of business, operate in a continuously changing marketplace, which prompts changes in their needs. If they are to survive they have to learn how to adapt and respond to these changes.

The introduction of HACCP in EU legislation has proved to be one of the most important stimuli prompting changes in the way food businesses operate. Although this can be viewed as a negative driver, forcing companies to carry out HACCP, it will inevitably speed up the progress of HACCP in the food industry (Khandke and Mayes, 1998).

However, legislation is not the only driving force to HACCP implementation. There are internal driving forces in addition to external forces that push the food industry towards the need to implement HACCP. The impetus behind a company's drive towards adoption of the HACCP system generally falls within one or more of the following categories:

External drivers

- Pressure from competitors
- Customer demands and expectations (ability to attract new and retain existing customers)
- Regulatory environment (government, enforcement authorities)
- Media
- International trade

Internal drivers

- Reduced product microbial counts (production of food safe for human consumption)
- Limitations of inspection and testing
- Derived benefits (expand market share, reduce scrap rate, reduce rework, increase productivity, reduce product defect rate, increase employee morale, minimize role ambiguity, better control of suppliers, improve existing system)

6. Management of HACCP Programmes

The requirement to install a HACCP system to a food business, according to Panisello *et al.* (1999), is influenced by a number of factors:

- Company size
- Company's customer(s) food safety requirements
- Product type produced
- Main processing operation used to process food

All HACCP systems, however, go through 4 phases, namely: **introduction**, **design**, **integration**, and **operation**. A careful consideration of what is involved in each phase could save the food business a lot of unnecessary and/or wasted effort, time and money. This preliminary strategic planning is of paramount importance because it is introduced as a new way of thinking and will form the basis upon which HACCP will be based.

First therefore, it is necessary to examine the important parameters associated with each phase of developing a HACCP system in the company (Easter *et al.*, 1994; Stier and Blumenthal, 1995; Khandke and Mayes, 1998; Taylor, 1999):

Introduction

- ***Good Manufacturing Practice (GMP) and HACCP***. Problems arise if HACCP is attempted before GMP is in place; every process step is associated with hazards which are, in reality, controlled through GMP
- ***Senior management leadership and commitment***. Leadership should generally be exercised on three different levels (Mintzberg, 1998): at the *individual* level, leaders mentor, coach, and motivate; at the *group* level, they build teams and resolve conflicts; at the *organizational* level, leaders build culture. All staff must be made aware of the importance of food safety and their role in achieving company goals. Management's primary function is to give the example when attempting such a cultural change

Design

- ***Teamwork***. Many companies are not prepared to invest on a group of people to develop a system; instead, they prefer simply to add it to the job description of one member of staff. This is a recipe for failure: a multidisciplinary team that has complete understanding of the business and manufacturing operations should carry out HACCP implementation. At this point many companies (especially small and medium size) seek outside help to deal with the creation of the plan because they lack a degree of technical sophistication, knowledge and expertise
- ***Variability in interpretation of HACCP terminology and principles***, such as the identification of too many CCPs & the adoption of minimal flow charts, is another issue for consideration. To tackle this problem, the adoption of the Codex Alimentarius Commission methodology (Codex Alimentarius, 1993) is recommended (because of its worldwide approval), together with the European Commission's consideration of the "farm to fork" principle.
- ***Training***. In order that operatives, supervisors and managers are able to implement the HACCP system it is essential that they be adequately trained. Consequently, training becomes the single most important element of any HACCP system

Integration

- ***Costs required include***: (a) a resource allocation to monitor the system and keep records, (b) capital costs for equipment or equipment modifications and plant reconstruction, (c) costs associated with the use of private consultants, and (d) costs of training. However, as Coleman and Kailasapathy (1995) emphasize, the cost of implementing HACCP should not just be considered in terms of the initial outlay for training, documentation, and outside technical resources. It should also take into account the potential increase in income, the decrease in customer complaints and therefore the decrease in staff needed to follow up complaints

- During ***integration*** there is a need to retain and motivate staff (in particular supervisory/managerial personnel) because the running of the programme depends on highly skilled individuals
- ***Time required for implementation***: It can take from 6 months to 2 years as it depends on a number of factors such as: company size, product type produced, complexity of the operation, number of personnel trained in HACCP, resources available to develop and implement the system, and company's customers and enforcement authorities food safety requirements

Operation

- HACCP systems demand much more intensive ***monitoring and documentation*** efforts than other systems. Probably the determining factor in the cost to a company is the frequency at which monitoring is needed (Coleman and Kailasapathy, 1995)
- Design of ***verification procedures***, to check if the system is working, is important because HACCP is a dynamic system and not a one-off paper exercise
- The ***effectiveness*** of HACCP programmes should be measured through the CCPs. Long term programme effectiveness should be reflected in changing trends in reported foodborne disease outbreaks (Unger, 1992)

Mazzocco (1996) states that internally driven management systems, such as HACCP, have three **key characteristics**:

- Heavy reliance on employee involvement and training
- Development of new measurement methods
- The continual change in processes

In the initial stages, the implementation of the HACCP system would require enormous resources in terms of qualified personnel, technical support facilities and financial inputs (particularly to ensure adequate training) (Motarzemi *et al.*, 1996). If a company is effective in implementing HACCP it will have, by default, created a thinking culture with operators and management who truly understand the meaning of process control. If it has failed to create this culture, HACCP is a fragile tool that will invariably fail (Adams, 1998).

How can, therefore, learners' attitudes towards food safety be changed and how could the HACCP concept be introduced to them? The changing process can be divided in three stages (Snyder, 1994):

Stage 1: Assessment of the existing situation (trainer)

- Define what is the current food operating system (input-process-output)
- What are the current food safety regulations and rules that drive the system?
- What is the correct knowledge for food safety in the system?
- Who is involved? What HACCP skills do they have?

Stage 2: Structured process information (trainer and trainee)

- What is my process?
- What hazards are associated with the process?
- At what stages of the process are these hazards likely to occur?
- What is the likelihood that these hazards will constitute a risk to my customers, and what is the severity of such risk?
- What must I do to prevent or control these hazards in order to ensure the safety of my customers?

Stage 3: Training for mastery of HACCP performance (trainer)

- Masters the task to be taught
- Prepares the trainee for instruction
- Presents the task
- Allows for a practice period
- Tests and follows through

Measurement of learning is notoriously imprecise. A lot depends upon the type of assessment used and how soon the measurement is made after the learning event. Some subjects are learned less efficiently and less permanently than are others and, of course, different individuals will learn at different rates. However, there is no doubt that a combined use of telling and showing is more effective than just telling (Johns, 1995). HACCP training, as Engel (1998) suggests, should always involve the principles:

- If all I do is HEAR – I will forget
- If I HEAR and SEE – I will remember
- If I HEAR, SEE and DO – I will understand

Learning to do something, like how to prepare food safe for human consumption, usually involves three distinct aspects:

- ***Purpose***. In order to learn a skill the trainee needs some understanding of what it is that is to be achieved
- ***Procedures***. Skills need some procedures that are necessary in order to complete them efficiently
- ***Practice***. All skills need practice so that they are completed correctly, quickly and automatically. This part is very important so that the movements are right from the start (unlearning of wrong movements can be difficult to rectify)

Thus, the first question that needs to be asked is: *"What is it that I want the trainees to learn?"* This, however, leads to the following interlinking questions: *"How can I get them to learn it?"* and *"How will I know when they have done so?"*

The challenge is to develop a food safety/HACCP training programme based on an effective learning sequence. The **Basic Teaching Model** (Reece and Walker, 1997) could be used to accomplish this task. This model sub-divides the teaching process into four components (objectives, entry behaviour, instructional techniques, assessment) allowing the trainer to plan and implement an effective learning sequence (Figure 1).

Objectives involve the determination of what the trainees should be able to do when they have completed a segment of instruction (for a trainer, knowing where he/she is taking the trainees means there is more likelihood of him/her getting them there).

Entry Behaviour describes the trainee's level of understanding of the topic before instruction begins.

Instructional Techniques describes the teaching process, which results in those changes in trainee behaviour called 'learning'.

Assessment consists of tests and observations that assessors use to determine how well the trainee has achieved the objectives (how much he or she has learned). It can also serve as an evaluation of the previous components of the model and provide information for necessary adjustments.

7. Strategy for the Successful Implementation of HACCP

In his review on behavioural management, Snyder (1994) suggests that HACCP is actually a process of behavioural management that undergoes continuous improvement. It strives to maintain today's company standards, while at the same time educating personnel to perform (i.e. behave) in a manner that will improve

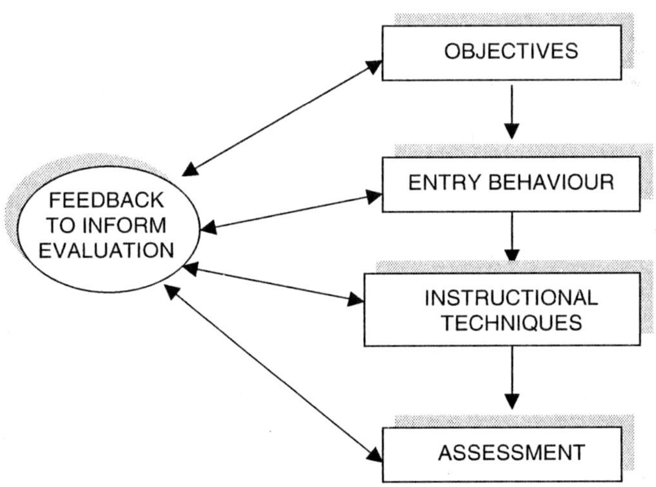

FIGURE 1. The Basic Teaching Model.

the company's competitive position tomorrow. He also adds that in a company there can be tremendous change and turmoil as HACCP is getting started and it might be difficult to manage.

The difficulty with culture change, as Adams (1998) observes, is that it requires real behaviour modification, whereby our assignment requires more training in human psychology than any form of food science.

By assuming that people are rational, logical beings, many of the recommendations of organizational transformation specialists tend to be of a quick-fix nature. Thus, in consequence of being only skin deep, the recommendations are all too often devoid of enduring influence (De Vries and Balazs, 1998).

In conclusion, we propose a model for visualizing the critical success factors and providing a basis for the development of a successful HACCP implementation strategy. The model is illustrated in Figure 2. **In this model**:

- HACCP is being perceived as an educational process
- External and internal drivers provide the foundation for the implementation of HACCP by food businesses
- The behaviour and attitude of the personnel in a company are recognized as the keys to the successful implementation of HACCP

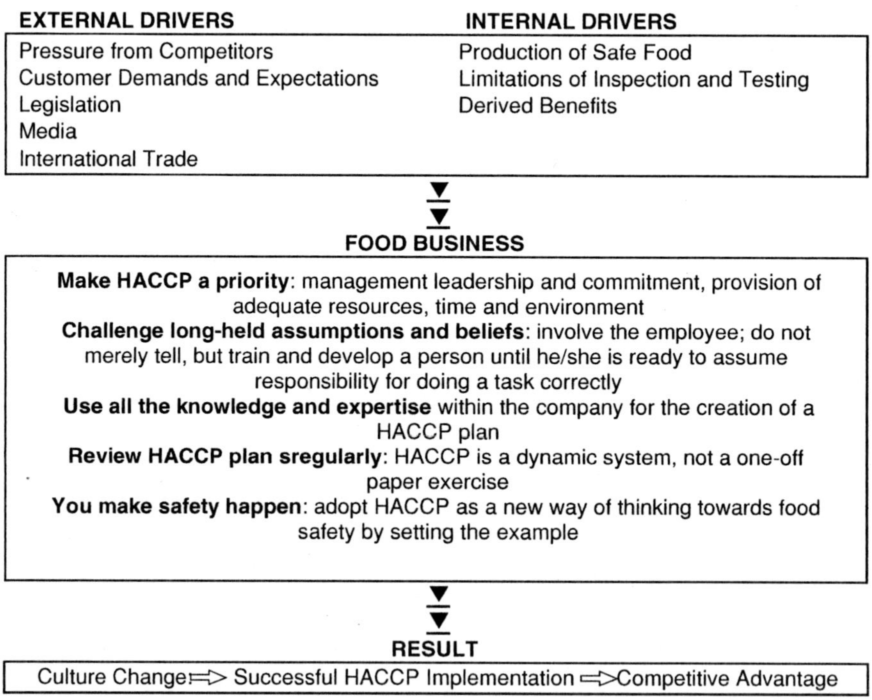

FIGURE 2. Strategy for the Successful Implementation of HACCP.

- The focus is on the actions necessary (from management and production staff) for the successful adoption of this new way of thinking
- By applying the insights derived from individual change processes to the domain of organizational transformation, it is possible to induce, facilitate, and even speed up, otherwise lengthy organizational intervention and change processes (De Vries and Balazs, 1998).

The proposed model, as practice has proved, offers many **advantages** to the food industry, because it can help their staff to:

- Assure the fitness of food for human consumption
- Prevent food contamination, microbial multiplication and survival
- Prevent foodborne illness
- Reassure customers and clients (if they see the evidence of effective food hygiene measures)
- Demonstrate *due diligence* (particularly if training records are kept)

8. References

Adams, C.E., 1998, Maximizing value from HACCP: managing food safety for the business, *Food Australia*, **50**(7):332-335.

Blanchfield, J.R., 1997, Verocytotoxin *E. coli* food poisoning, *Keynote Newsletter of the IFST*, March 1997.

Codex Alimentarius, 1993, *Guidelines for the Application of the Hazard Analysis Critical Control Point (HACCP) System* [Volume 1, Supplement 1], FAO and WHO, Rome.

Coleman, P. and Kailasapathy, K., 1995, HACCP-its role in dairy factories and the tangible benefits gained through its implementation, *The Australian Journal of Dairy Technology*, **50**(2):74-78.

Council Directive 93/43/EEC (June 14, 1993) on the hygiene of foodstuffs, *Official Journal of the European Communities*, July 19, 1993, **No. L 175/1**.

De Vries, M.K. and Balazs, K., 1998, Beyond the quick fix: the psychodynamics of organizational transformation and change, *European Management Journal*, **16**(5):611-622.

Easter, M.C., Mortimore, S.E., and Sperber, W.H., 1994, The role of HACCP in the management of food safety and quality, *Journal of the Society of Dairy Technology*, **47**(2): 42-43.

Ehiri, J.E. and Morris, G.P., 1995, HACCP implementation in food businesses: the need for a flexible approach, *The Journal of the Royal Society of Health*, **115**(4):249-253.

Ehiri, J.E., Morris, G.P., and McEwen, J., 1997, Evaluation of a food hygiene training course in Scotland, *Food Control*, **8**(3):137-147.

Engel, D., 1998, Teaching HACCP-theory and practice from the trainer's point of view, *Food Control*, **9**(2-3):137-139.

Griffiths, S. and Wallace, J. (eds.), 1998, *Consuming Passions: Food in the Age of Anxiety*, Mandolin, Manchester.

Henson, S., Holt, G., and Northen, J., 1999, Costs and benefits of implementing HACCP in the UK dairy processing industry, *Food Control*, **10**(2):99-106.

Johns, N., 1995, *Managing Food Hygiene*, Macmillan Press Ltd, Basingstoke.

Jouve, J.L., 1994, HACCP as applied in the EEC, *Food Control*, **5**(3):181-186.

Khandke, S.S. and Mayes, T., 1998, HACCP implementation: a practical guide to the implementation of the HACCP plan, *Food Control*, **9**(2-3):103-109.

Kirby, R., 1994, HACCP in practice, *Food Control*, **5**(4):230-236.

Knowles, M., 1984, *The Adult Learner: A Neglected Species*, Gulf Publishing, Oxford.

Lachlan, H., 1998, Food poisoning cases are on the increase, *Environmental Health News*, **13**(44):7.

Mazzocco, M.A., 1996, HACCP as a business management tool, *American Journal of Agricultural Economics*, **78**(3):770-774.

Mintzberg, H., 1998, Covert leadership: notes on managing professionals, *Harvard Business Review*, **76**(6):140-147.

Morrison, Y., 1999, Who cares about HACCP? *Environmental Health Journal*, **107**(3): 76-78.

Mortimore, S. and Wallace, C., 1998, *HACCP – A Practical Approach, 2nd edition*, Aspen Publishers, Inc., Maryland.

Motarjemi, Y., Kaferstein, F., Moy, G., Miyagawa, S., and Miyagishima, K., 1996, Importance of HACCP for public health and development. The role of the WHO, *Food Control*, **7**(2):77-85.

Panisello, P.J., Quantick, P.C., and Knowles, M.J., 1999, Towards the implementation of HACCP: results of a UK regional survey, *Food Control*, **10**(2):87-98.

Peters, R., 1995, HACCP: planned commitment or casual lip service? *Food Australia*, **47**(6):266.

Reece, I. and Walker, S., 1997, *Teaching, Training and Learning. A Practical Guide, 3rd edition*, Business Education Publishers Limited, Sunderland.

Rogers, A., 1989, *Teaching Adults*, Open University Press, Maidenhead.

Snyder, O.P., Jr., 1994, Achieving successful total quality through behavioural management, *Food Technology*, **48**(9):144, 146-151.

Stier, R.F. and Blumenthal, M.M., 1995, Will HACCP be carrot or stick? *Dairy, Food and Environmental Sanitation*, **15**(10):616-620.

Taylor, E., 1999, Securing public health through the application of HACCP principles: a reflection on problems experienced in the UK food industry, *Food Science and Technology Today*, **13**(2):102-104.

Unger, H., 1992, Time for total testing, *Food Processing UK*, **61**(10):35-36.

13
Cleaning and Disinfection Procedures in the Food Industry General Aspects and Practical Applications

GIULIANO SANSEBASTIANO, ROBERTA ZONI, AND LAURA BIGLIARDI[1]

In the field of general prevention of infectious diseases, one of the most important issues concerns disinfection and sterilization, which both play a fundamental role in reducing the risk of infection. This is also true for the food industry, where these processes are used to produce the same outcomes.

The different phases of cleaning and disinfection process of machines and of accessory parts (tanks, tubes etc.) with which foods come in contact are generally achieved by both physical and chemical means. These processes should result in cleaning, disinfection and removal of foreign bodies from with the structures of the equipment being cleaned.

1. Definitions

Cleaning/cleansing: a series of combined operations aiming at the elimination of dirt from the surface areas of equipment or from materials, through the use of chemical and physical means;

Disinfection: elimination on a given substrate of pathogen microorganisms, viruses and bacteria, in a vegetative form through the use of chemical or physical means.

Sterilization: elimination on a given substrate of all the microorganisms including bacterial spores through the use of chemical or physical means.

Sanitization: reduction of the contaminating bacterial load present on a given substrate in order to reach safety levels for human health.

[1]Department of Public Health Sec. Hygiene University of Parma. Parma Italy 43100

2. Introduction

First of all, it should be borne in mind that there are no standard procedures for cleaning and sanitiziation. As a consequence the procedures need to be defined for each industrial situation.

Each facility must be considered as unique, and the processes have to be adapted to suit the features of the processing plant, the type of food being processed, and the type of processing carried out. The cleaning process has to take into account those factors which affect the type of dirt that is accumulated and the hygiene standard that need to be achieved.

The first step is the choice of the detergent and the disinfectant. This choice has to take into account: the dirt to be eliminated; the materials of the surface to be treated; the hardness of the water which may affect the efficiency of the deterging and in some cases also of the disinfection; the hygiene standards that need to be achieved.

Another important parameter that needs to be defined is the frequency with which cleaning will be carried out. This will depend mainly on the speed with which the system gets dirty and the hazards presented by the dirt that accumulates; the greater these two factors are, the higher the frequency with which the cleaning and sanitizing procedures will need to be carried out.

Furthermore, production must be stopped to carry out the cleaning of systems. Cessation of production will necessitate the planning of operations so that there is a limited amount of interference with output without, however, compromising the quality of hygiene in the processing procedures.

Finally, routine procedures should be backed by more thorough interventions of cleaning and disinfection, to be carried out every now and then when production is stopped for longer periods of time owing to breakdowns or special maintenance to the system.

To ensure that the best possible cleaning procedures will be carried out, continuous monitoring systems should be set up to check the effectiveness and correctness of the operations; the monitoring systems should collect and record a series of data regularly.

3. Dirt Removal

'Dirt' is any unwanted material which remains on the surface, whether or not it comes into contact with the foodstuffs. Any dirt present in equipment associated with the food industry is generally formed by complex substances which may differ in size (macroscopic or microscopic), nature and origin. The substances are generally formed by food residues, chemical compounds, microorganisms, dust of environmental origin, particulate material deriving from the foodstuff or system, etc.

The main sources of dirt found in the foodstuff that is being processed are the solutions used for deterging and sanitizing, the water used, the external environment, and the personnel coming into contact with the foodstuff.

The ease of removal of the dirt is dependent on its solubility; the more soluble the material the easier its removal. On the basis of this characteristic, one can categorise dirt according to its solubility in water, acids or alkali and emulsions. Inert particulate material would remain insoluble.

Dirt composition is hardly ever simple – indeed, it is frequently the result of the combination of several types of substances.

In general, it is possible to distinguish the following:

Greasy dirt: This is caused by the foodstuffs that are being processed. Once emulsified, it is possible to remove the dirt with hot water even though the use of high temperatures may cause polymerization which would then make removal even harder. Complete elimination of greasy dirt requires the use of alkaline solutions with emulsification or saponification properties.

Protein dirt: this is the most difficult to eliminate owing to its strong adhesive properties. The greater or lesser difficulty encountered in removing proteins lies also in their structural complexity and in their tendency to denature and form precipitates at high temperatures (>60°C). It is generally necessary to use strongly alkaline solutions with proteolytic properties and to which humectants have been added to increase solubility.

Dirt with carbohydrate-base: simple sugars can be removed simply with water, while mild detergents remove starch residues. If the latter are present in association with protein dirt, then alkaline solutions need to be used. In this case, too, and in the presence of proteins high temperatures cause caramelizing effects which make dirt removal difficult.

Dirt with inorganic compound-base: most inorganic salts are hydro-soluble but some can, under special conditions, form precipitates which are difficult to remove. For instance, as a result of heat or in alkaline pH conditions, calcium and magnesium salts form insoluble compounds which precipitate and give rise to persistent incrustations. These precipitates can compromise the hygienic level of the system as they facilitate the adhesion of other dirt of microbial nature. They can also cause corrosion as well as interfere in the efficacy of certain detergents and disinfectants. Once formed, incrustations resulting from inorganic precipitates can only be removed by using acid solutions.

Grease and lubricating oils: these are used in the conveyor belts of certain plants to transport foodstuffs or their containers along the processing line. They can only be removed with water once they have been emulsified with suitable detergents.

Other insoluble substances: inert dirt such as sand, clay or fine metal particles can be removed by using a surfactant detergent.

Apart from the nature and solubility of the dirt, when defining the sanitizing procedure for an industrial plant other factors need to be taken into account as they may affect their efficiency:

Contact time: the longer the contact time between detergent and surface area to be cleaned, the more effective the removal of dirt. Generally, as this parameter increases, it is often possible to reduce the concentration of detergents used. To increase the contact time for vertical or sloping surfaces it is suitable to use detergents formulated as foams or gel, which adhere better.

Temperature: the increase in this parameter may, within certain limits, favour solubility of the dirt and allow easier removal even though, as has already been stated that it is necessary to take into account the fact that high temperatures may impede removal. The increase in this physical parameter mainly affects the deterging action of compounds used for cleaning, making the latter faster and more effective.

Detergent concentration: this depends on the type of detergent used (each compound presents an optimal concentration for use) but it also needs to be adjusted to the quantity and quality of the dirt and to the characteristics of the system, which can be damaged if the concentration is excessively high. Moreover, excessive concentrations are not only costly but also make subsequent rinsing of the surface areas undergoing treatment difficult.

Mechanical strength: removal of dirt is more effective if the action of the detergent is accompanied by a mechanical action carried out by equipment such as brushes (provided they do not scratch the surface areas), or turbulence or pressure in the solutions used.

Removal of the dirt is carried out in three stages. First of all, it is necessary to overcome the forces which keep the dirt adherent to the surface areas; this can be achieved by a mechanical action, a chemical action and an emulsifying action, which may be operated singly or together to produce a synergistic effect. Once the dirt has been detached, it must then be dispersed in the detergent solution, and this can achieved by dissolution, or by dispersion after it has been reduced to particles or fine drops by the synergistic action of the detergent and the mechanical force used. Finally, the re-deposition of the dirt on the surface needs to be prevented, and this is achieved by keeping it in solution/suspension (turbulence and speed of flow of the means), and its removal and dilution in the new detergent solution or in the rinsing water.

4. The Procedures

The cleaning and sanitizing systems procedures intended for use in food processing plants consist of the following:

1. Pre-washing to remove the gross dirt
2. Cleaning with water and detergents
3. Intermediate rinsing with water
4. Disinfection or sanitizing
5. Rinsing with water

4.1. Pre-washing

This involves removal of the dirt which is principally formed by food residues. It allows removal of a considerable amount of dirt, thereby making subsequent deterging easier. Moreover, it allows a limited amount of detergent to be used or to use less aggressive forms of cleaning.

In order to ensure the best possible effect, it is important to carry out the operation promptly in order to prevent dehydration and a greater adhesion of the dirt caused by its length of stay on the surface areas.

This operation can be carried out with cold or hot water (<50°C to avoid denaturation of the proteins or precipitation of the magnesium and calcium salts present in the water). If the dirt is resistant, it is possible to use brushes or water under pressure in order to more effectively remove any residues. If dust or dry food residues need to be removed, it is also possible to use suction, which is preferable to compressed air which tends to circulate in the surrounding environment or in the mechanical parts near the dirt that has been removed.

4.2. Cleaning with water and detergents

Water is undoubtedly the most widely used method for cleaning, capable of removing even the finest residues of foodstuffs after processing. The water used has to be drinkable, clear and colourless with a low content of mineral salts. In fact, hard water can interfere with other compounds used in cleaning thereby reducing their efficacy.

Adding detergents to water to lower the surface tension of the water helps the detachment of the residues and their removal particularly fats, oil and greases.

It is not necessary for detergents to have disinfecting activities although certain compounds may possess both characteristic (detergent and disinfectant). However, it should be remembered that besides cleaning the surfaces they come in contact with, detergents reduce the total microbial load, as has been reported in various studies (Holah., 2003). This reduction of bacteria number present favours the subsequent process of disinfection.

4.3. Intermediate rinsing with water

Once the detergent stage has been completed, a rinsing stage should take place to remove the dirt definitively and to remove the detergent from the surface areas of the system. This step can be carried out simply by using cold or hot water, or first rinsing with water followed by a second rinsing with an acid solution to neutralize the alkaline residues and remove any inorganic salts that may still be present. Moreover, the effectiveness of this step in removing any organic residues is determinant in order to optimise the subsequent disinfectant procedure.

4.4. Disinfection or sanitizing

This stage is often necessary in the food processing industry when the aim is to reduce the microbial flora contaminating the surface areas which come into contact with the food to acceptable levels and to eliminate pathogenic microorganisms. The removal of organisms is required for the sake of the integrity of the finished product (spoilage microbial flora) and to render it free of disease causing microorganisms thus rendering it safe for all. The procedure mainly requires the use of chemical substances with an oxidizing effect, and its efficacy is greatly conditioned by the effectiveness of the detergent treatment which precedes it.

4.5. Final rinsing with water

This is generally carried out with cold water, even though hot water (<80°) may be used in systems where a high level of hygiene is required as a further intervention for physical decontamination of the surface areas. The main aim of the final rinsing is, nevertheless, the complete removal of the residues of chemical disinfectants from the surface areas in contact with the foodstuffs.

5. Detergents

5.1. Inorganic alkalis

Sodium hydroxide is a material which possesses excellent dissolution and saponification properties together with good bactericidal action and for this reason is widely used to remove dirt (Holah, 2003); it is active both on protein and greases.

Its main limit is its aggressiveness. If handled incorrectly it may corrode metals and if it comes into contact with skin it may cause severe chemical burns. Sodium hydroxide causes the precipitation of magnesium and calcium salts in water. It should, therefore, either be used with soft water or with hard water to which softening or chelating compounds have been added (Holah., 2003).

Sodium metasilicate is another alkali which is less corrosive and can actually eliminate the caustic activity of other alkaline compounds; it is therefore often added to caustic alkalis. This compound also has a good dispersing and emulsifying effect and is easily rinsed away.

Two other alkalis with a good saponification action for grease, and deterging for protein dirt, are *orthosilicate* and *sodium sesquisilicate*; however, they are corrosive versus aluminium.

Sodium bicarbonate and *sodium orthosulphate* are non caustic alkalis used in detergent formula. The former is a mild detergent, rather corrosive and it precipitates magnesium and calcium salts; however, it is cheap and can stabilize the pH.

The latter, instead, has a good capacity to dissolve dirt and to emulsify and saponify fats and it can soften water by precipitating magnesium and calcium salts as flakes rather than as incrustations.

There are also weak, non caustic alkalis which have certain corrosive power towards metals, with lower detergent properties on fats and proteins; they are used for cleaning surface areas which are not very dirty.

5.2. Inorganic and organic acids

The use of certain strong inorganic acids, in particular *nitric acid* but also *hydrochloric acid, hydrofluoric* and *sulphuric acid*, is limited to certain situations, such as the dairy industries, where it is necessary to remove incrustations formed during the production process (e.g. milkstone in pasteurizers). Where the problem of inorganic saline precipitates is less important, weaker acids are used such as *phosphoric acid* and *sulfamic acid*, which are less corrosive. In particular, the former can be combined with surfactants and this allows removal of all the dirt, including incrustations, in one step. Organic acids are weaker and therefore less dangerous to handle than inorganic acids, and they also have a bacteriostatic activity. The most common are *gluconic acid, hydroacetic acid, citric acid* and *tartaric acid*. However, they are not widely used because they are less effective and more costly.

5.3. *Surfactants*

They are able to reduce the superficial tension of water and favour humectant effect on the dirt, and promote emulsion of fatty substances. According to the electric charge they present in solution, surfactant agents can be divided into:

1. Anionic detergents if negative charges prevail
2. Cationic detergents if positive charges prevail
3. Non ionic detergents if they do not dissociate in solution.
4. Amphotheric detergents if the type of predominant charge depends on the pH of solution in which they are found.

They are not corrosive or irritant and are easily removed because they are easily dissolved in water. With certain exceptions (soaps), they are not affected by the hardness of the water and many of them are stable both in acid and in alkaline solutions. Anionic detergents and non ionic detergents do not present bactericidal properties, unlike cationic ones which are used as disinfectants since their detergent capacity is more limited.

Many surfactants produce considerable amounts of foam, especially when the cleaning procedure requires the use of turbulence or pressure; that is why they are used together with anti-foaming agents.

5.3.1. Anionic surfactants

Anionic surfactants include a very large group of compounds, including traditional soaps, which are not used because they have a more limited detergent power. The main anionic surfactants used in the food processing industry are those known to have excellent detergent capacity, these include *alkylsulphates*, *alkylbenzen-sulfonates* and the group of *ethoxylate sulph*ates; they are usually sodium compounds, less frequently potassium compounds. They cannot be combined with cationic detergents because they form precipitates.

5.3.2. Non ionic surfactants

Non ionic surfactants are polymers of *ethylene oxide*; they do not dissociate in solution and can therefore be used both with anionic surfactants and with cationic ones.

They present strong emulsifying capacities, are not affected by the hardness of the water, and present a variable foaming power. Although very soluble in water, some tend to separate from the aqueous solution above a certain temperature (point of turbidity), causing the solution to become turbid and losing their foaming properties. This phenomenon is exploited in a practical way in cleaning operations carried out on systems by using these detergents at a temperature that is higher than the point of turbidity when they need to be distributed under pressure on the systems; the subsequent cooling of the solution restores the solubility and therefore the detergent action.

5.3.3. Amphotheric surfactants

Amphotheric surfactants can be present in solution both in an anionic form and in a cationic one, according to the pH value. They present a relatively good emulsifying capacity, and they are stable both in acids and in alkalis; they are affected by the hardness of the water even though to a lesser extent than anionic surfactants. Furthermore, they also possess a good disinfectant effect on Gram+ and Gram-bacteria and on actinomycetes. This characteristic allows the combination of the detergent stage with the disinfectant stage, when a high level of hygiene is not required, thereby simplifying the sanitization procedure.

5.4. Water conditioners

These substances are added to detergents to prevent or limit the precipitation of magnesium and calcium salts in water and the subsequent formation of incrustations. They may be organic or inorganic compounds; of the inorganic ones, the most frequently used are sodium polyphosphates, which often have good emulsifying properties. They present good dissolution and dispersion capacities and enhance the rinsing properties of the detergents to which they have been added. The less costly and most widely used is *tetrasodium pyrophosphate*, which is more effective than calcium towards magnesium, while *sodium tripolyphosphate* and *sodium tetraphosphate* remove calcium and magnesium and form compounds rather than precipitating phosphates. It is actually better to use a softening treatment for the water than to use these substances, whose effects may be insufficient in the case of particularly hard water.

Polyphosphates present a different degree of instability and tend to convert to orthophosphates, losing their capacity as softeners. The most instable of all is *sodium esametaphosphate*, a compound which can effectively eliminate calcium salts.

The main organic water conditioners or chelating agents are *EDTA* (*ethylendiaminotetracetic acid*), *NTA* (*nitrilotriacetic acid*), used as potassium or sodium salts, and sodium salts of gluconic and heptonic acids. Although rather costly, they are widely used because of their high solubility in liquid detergents.

As we have already pointed out, the detergents in use are mixtures of several compounds, of those mentioned. A detergent which meets requirements of effectiveness as described above for an ideal product will contain, for instance, a first compound which is a surfactant with physical activity on dirt; a second detergent compound with chemical activity; and possibly a corrector for water hardness.

Besides guaranteeing good cleaning, detergents should also present limited costs and be used in concentrations which allow the required level of hygiene to be attained with the lowest cost. At a final analysis, the proof of the effectiveness of a detergent is measured on the basis of the standard of hygiene achieved, of the time taken and the costs required in order to attain that standard.

6. Disinfectants

The most widely used disinfectants in the food industry are the halogens (*hypochlorous acid, chlorine dioxide* and *iodine*), the peroxides (including *hydrogen peroxide* and *peracetic acid*) and the surfactants which include cationic surfactants with prevalently disinfectant characteristics and amphotheric surfactants which are discreet disinfectants.

In general, the aim of disinfectants is to inactivate non spore forming pathogenic microorganisms, to reduce total bacterial load. To achieve this, the disinfectants must possess a broad spectrum of action.

An ideal disinfectant should not be toxic for man, should remain stable under the working conditions in which it is used, should be soluble in water, should not be corrosive or toxic or irritant and should provide a good cost-benefit ratio.

The effectiveness of disinfectants can be affected by various factors such as temperature, pH, concentration and contact time, and presence of organic and inorganic substances.

As regards temperature, an increase generally corresponds to an increase in the speed of reaction of the disinfectants; disinfection is usually carried out with neutral pH even though an acid pH improves the effectiveness of the disinfectants with oxidizing properties; furthermore, for a correct disinfection suitable combinations of concentration-contact time of the disinfectant should be taken into account.

In fact, in order to be effective a disinfectant must be allowed to act for a certain period of time, and obviously by increasing its concentration its effectiveness also increases, bearing in mind, however, that if concentrations are too high they may be toxic for operators and damage equipment.

The efficacy of disinfectants is also affected by the presence of organic and inorganic substances on the surface to be disinfected since they partially consume the disinfectants as they are oxidized by the latter, and they partially bind with them producing by-products which may be toxic for man and require longer times for their action to occur.

6.1. Halogens

6.1.1. Hypochlorous acid

Hypochlorous acid has a wide range of action as it is effective against vegetative bacterial cells and spores, as well as viruses. It provides a good cost-benefit ratio, is easy to use and even when dissolved in hard water it does not form compounds with the salts present and it maintains its oxidant properties.

The main disadvantage of using hypochlorous acid is that, in the presence of organic compounds, it binds with the latter determining the formation of by-products – DBPs (disinfection by products) – which may be toxic to man.

These by-products include trihalomethanes (THMs), whose principal compound is chloroform; these are the by-products most frequently encountered when disinfecting with hypochlorous acid, and they take on the role of indicators of the possible presence of other chlororganics (Wallace., 1997).

Hypochlorous acid is formed when adding chlorine gas to water, or more commonly chlorine salts, especially calcium hypochlorite or sodium hypochlorite (see Table 1)

The resulting hypochlorous acid is in equilibrium with its dissociated form according to the reaction:

$$HClO + H_2O \leftrightarrow H_3O^+ + ClO^-$$

The prevalence of one or the other form depends mainly on the pH. In fact, when the pH is neutral equal amounts of undissociated and dissociated acid are present; when solutions are more acid, the undissociated HClO form is present, whereas in more alkaline solutions the dissociated form ClO^- is prevalent. (Figure 1)

Disinfection with hypochlorous acid is based on a mechanism of oxidation and it is believed that the molecular oxygen liberated by the acid acts both on the key enzymes in the bacterial metabolism and on the nucleic acid of bacteria and viruses.

Both the undissociated form of hypochlorous acid (HClO) and the dissociated form (ClO^-) have this oxidizing power although it is the undissociated form of hypochlorous acid which has a greater oxidizing power.

Experimental studies have highlighted the greater effectiveness of hypochlorous acid as compared to its dissociated form; in fact, in order to obtain a 99.9% inactivation of Poliovirus 1, at equal temperatures (25-28°C) and acid concentration (0,2-0,3 mg/l), an increasingly greater contact time is required (from a few minutes to half an hour) as the pH of the solution increases (from pH 6 to pH 10), thereby favouring the prevalence of the hypochlorite ion.

In industry, disinfection is carried out at neutral or slightly acid pH levels, taking care not to excessively acidify the aqueous solution in which the hypochlorous acid is dissolved since at pH values lower than 4 highly toxic chlorine gas is liberated.

Apart from the pH value, the other factors affecting disinfection with hypochlorous acid include temperature and the chemical composition of the material: lower temperatures cause a slowing down of the oxidizing effect of the

TABLE 1. Reactions of chlorine and hypochlorite salts in water (Baker, 1959)

$Cl_2 + H_2O \leftrightarrow HClO + H^+ + Cl^-$
$NaOCl + H_2O \leftrightarrow HClO + Na^+ + OH^-$
$Ca(OCl)_2 + 2\ H_2O \leftrightarrow 2\ HClO + Ca^{2+} + 2\ OH^-$

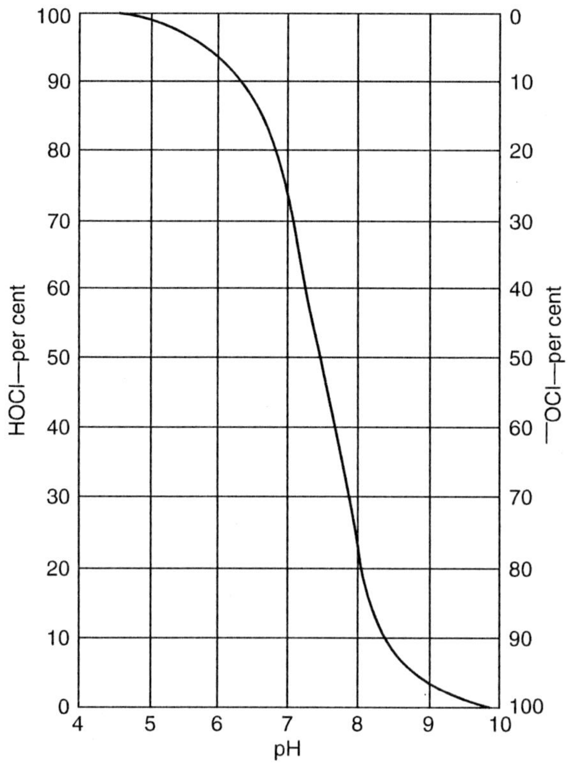

FIGURE 1. Relationship among HClO, ClO^- and pH.

disinfectant, which consequently requires a longer contact time; organic and inorganic substances where disinfection is being carried out react with hypochlorous acid and decrease the disinfectant activity. In fact, if ammonia or ammonium salts are present during chlorination of the waters, hypochlorous acid may combine with these and give rise to reaction by products called chloramines (see Table 2).

When active chlorine is combined in the form of chloramines, it maintains its microbicidal effect but requires longer contact times and this goes to the detriment of industrial disinfection where it is impossible to stop an entire processing line for long periods of time.

The most widely used solutions of hypochlorous acid in the industrial sector are those with sodium hypochlorite salts, with active chlorine equal to 10-14%, or calcium hypochlorite salts (e.g. $Ca(OCl)_2$), which contain 30% active chlorine. The pH of these commercial preparations is approximately 12;. At this pH value the dissociated form of the acid prevails, it is known to be less reactive and can therefore remain stable for a longer period of time; however, it is advisable to check the strength of the solution before use as the hypochlorite ion gradually decreases as time passes and loses oxygen.

TABLE 2. Reaction of hypoclorite acid with NH group

$NH_4^+ + HClO \leftrightarrow NH_2Cl + H_2O + H^+$;	NH_2Cl monochloroamine
$NH_2Cl + HClO \leftrightarrow NH_2Cl_2 + H_2O$;	$NHCl_2$ dichloroamine
$NHCl_2 + HClO \leftrightarrow NCl_3 + H_2O$;	NCl_3 tricholoroamine

Several concentrations of hypochlorite solutions can be used; they vary in accordance to what needs to be disinfected (see Table 3).

6.1.2. Chlorine dioxide

Chlorine dioxide is a oxidizing disinfectant often employed as an alternative to hypochlorous acid. This compound is very stable, highly inflammable, highly toxic, and consequently requires preparation at the time of use and in a closed circuit system. Chlorine dioxide can be formed by the reaction between a strong acid and chlorine, However, chlorine dioxide manufacturers preferably use sodium chlorite instead of the highly toxic chlorine (see Table 4).

Sodium chlorite is a commercial product and as such it is not pure; it is therefore necessary to use excess acid in order to make sure that all the chlorite reacts; in fact an incomplete reaction in the formation of chlorine dioxide may leave a toxic residue of chlorite (Aieta E.M. et al., 1986).

The disadvantage of the use of chlorine dioxide is the presence of chlorite and chlorate.

The advantages of the use of chlorine dioxide as compared to sodium hypochlorite are due to the fact that chlorine dioxide is less affected by the presence of organic substances; moreover, its higher oxidizing power -as compared to hypochlorous acid -allows it to oxidize any organic compounds present without producing chloro-residues or volatile compounds such as THMs (even though we cannot rule out its involvement in the formation of other non volatile compounds).

TABLE 3. Example of appliance of hypochlorite

	Disinfection with Hypochlorite
Food processing industries	In general: 50-200 ppm for 3-30 minutes' contact time (Forsyte S.J, 1998)
	Emergency intervention for heavy pollution: storage with 1 g/l over night
Waters	0,3-0,5 ppm up to 30-60 minutes' contact time (Sobsey M.D, 1989, WHO, 1996)

TABLE 4. Chlorine dioxide preparation

$Cl_2 + 2NaClO_2 \leftrightarrow 2NaCl + 2ClO_2$

$4HCl + 5NaClO_2 \leftrightarrow 5NaCl + 4ClO_2 + 2H_2O$

A further advantage lies in the fact that it is scarcely affected by the pH value in its action mechanism and remains stable as a compound in a pH range between 5 and 9.

Chlorine dioxide is applied in chlorination of waters destined for human consumption and in the disinfection of waters in swimming pools. It could be also used in food industry to prevent bacterial biofilm formation; in this case it is important to control chlorite residues.

6.1.3. Iodine

Iodine is a disinfectant with a wide range of action since it affects both Gram+ and Gram– bacteria, as well as actinomycetes and has sporicidal and virucidal effects when its concentration and contact times are suitably increased.

There are three main commercial solutions: Lugol solution, iodine tincture, and iodophores (see Table 5).

Iodophores are compounds in which iodine is the true disinfectant while the surfactants act as carriers for the iodine, increasing its solubility in water (which is generally low) and providing a reserve for it.

They are generally preferred to other solutions because they do not leave any colouring after disinfection and are less irritant to the skin; when iodine is dissolved in water it gives rise to very complex reactions in which several ions are formed, including I^-, OI^-, IO_3^-, I_3^-, but it is the molecular iodine I_2 which is active in disinfection processes. The procedure whereby molecular iodine acts in microorganisms has not yet been fully understood; however, it is thought that it acts in the NH binding of amino acids and nucleotides, thereby modifying the structure of the proteins; it can also affect the oxidation of SH groups of amino acids and form S-S disulphuric bridges which inhibit the protein synthesis.

Commercial solutions containing iodine are ready for use, do not need to be diluted and have a 9-12% iodine concentration.

They are not widely used in the food processing industry because they leave a typical brown colouring on the disinfected surfaces and are very expensive products; that is why iodophores are often preferred to solutions containing iodine, for the advantages mentioned above.

Iodophores are used also in industrial systems for cleaning and disinfection systems such as Cleaning-In-Place (CIP) although they need to be used in conjunction with non ionic surfactants to limit the formation of foam by the iodine.

Iodophores are also used for disinfection of the skin and in hospitals for the disinfection of medical instruments (see Table 6)

TABLE 5. Iodine solution

Lugol solution	5% iodine + potassium iodide
Weak iodine tincture	2% iodine in alcoholic solution
Strong iodine tincture	7% iodine in alcoholic solution
Iodophores	Non ionic surfactants + iodine

TABLE 6. Example of appliance of Iodine

In manual system (CIP), dairy and brewering industry	10–100 ppm at 50°C (Forsyte , 1998)
In water destined for human consumption	8 ppm for 10 minutes' contact eliminates pathogen present (Block S.S., 2001)
On the skin or surfaces	5% tincture for 60″ and 7% for 15″ eliminates 90% of bacteria present (Goodman and Gilman's, 1990)

6.2. *Peroxides*

6.2.1. Hydrogen peroxide (HP)

Hydrogen peroxide is normally found in the human organism both in the mucosa (e.g. in the oral cavity) where it has a microbicidal effect which increases the a-specific defence role of the mucosa versus pathogen microorganisms, and in the phagocytes, the cells of the immune system, which are also destined to an a-specific defence role.

HP is produced in the cells through the reduction of the molecular oxygen

$$O_2 + e^- \rightarrow H_2O + OH\cdot$$

$$O_2 + 2e^- + 2H+ \rightarrow H_2O_2$$

The action mechanism is based on oxidizing reactions even though hydrogen peroxide has an indirect oxidizing effect since it can react with the chloride ion giving hypochlorite ion as a disinfectant, or it reacts with superoxide ion and gives an hydroxyl radical which also has a microbicidal action which damages the membrane lipids as well as the DNA.

$$Cl^- + H_2O_2 \rightarrow OCl^- + H_2O$$

$$O_2^- + H_2O_2 \rightarrow OH^{\cdot} + OH^- + O_2$$

Since H_2O_2 is able to damage bacterial cellular membranes and is naturally produced in human cells, it needs to be promptly detoxified by those same cells in order to prevent the production of an excessively high concentration which could cause damage.

Detoxification occurs through the enzymes, such as catalase and peroxidase, which transform H_2O_2 into H_2O and oxygen; it also occurs intracellularly in pathogenic microorganisms thereby causing a reduced disinfectant effect of hydrogen peroxide if it comes into contact with them.

Very high concentrations of peroxide are generally used, even 10-20 g/l, owing to its possible inactivation on the part of pathogens, and to the ease with which peroxide is disproportioned in water and oxygen.

Disinfection with peroxide occurs in a pH range which varies between 5 and 8 and at high temperatures, especially if one wishes to obtain a sporicidal effect, which occurs only at temperatures higher than 60°C.

Hydrogen peroxide is used in the food industry for aseptic packaging of milk and fruit juices; the container is disinfected prior to assembly to contain the product, and high temperatures are used together with high concentrations of disinfectant.

It is also used for the sterilization of metal tanks, e.g. in dairy industry where the metals increase the kinetics of oxidation reaction of the HP since they act as catalysers.

The mechanism of action of hydrogen peroxide can, in fact, be enhanced both by the presence of metals and by UV rays (which can be used together with peroxide especially in water treatment), as well as by ultrasounds; the UV rays often combined with peroxide in aseptic packaging, e.g. of UHT milk.

Often in industrial disinfection, especially in the case of joints or valves which are difficult to reach for a correct microbial inactivation, H_2O_2 is used in its vapour phase and at low concentrations for very short times (see Table 7).

Hydrogen peroxide is also used in the medical field as an antiseptic both to arrest microbial growth and to cleanse wounds; it is also used as a disinfectant for medical and dentistry instruments.

6.2.2. Peracetic acid (PAA)

Peracetic acid is a disinfectant with a wide range of action since it can act on Gram+ and Gram– bacteria, on moulds and yeast, and it is active on spores and viruses at environmental temperatures.

When compared to other disinfectants such as chlorine, it has the advantage of not being particularly affected by the presence of organic substances and not forming toxic reaction by-products except in very low concentrations and therefore negligible.

It is soluble in water and in fat, and since it is not inactivated by cellular enzymes and therefore by bacteria, it is extremely effective in its microbicidal activity.

In the various commercial solutions available on the market, peracetic acid derives from the equilibrium reaction between 15-20% acetic acid and hydrogen peroxide, according to the following reaction:

$$CH_3COOH + H_2O_2 \leftrightarrow CH_3COOOH + H_2O$$

All commercial solutions need to be checked at the time of use, and since they are very concentrated they need to be diluted to the required working concentration. Like the other disinfectants described so far, peracetic acid is an oxidant and

TABLE 7. Example of appliance of H_2O_2

	Disinfection with H_2O_2
H_2O_2 as a solution	In aseptic packaging: temperature up to 80°C, concentration 35%, for 3-9 seconds' contact (Leaper , 1994)
H_2O_2 in vapour form	To sterilize medical instruments: concentration 30% H2O2 in solution, at 35°C, to obtain 4mg/l H_2O_2 in vapour form (Block , 2001)

is reduced with subsequent liberation of active oxygen (O_2); there is limited data available on the microbicidal effects of PAA, but it is thought that it acts on sulfhydryl groups of proteins and amino acids as well as on the cellular chemiosmotic mechanisms thereby determining the denaturation of bacterial and viral proteins or the breakage of their external lining.

Peracetic acid has a greater oxidizing capacity at an acid pH value although for practical purposes disinfection is carried out at neutral pH values; it is therefore necessary to use increasingly higher concentrations of disinfectant the further one is from the initial optimal acidity situation, in order to guarantee the effectiveness of peracetic acid.

Various studies carried out under experimental conditions on the bactericidal and sporicidal effects of the acid have underlined its effectiveness against bacteria at concentrations which are generally lower than 100 mg/l with contact times of 5 minutes; much higher concentrations are required in order to obtain inactivation of spores, from 500 to 30000 ppm for contact times ranging from15 minutes to 15 seconds at environmental temperatures (Baldry., 1983).

Its virucidal action has not been thoroughly studied, but it appears to occur at higher concentrations and with longer contact times when compared to those required for the inactivation of bacteria. For the inactivation of the Poliovirus present in demineralised water, for instance, it has been noted that concentrations of 750 to 1500 ppm are required with a minimum contact time of 15 minutes (Baldry et al., 1991).

There are numerous practical applications of PAA; commercial preparations have a 5 to 35% content of peracetic acid, in the presence of hydrogen peroxide and acetic acid.

Peracetic acid is also widely used in the food processing industry, where it is used as a disinfectant and to sterilize stainless steel and glass tanks, piping and tubes, tank trucks, cisterns, cistern freight cars for food products, bottling lines for glass or plastic bottles, pasteurizing lines and Tetrapak containers for the aseptic packaging of milk and fruit juices.

It is widely used also in the CIP (cleaning in place) stage of disinfection.

0.1 to 0.5% peracetic acid is also used to eliminate, or at least reduce, the formation of biological matrixes, made up prevalently of bacteria.

Peracetic acid can also be used for tertiary treatment of sewage where its bactericidal capacity has been noted in reducing or totally inactivating the bacterial indicators of faecal contamination (total and faecal coliforms and faecal streptococci).

Its virucidal effects on sewage are lower at the same concentrations of PAA used for bacterial inactivation (see Table 8)

6.3. *Surfactants*

6.3.1. Cationic and amphotheric surfactants

Cationic surfactants, or quaternary ammonium compounds (QACs), are ammonium salts whose general formula is $(NR_4)^+X^-$ where R represents the hydrogen

TABLE 8. Example of appliance of peracetic acid

	Disinfection with PAA
Food area	
Disinfection of plastic containers	Solutions with 0.1% PAA and 20% HP eliminate *Bacillus subtilis* spores in 12 seconds, at 65°C. (Dally H., 1980)
Soft drinks industry	2500 ppm PAA in 30 minutes and 5000 ppm in 15 minutes inactivate all bacteria, yeasts and fungi
Sewage	500 ppm PAA, for 30 minutes' contact guarantee total inactivation of bacteria

atoms which can all be replaced by alkyl or aryl groups, and X^- represents the anion which is usually represented by chloride or by bromide.

The hydrophobic part which interacts with the molecules of fat or oil is represented by the cationic portion $(NR_4)^+$, while the anion X^- increases the solubility of the surfactant in an aqueous solution.

Cationic surfactants have good disinfectant effects although they are not very active against *P. aeruginosa*, a microorganism which is very frequently found in the environment (Adair et al, 1969). Cationic surfactants known to lack activity towards bacterial spores (Mackinnon, 1974) and *Mycobacterium tuberculosis* (Smith, 1968), both of which are extremely resistant against the effects of disinfectants. Their use on surface areas, e.g. on a working table, requires a thin layer to remain attached to the surface thereby preventing the re-growth of any vegetative bacterial forms which have not been inactivated by disinfection.

Cationic surfactants are effective at a wide range of pH values although they appear to be more effective under slightly alkaline conditions; moreover, they are not affected by the presence of organic substances, they are not corrosive even though they may damage certain types of rubber, and they are not irritant to the skin except at very high concentrations; that is why they can be used by personnel without the need for any special precautions.

They have the disadvantage of a reduced action if dissolved in hard water and they form a lot of foam, which requires thorough rinsing after use and which prevents their use in CIP systems.

The QACs are generally used at concentrations which vary from 50 to 500 ppm; temperatures of 40°C are required for contact times ranging between 1 and 30 minutes to ensure a good disinfectant effect.

Amphotheric surfactants can exist both in cationic and in anionic forms in relation to the pH. In fact, they have the general formula of an aminoacid and if the positive charge of the ammonium ion prevails, they are found in the cationic form, vice versa if the negative charge of the carboxylic ion prevails they take on the anionic form.

These surfactants have a good detergent effect, they are good emulsifiers, they are stable both under acid pH and alkaline pH, they maintain a discreet activity even when dissolved in hard water and they are not affected by the presence of organic substances.

Certain amphotheric surfactants are not only good detergents but also discreet disinfectants; in fact, they are active versus Gram+ and Gram–, they are also effective against resistant mycobacteria, but they are not sporicidal.

They are used in industries for meat processing, milk and its by products, drinks (mineral water, soft drinks, beer, etc.).

There are many amphotheric detergents available commercially, like Dodecil-beta-alanina, Dodecil-beta-aminobutyric acid and Dodecil-di(aminoetyl)-glycine; the last is the active compound in a series of products named TEGO (a trade name) which are used in the disinfection of equipment and machinery in food industries, and for the disinfection of hands as they leave a thin layer which ensures an antibacterial effect for many hours.

There is currently also a thermal aerosol unit which is commercially available, which can be used with the same products for quick, effective disinfection treatments in closed environments whose volume ranges from 100 to 8000 m^3.

7. Main Cleaning and Disinfection Methods: Manual, COP, CIP

There are different types of interventions which are carried out according to how the cleaning procedure occurs. Mechanical cleaning requires the complete disassembly of the equipment to be cleaned and to be inspected by qualified staff. This procedure is time-consuming and has considerable economic costs both in terms of labour and of time. Moreover, the regular disassembly and assembly of the parts may lead to the gradual deterioration of the machinery as well as to the risk of re-contamination during the assembly stage.

The Cleaning out place (COP) requires disassembly only of the parts in a system which then undergoes cleaning in special pressure containers where hot solutions are used, together with turbulence and the pressure of the washing liquids.

The most advantageous is the *Cleaning in place* (*CIP*): this procedure is widely used especially for cleaning systems for the processing of liquid foodstuffs characterised by a type of dirt that is easily removed (dairy industries, industries for the production of mineral water and soft drinks). It is suitable mainly for the cleaning of system with closed parts (piping, tanks, pumps, valves, thermal exchangers, homogenizers ...) which cannot immediately be reached. Its main characteristic, on which its name depends, is given by the fact that no disassembly of the mechanical parts of the system is required. The procedures of sanitization attained in CIP follow the steps described above: pre-washing, deterging, rinsing, disinfecting, and final rinsing.

The solution of the detergent as well as the water used in the pre-washing and in the rinsing are all taken from special feed tanks, which are found within the circuit of the system and drain externally to it. The action of the detergent can be enhanced by modulating not only its concentration but also the temperature and the speed of circulation (contact time). The mechanical strength generated by the flow of liquids from inside the circuit, or the pressure produced by special spray nozzles favours the elimination of the dirt. In particular, spray nozzles with different characteristics (ball nozzles, rotating nozzles ...) are useful for cleaning tanks and cisterns since they allow all the parts of the internal walls to be reached without needing to fill the whole container.

There are three fundamental CIP systems: single-use, re-usable and multi-use. In the first case, the detergent solutions are used only once then disposed of; this type of CIP is suitable mainly for the cleaning of small systems or parts of systems with a high degree of dirt.

The re-usable and multi-use systems, instead, require repeated use of the detergent solution and of the rinsing water collected in special separate tanks. This allows limited consumption of the detergent as well as a limited production of sewage, but the basic condition that allows the use of this procedure is that the amount of dirt present in the system must be limited. Re-usable procedures often require automatic monitoring not only of the various stages of sanitization and of the parameters being used, but also of the strength of the detergent solutions used in order to establish when they can be recycled and when they should be definitively discarded. The rinsing water is generally used for removal of gross dirt in the subsequent cleaning procedure. These systems are suitable for complex systems and large systems, or for systems requiring frequent sanitization procedures characterized by a limited deposit of dirt and for complex installations.

The advantages which have favoured the widespread use of this system of cleaning and sanitization as an alternative to manual cleaning, in spite of the high costs required for its installation, are:

1. reduced maintenance costs;
2. optimizing the cleaning process;
3. strict application of the procedures with subsequent reduced variability of the execution and increased standard of hygiene for the system;
4. reduced duration of stopping times in production and the possibility of increasing the frequency of the cleaning procedures;
5. eliminating the risk of damage due to disassembly of the mechanical parts to be cleaned;
6. eliminating the risk of re-contamination of the equipment in the assembly stage;
7. less risk for the personnel involved in the handling of dangerous substances or in access to the parts of the system for cleaning (e.g. tanks).

The cleaning and disinfecting procedure of a food-processing system is an important issue in guaranteeing the safeness of the products destined for consumption, and should therefore be accurately defined and undergo regular monitoring as regards its effectiveness and the correctness of its execution. This means that suitable monitoring and control instruments need to be available.

That is why food industries have written procedures which list in significant detail all the indications for the correct execution of cleaning and disinfection. In the case of cleaning, the procedures adopted in the production areas as well as in other areas (storage, changing rooms, toilets, yards ...) must also be clearly listed. The essential indications that need to appear in a cleaning and sanitizing schedule include:

1. what needs to be cleaned
2. how to clean it
3. what to use (equipment and products; for the latter, it is necessary to indicate the concentration, the temperature of use, contact time and any hazards in the use of the compound)
4. when and how often the treatment should be carried out
5. who should carry it out (personnel in the industry or external personnel; if it is necessary to disassemble the parts, is a technician required)
6. who is in charge of checking the procedure.

Of course, points 5 and 6 introduce a new type of problem connected to the cleaning and sanitizing processes: the personnel involved.

People who need to execute the procedure require a suitable training course that covers principles of general hygiene in the food industry specific hygiene precautions required for the product being processed together with specific aspects of standard cleaning procedures which will be carried out in their facility and regulations regarding personal hygiene and behaviour in the plant.

The person in charge will need to carry out regular checks on the procedure as regards correctness of execution and as regards results. In this case, too, the (supervisory) personnel will have to be suitably trained.

As for checking the results, when cleaning concerns the environment the aim of the treatment and the elimination of dust and residues from all the surface areas treated and its result are checked with regular inspections in which data are recorded on special forms as well as any corrective actions identified.

Expected results from a sanitizing procedure of the systems are more complex and involve:

1. absence of organic and inorganic residues
2. absence of residues of the detergents and disinfectants used
3. absence of pathogen micro organisms
4. presence of micro organisms altering the product to levels lower than the standard value adopted (e.g. yeasts <5cfu/dm^2)
5. presence of micro organisms not altering the product to levels lower than the standard value adopted (e.g. CBT<100cfu/dm^2)

In this case, controls involve both sensorial inspection (visual check on the absence of residues, olfactory check on the absence of smells, and tactile check on the total elimination of traces of fat and of microincrustations) and, in the case of microbial dirt, the execution of tests on the treated surfaces.

The most frequently used tests are microbiological tests (buffers, washing, RODAC plates and cellulose nitrate membrane) and specific tests carried out on the ATP levels with a bio-luminometer or on proteins by means of colorimetric reactions.

In this case, too, results of the inspections must be carefully recorded on special forms.

The aim of recording these data is not only to monitor the effectiveness of the procedures but also to have available regular monitoring data over time so as to quickly identify the effectiveness of the procedures and to introduce any corrective measures required in order to recover the hygiene level of the system.

7.1. Some practical applications

As practical applications we considered two industrial plant types, one relating mineral water and soft drink production, the other relating fruit juices production, with or without probiotics addition. The data are furnished by Norda SpA, Masanti factory (Bedonia – Parma) and by Berchi SpA and Wab of Parma.

7.1.1. Mineral water and soft-drink bottling

It is a question of a classic bottling plant of mineral water and soft drink production with PET and glass packing.

The cleaning and disinfection of the equipment are carried out by CIP with different systems relating to production phases.

Cleaning and disinfection procedures are different in relation to different type of production. Generally the phases provide an alkaline washing and disinfection. The treatment with acid is used to dissolve limestone deposits on which microorganisms can settle, colonise and consequently form biofilms; for such a scenario Peracetic acid (0.5%) is used as a disinfectant.

Below the features of the cleaning and disinfection plant (see Table 9).

There are two large tanks containing 3000 litres (Figure 2) in which the detergent solution, generally NaOH is prepared as well as the acid solution to eliminate any deposits. The disinfectant at the required concentration is placed in one of the two tanks. The equipment also includes a manually-operated system to supply alkaline and acid detergent and disinfectant to the various sections of the bottling machine, such as the saturator for CO_2, the filling machine and the soft-drinks container. An electronic procedure which automatically controls the various stages of the operations is also installed. Sanitization operations of the machine include various stages (see Table 10).

The first wash is performed with warm water at 30°C for approx. 10 minutes. The water used is generally the same mineral water that is then bottled. The second stage consists in NaOH treatment at a concentration of 1,5-2% and high temperature

TABLE 9. Equipment characteristics

2 steel tanks of 3000 lt.
1 tank to prepare caustic solution
1 tank to prepare acid solution and disinfectant
1 manual selector for products distribution to different parts, saturators, fillers, soft-drinks tanks
1 electric command panel with keyboard for different sanitization phases management

FIGURE 2. CIP system with two tanks. (by courtesy of Norda spa Parma).

–80°C – with recirculation for approx. 20 minutes. This recirculation ensures a good cleaning of all the machine parts -in particular the filling nozzles - and a reduction of the total bacterial load. The third stage consists of recirculation for approx.15 minutes at room temperature of an acid solution to eliminate deposits of mineral salts. In this case phosphorus based acids are not used to avoid polluting surface waters with Phosphorus and Nitrogen. Next there is rinsing with tepid water at 30°C for approx. 10 minutes. In the following stage disinfection is carried out with peracetic acid at a concentration of 0.5% and at 40°C; the disinfectant recirculation lasts 20 minutes. The final rinsing of the whole machine uses mineral

TABLE 10. Sanitization different phases of bottling line 1

Initial pre-rinse	tepid water	30°C	10 minutes
Washing with soda	NaOH 1,5-2%	80°C	20 minutes
Descaling treatment	acid solution ECO*		15 minutes
General rinse	tepid water	30°C	10 minutes
Disinfection	peracetic acid 0,5%	40°C	20 minutes
Final rinse	tepid water	30°C	10 minutes

*The use of acid is limited to avoid discharge of Phosphorous and Nitrogen in surface water

water with recirculation for 10 minutes at 30°C. The frequency is monthly; however it should be carried out every time manual work is carried out on the machine and after a longer than 3 days stop of the same machine. In case of longer than 4 hours stop it's possible to carry out cold sanitization.

In the case of cold sanitization a disinfection treatment is performed with peracetic acid at a lower concentration, 0.1%, with recirculation for approx. 20 minutes; this is followed by cold water rinsing for 15 minutes (see Table 11). At each working turn one goes on an external block cleaning (rinser, filler, capping machines).

The external cleaning of the machine is also important, and the machine is opened. The procedure includes washing with detergent at 3% concentration for 20 minutes, an intermediate rinsing and a treatment with 2% acid acting for 20 minutes, an intermediate rinsing and treatment with 2% acid for 15 minutes and a final rinsing with water (see Table 12). Bacteriological test confirmed the right operation system of sanitization.

During output the empty bottles are disinfected with peracetic acid at 0.1% with a quick spray (2-3 seconds contact time) and later rinsed with the mineral water before filling.

7.1.2. Probiotic products conditioner plant

It's a question of a bottling plant in clean room with filtered air through HEPA filters 0,3 μ to keep the most possible an uncontaminated environment in the filling machine zone.

It concerns a mono-block set up by sterilizing, rinsing, filling and capping machines in the ultra-clean configuration; the plant is completed by a unit of dosage, heating and recycling of the sterilized product used in production; it's also furnished filtered water according to the FDA specifications.

In this kind of machine are provided cleaning and disinfection phases and of filling machine sterilization too. The cleaning system is a CIP combined with a SIP vacuum sterilization system of the filling machine (Figures 4, 5).

TABLE 11. Cold sanitization

Starting after sanitization		
Disinfecting treatment	peracetic acid 0,1%	20 minutes
Rinse	cold water	15 minutes
Filler machine potting		
Disinfecting treatment	peracetic acid 0,1%	10 minutes
Rinse	cold water	15 minutes

TABLE 12. External sanitizing

Washing	Hypofoam 3%	20 minutes
Rinse	water	
Acid treatment	Acifoam 2%	15 minutes
Rinse	water	

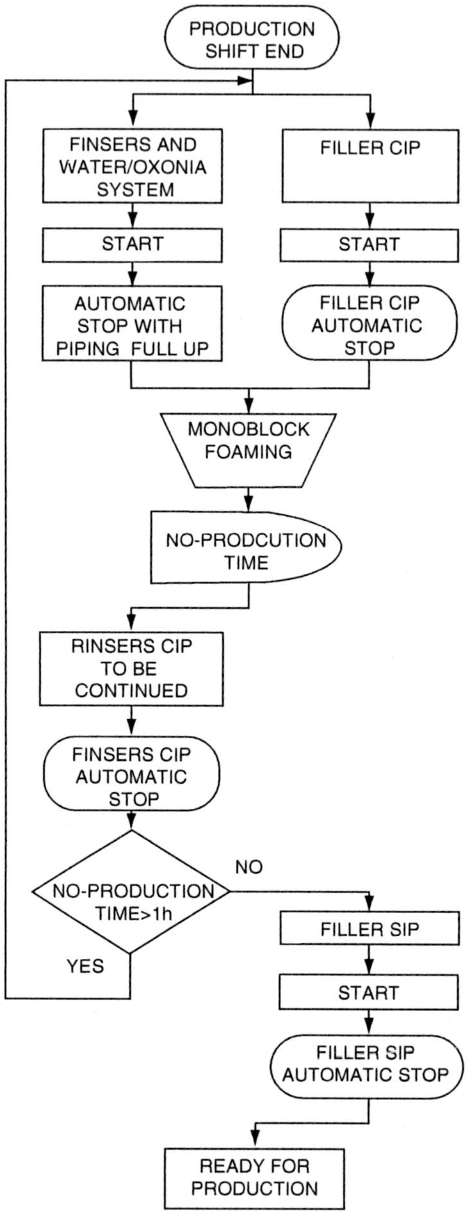

FIGURE 3. Flow chart of SIP/CIP combined system.

Two CIP procedures are implemented in the plant which simultaneously target different sections of the machine. The first procedure (see Table 13) concerns sanitization of the turrets involved in treatment of the bottles and their rinsing.

FIGURE 4. SIP/CIP system. CIP side (by courtesy of Berchi spa Parm.

FIGURE 5. SIP/CIP system. SIP side(by courtesy of Berchi spa Parma).

TABLE 13. CIP-Rinsers and water/PA mix system CIP procedure

Action	Fluid	Duration	Note
Tank and piping empting	Water, Water/PA mix	~10 min	
CIP preparation	Water/PA mix	~10 min	
1st Treatment sanitation	Water/PA mix	~2 min	
2nd Treatment sanitation	Water/PA mix	~10 min + Wait + ~10 min	filling pipes sanitation is carried out by spraying nozzles. "downstream" flushing of manual drain and filters valves, (operator confirmation request needed). step stopped prior to the time of the new production shift.
Tank and piping empting	Water/PA mix		
1st Treatment washing	Sterile water		
2nd Treatment washing	Sterile water		Flushing of the manual draining and filters valves, in "downstream" order, not simultaneous (with operator confirmation request). The filling pipes are not washed; they will be in contact with steam later contact with steam later.
Tank and piping empting	Sterile water		

Water/PA mix = solution at 0,2% at 20°C; Sterile Water: as for FDA specifications, by the means of the Water filtering system

During the first phase the tanks and tubes are emptied; next the disinfectant solution, peracetic acid at 0.2%, is prepared and left for the first treatment in bottle turret for 2 minutes at 20°C; in the second treatment of the rinsing turret peracetic acid at 0.2% is used for 10 minutes and after a short interval for other 10 minutes always at 20°C. In the subsequent phases the tubes and tanks are emptied once again and both turrets are rinsed with sterile water. Sterile water is obtained by filtration on 0.2 μ cartridge filters the second procedure (see Table 14) is performed on the filling machine.

During the first phase the tubes and tank are emptied; the filling nozzles are closed (to allow recirculation of the sanitizing product). Next, the tubes and tank are cleaned with detergent, NaOH 2% at 80°C, by recirculation. In the following phase rinsing is carried out with water from the mains. Before the final disinfection of the filling machine the whole unit is washed externally with detergent (see Table15) and disinfected with chlorine based products. The final disinfection of the filling machine is carried out with steam at 115-120°C for 15 minutes (see Table16); this last phase involves cooling of the tank and tubes with sterile water.

TABLE 14. Filler CIP Procedure

Action	Fluid	Duration	Note
Tank and piping empting	Product		
Dummy bottles insertion			
Tank and piping cleaning	Water/NaOH	≅10 min	
Tank and piping washing	Water	≅5 min	
Monoblock washing	Water	≅10 min	
Disassembling dummy bottles			

Water/NaOH: Solution of water and NaOH at 2% at 80°C

TABLE 15. Monoblock Foaming Procedure

Action	Fluid	Duration	Note
Underired bodies removal			Bottles, caps, etc.
Foaming	Acid detergents	≅10 min	
Washing	Water		
Disinfection	100-500 mg/l of NaClO	≅5 min	

Part to be foamed and washed: a)rinsing turrets; b)starwheels and guides, basement surface; c)filler from below the flowmeters to the basement, inner surface of the monoblock walls; d)caps channels from below the UV device to the basement, capping pockets and heads

TABLE 16. Filler SIP Procedure

Action	Fluid	Duration	Note
Tank and piping steaming	Steam	~15 min	
Tank and piping cooling	Water		

Steam: 115-120°C; Sterile water: as per FDA specifications, by the means of the Water/PA mix filtering system

The flow-chart (Figure 3) shows the various sanitization procedures according to the type of production. The two sanitization procedures start at the same time at the end of production. All the phases described are completed unless production is interrupted. If the interruption of production lasts over 1 hour the system starts from the beginning again without completing the final sterilization phase of the filling machine.

8. References

Adair, F., Septic, S., and Gelrer, J., 1969, Resistance of Pseudomonas to Quaternary Ammonium Compounds.1: Growth in benzalkonium chloride solution –*Applied Microbiology* **18**:299-302.

Aieta, E.M. and Bery D.J., 1986, A Review of Chlorine Dioxide in Drinking Water Treatment, *J.Am. Water Works Assoc.*, **57**:62-72.

Baker R.J., 1959, Types and significance of chlorine residues. *J. Am. Water Works Assoc.*, **51**:1185-1190.

Baldry, M.G.C., 1983, The bactericidal, fungicidal and sporicidal properties of hydrogen peroxide and peracetic acid *J. Appl. Bacterial.*, **54**:417-423.

Baldry, M.G.C., French, M.S., and Slater D., 1991, The activity of peracetic acid on sewage indication Bacteria and viruses. *Water Science and Technology*, **24**(2):353-357

Block, S.S., 2001, *Disinfection, Sterilization and Preservation.* 5rd ed. Philadalphia. Lippincot Williams and Wilkins.

Dally, H. (inventor), 1980 – Sterilization of articles – Patent N. GB 1570492

Forsythe, S.J. and Hayes P.R., 1998, Cleaning and disinfection: methods, In: Forsythe SJ., Hayes PR eds. *Food Hygiene, Microbiology and HACCP.* 3rd Maryland Aspen Publishers Inc.

Goodman and Gilman's, 1990, *The Pharmacological Basis of Therapeutics.* 8rd ed. New York. Pergamon Press Inc.

Holah, J.T., 2003, Cleaning and disinfection, In: Lelieved, H.L.M., Mostert, M.A., Holah, J., White, B. eds., *Hygiene in Food processing.* Cambridge, Woodhead Publishing Limited.

Leaper, S., 1984, Comparison of the resistance to hydrogen peroxide of wet and dry spores of *Bacillus subtilis* SA22, *J. Food Technol.*, **19**:355.

Mackinnon, I.H., 1974, "The use of inactivators in the evaluation of disinfectants". *Journal of Hygiene*, **73**:189-195.

Smith, C.R., 1968, "Mycobactericidial agents". In: Lawrence CA, Block SS, eds. *Disinfection, Sterilization and preservation.* Philadelphia: Lea & Febiger.

Sobsey, M.D., 1989, "Inactivation of health-related micro-organisms in water by disinfection process". *Water Science and Technology*, **21**:179-195

Wallace, L.A., 1997, Human exposure and body burden for chloroform and other trihalomethanes. *Crit. Rev. Environ. Sci. Technol.*, **27**(2):113-194.

World Health Organization, 1996, *Guidelines for drinking water quality.* Geneva: WHO.

14
Ensuring Biosafety through Monitoring of GMO in Food with Modern Analytical Techniques, a Case Study

AURORA RIZZI[1], CLAUDIA SORLINI, SAVERIO MANNINO, AND DANIELE DAFFONCHIO

1. Introduction

For at least 10000 years, crop cultivars have been modified from their original wild state by domestication, selection and controlled breeding, to become more pest resistant, to produce higher yields or to produce a better or different quality of product. In recent decades, the application of recombinant DNA technologies, including genetic cloning and transformation, has permitted the introduction of exogenous genes into unrelated species across species barriers. Thus, the term genetically modified has been introduced to describe an organism in which the genetic material has been altered in a way that does not occur under natural conditions of cross-breeding or natural recombination (Article 2, Directive 2001/18/EC).

Sixty-seven million hectares of genetically modified (GM) plants were globally grown in 2003 and 99 % of global transgenic crop area was grown by six leading countries: USA, Argentina, Canada, Brazil, China and South Africa. The principal commercialised GM crops were soybean, maize, cotton and canola (James, 2003). The initial objective for developing GM plants was to improve crop protection; genetic modifications of GM plants performed up to now has mainly concerned the following traits: herbicide tolerance, Bt-derived (*Bacillus thuringiensis*) insect resistance, virus resistance, fungal resistance, male sterility/fertility restoration. The production of "next generation" transgenic crops is intended to generate products with enhanced nutritional value, durability, as well as those termed functional foods. The focus and emphasis is on applications with

[1]Department of Food Science and Microbiology, Via Celoria 2, 20133, Milan, Italy. Tel: +39-0250316730, fax:+39-0250316694, e-mail:aurora.rizzi@unimi.it

obvious benefits to consumers, an even more ambitious deliverable is the potential contribution to the solution of malnutrition in developing countries.

Though the benefits of this new technology has led to increased crop productivity, increased stability of production and some say a more sustainable agriculture and environment, a number of concerns have been expressed regarding its safety to both human health and the environment. The issues debated: allergenicity of GM foods to humans, horizontal gene transfer, outcrossing, loss of biodiversity, effects on non-target organisms, increased use of chemicals (Dale et al., 2002). The major concerns regarding foods risks are principally allergic reactions that new expressed proteins may provoke or the potential alterations of metabolism of GMO caused by genetic engineering (Bindslev-Jensen, 1998; Mayeno and Gleich, 1994).

Although cases of allergic effects (or reactions) have occurred, e.g. allergenicity caused by the expression of brazil nut transgene in soybeans, new regulations provide for the marketing of GM food only after a severe risk assessment has been passed. Another negative effect for human health may derive from the transfer of transgene or, of other parts of construct, to the cells of the body or to intestinal or food-associated bacteria (van den Eede et al., 2004). This would be particularly relevant if antibiotic resistance genes, used in creating GMOs, were to be transferred.

The risk associated with an intense agriculture of GM crops consists of the movement of genes from GM plants into conventional crops or related species in the wild which may cause loss of varieties. This risk is real as several countries have adopted strategies to reduce mixing, including a clear separation of the fields within which GM and conventional crops are grown. Another risk is represented by the possible effects on non-target organisms: for example, it has been shown that pollen of Bt-corn threatens Monarch butterfly larvae and aphids which have fed on Bt crops may affect ladybugs.

On the other hand, a certain concern has been expressed about the undesirable level of control of agriculture production by a few chemical companies; this impacts on farmers who become dependent from chemical industry for GM seeds supply as well as for chemicals in the case of herbicide-tolerant GM crops.

The global market value of GM crops, which is based on the sale price of transgenic seed plus any technology fees that apply, was estimated in 2003, to be \$4.50 to \$4.75 billion and is projected at \$5 billion or more, for 2005 (James, 2003).

By virtue of the precaution principle and of the possibility for consumers to make a informed choice between GM or not-GM food, a series of regulations have been introduced in Europe.

2. European Union (EU) Legislation

The use of genetically modified organisms (GMOs), their release into the environment, cultivation, importation and, particularly, their utilisation as food or food ingredients, is regulated in the European Union by a set of strict procedures.

Council Directive 2001/18/EC of the European Parliament and of the Council of 12 March 2001 (European Commission, 2001), repealing council Directive 90/220/EEC (European Commission, 1990), concerns the deliberate release of into the environment of GMOs. Directive 2001/18/EC deals with both small-scale field trials (releases carried out for experimental purposes, dealt with in part B of the Directive) and the marketing provisions of GMOs (dealt with in part C of the Directive).

As of the 18th April 2004, new regulations, Regulation (EC) No 1829/2003 (European Commission, 2003a) and Regulation (EC) No 1830/2003 (European Commission, 2003b) came into force. These regulations aim to harmonise national rules on GM food and feed within the EU, and relate to the authorization of GM food and feed, their labelling and traceability. Regulation (EC) No. 1829/2003 repeals previous regulations, namely Regulations (EC) No. 1139/98 (European Commission, 1998), (EC) No. 49/2000 (European Commission, 2000a) and (EC) No. 50/2000 (European Commission, 2000b), and amends Regulation (EC) No. 258/97 (European Commission, 1997) and 2001/18/EC. Regulation (EC) 258/97 dealt with the placing on the market within EC of novel foods and novel foods ingredients, including products containing, consisting of or produced from GMOs. The procedure applied to foods or food ingredients which on the basis of the scientific evidence or on the basis of an opinion delivered by one of the competent bodies were substantially equivalent to existing foods or food ingredients as regards their composition, nutritional value, metabolism, intended use and the level of undesirable substances contained therein. Regulation (EC) No 1139/98 was adopted to overcome the placing on the market of two GMOs (Roundup Ready Soybean and Maximizer maize) before the Novel Foods Regulation came into force; and was later amended by the so-called "threshold regulation" (Regulation (EC) No. 49/2000) setting a 1% threshold for mandatory labelling, extended also to additives and flavouring (Regulation (EC) No. 50/2000).

Regulation (EC) No 1829/2003 establishes a common EU authorisation procedure for GMO foods. Applications for authorization to place a product on the market are to be made to a national competent authority and forwarded to the European Food safety Authority (EFSA), providing the following information: name and address of applicant, detail of the food and its "transformation event", safety assessment studies, methods of detection and whether the food could give rise to ethical or religious concerns. The product goes through the approval procedure between the European Commission and Member states before being entered into the Register of GM food and feed. The authorization is valid for ten years on a renewable basis, on application to the European Commission.

Regulation (EC) No. 1829/2003 requires also that all food and feed containing, consisting of or produced from GMOs must be labelled as genetically modified (Article 13). In addition, further information must be provided in cases where the food differs from its conventional counterpart in nutritional value, composition, use, or health implications, or if it may give rise to ethical or religious concerns (Article 13(2)). For both food and feed a threshold is set, at 0.9% of the food ingredient, for an allowable presence of 'adventitious or technically unavoidable' traces of GMOs. Food or feed with traces of GMOs below the threshold do not have to meet the requirements of this Regulation. Operators must be in a position to demonstrate that they have taken appropriate steps to avoid the presence of GMOs in order to establish that the presence of the material is adventitious or technically unavoidable. Threshold reduces to 0.5% for GMOs that are not approved in the EU, but which have received a favourable opinion (favourable risk evaluation). These thresholds apply to each food ingredient considered individually or to food/feed consisting of a single ingredient in the food/feed (an example is shown in Figure1).

Regulation 1830/2003 provides a framework for products consisting of or containing GMOs, and food and feed produced from GMOs. It sets the traceability requirement for all handlers ("operators") to ensure that the following information is transmitted in writing at all stages of placing on the market: a) that it contains or consists of GMOs; b) the unique identifier(s) (codes) assigned to GMOs. The purpose of having procedures allowing traceability of GMOs is to guarantee consumer's choice, to facilitate monitoring, risk management, and possible withdrawal of products in case of unexpected adverse effects to human or animal health or to the environment. Member States are instructed to appropriately penalise non-compliance (Article 11) and also to establish appropriate inspection mechanisms for checking compliance (Article 9). The European Commission is to develop guidance on sampling and detection for verification of compliance, and also 'systems for the development and assignment of unique identifiers for GMOs'.

In 2000 the Commission signed the Cartagena Protocol on Biosafety (Convention on Biological Diversity, 2000), ruling the international trade of GMOs. This Protocol was created to try to minimise the potential risks to genetic diversity posed by living modified organisms (LMOs) resulting from modern biotechnology. It includes guidelines for the "safe transfer, handling and use of living modified organisms" (Article 1). It also established a procedure of

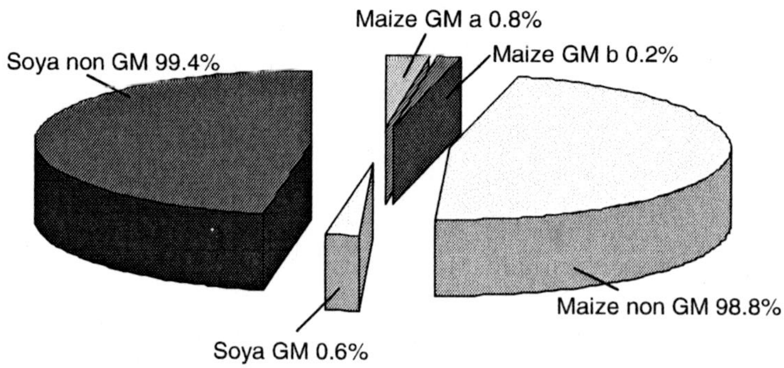

FIGURE 1. Example of foodstuff containing maize and soy: labelling is required only for GM maize.

'Advanced Informed Agreement' (AIA) for any transboundary movement of LMOs. For this exporting parties must give written notice and gain consent from the importing party prior to the first movement. After 50 ratifications, the Protocol entered into force on September 2003. As of March 2004 the Protocol has 92 states parties.

A central theme in the legislation cited consists of the availability of methods of analysis that are sound, precise and robust. This calls for research activity but also for harmonisation/standardization of procedures and performances within European control laboratories. The "Community Reference Laboratory (CRL)" (CRL, 2004) established in Regulation 1829/2003 (Article 32), is the Commission's Joint Research Centre (JRC). The JRC, in collaboration with a consortium of national reference laboratories, called ENGL (European Network of GMO Laboratories) (ENGL, 2003) has the core task of evaluation and validation of quantitative detection methods proposed by applicants as part of a Commission authorisation procedure.

3. Methods for Detection of GMO In Food

The need to identify the specific GMO and to quantify the amount of GMOs in different food products demanded by the current EU legislation, has generated the requirement of sensitive and reliable detection methods applicable to both raw materials and processed products. Two scientific approaches for detecting genetic modification can be distinguished: DNA-based methods and protein-based methods. Currently available methods are based on the detection of specific DNA fragments by polymerase chain reaction (PCR) or methods based on the detection of new expressed protein by enzyme-linked immunosorbent analysis (ELISA). Besides PCR and ELISA, alternative methods, some based on already available techniques (western blot, southern blot, mass spectrometry, chromatography, near infrared spectroscopy, surface plasmon resonance), and some novel ones (DNA microarrays) are being introduced. Here only PCR and ELISA methods will be dealt with as they are the most frequently used and have received much attention for validation.

3.1. ELISA (enzyme linked immunosorbent assay)

Immunological techniques, currently used in diagnostic clinic, are now routinely used for the rapid detection of protein in novel food. Methods focusing on separation of proteins, such as Western Blot, are difficult for routine testing and time-consuming. A variety of immunoassays formats have been developed to allow measurement of the primary binding reaction between antibody and its target antigen, the most common being sandwich ELISA, named in this way because the antigen (novel protein) binds both the immunosorbent (antibody bound to a solid support) and the labelled antibody. A sandwich ELISA assay with monoclonal antibodies against the protein CP4 EPSPS (5-enolpyruvylshikimate-3-phosphate synthase) in Roundup Ready (RR) soybean was tested and validated in a ring study coordinated by JRC of ISPRA, involving 38 laboratories (Lipp et al., 2000). The results were expressed in units of weight percentage for GMO in mixture with non-GMO of the same species. Assay data were given for a semi quantitative assay as a ratio between false positive and false negative samples, for a quantitative assay as an absolute error. Any sample containing RR soybean <2% was identified as below <2% with a confidence level of 99%, quantitative use of the assay measured on samples containing 2% RR soybean resulted in a repeatability and reproducibility of 7% and 10%, respectively, and detection limit of approximately 0.35% RR soybean. The method is applicable to samples where little or no treatment has been carried out during processing and thus the protein is not significantly degraded or denatured. The use of protein-based methods is limited to raw and partially processed materials. For a reliable quantification in complex matrices it is important to control the recovery of extraction and the matrix effect. Recovery of extraction can be quantified by spiking the samples under analysis with known quantities extracts of non-GMO containing materials at several levels across a quantitative range and measuring the recovery, thus evaluating interference, if any exists.

Furthers drawbacks linked to quantification are: expression below levels of quantification, expression only in specific part of the plant (tissue-specificity) or at different levels in correspondence of different plant parts or of different stages of plant development, limitation of quantification to only one-taxon. Nevertheless, advantages of ELISA methods consist in high degree of automation and high throughput of samples. Moreover, field variants, such as lateral flow strips kits, suitable for semi-quantitative test, are also available.

3.2. PCR-based methods

High sensitivity and specificity of PCR-based methods make them, at present, the most common tool in GMO detection (Holst-Jensen et al., 2003). A prerequisite for GMO detection by PCR is at least the partial knowledge of the target gene sequence in order to design the two primers that will function as the trigger for the polymerisation reaction leading to the amplification of a DNA fragment between the primers. PCR primers can be directed to different sequences of the introduced DNA. Generally, the gene construct is composed of several elements: 1) gene/s of interest, 2) regulatory sequences, most used are the 35S promoter (P-35S) of Cauliflower Mosaic Virus (CaMV) and the *Agrobacterium tumefaciens* nopaline synthase terminator (T-Nos), 3) marker genes, e.g. genes conferring resistance to antibiotics like neomycine-3'-phosphotransferase (*nptII*), conferring resistance to kanamycin (marker will be avoided by legislation from 31 December 2004 according to Directive 2001/18/EC) 4) DNA from the cloning vector. At least four categories of GMO tests, corresponding to different level of specificity, can be recognized (Holst-Jensen et al., 2003): screening, gene-specific, construct-specific and event-specific methods. Screening methods target the most common sequences present in GMOs currently on the market, e.g. P-35S, T-Nos, *nptII*, gene-specific methods target the gene of interest, construct-specific methods target junctions between adjacent elements of the construct, for example a region spanning the promoter and the gene of interest. Event-specific methods are "line specific" (transformation event specific) because the target is the junction between the host genome and the inserted DNA.

In conventional PCR, after amplification, PCR products are usually subjected to agarose gel electrophoresis to separate fragments according to size and then visualized by staining with an intercalating dye, e.g. ethidium bromide. Different methods can then be used to confirm the PCR results: restriction analysis, sequencing, hybridisation, nested PCR.

Compliance with the 0.9% mandatory labelling, makes GMO quantification a crucial step in GMO analysis. This can not be addressed with qualitative PCR because the amount of PCR product generated by amplification is poor indicative of the initial target concentration due to the "plateau phase" reached by conventional end-point measurement. At present, to overcome this problem, two PCR techniques are being used: quantitative-competitive PCR (QC-PCR) and, most commonly, real time PCR (RT-PCR).

QC-PCR is based on the co-amplification of the target DNA with an internal standard which, has the same sites of primer binding respect to the target, but is distinguishable from the target by size difference. By running a series of experiments in which each sample is co-amplified with varying amounts of the competitor, it is possible to determine the initial amount of template DNA. As long as amplification efficiencies of the target and the competitor are the same, if the resulting PCR products are of equal intensity, the initial concentration of target and internal standard are equal (equivalence point).

This technique has the advantage to be applied in laboratories that already use qualitative PCR without the need of specialised equipment, however drawbacks are: difficulty to obtain standardization, time-consuming, significant risk of carry-over contamination due to extensive handling and pipetting.

In the RT-PCR system, the increment of PCR product is measured by monitoring the increase of fluorescent signal emitted by specific fluorescent probes present in the amplification reaction. Fluorescent signals emitted from irradiated samples are detected by a CCD (charged coupled devise) camera and converted into quantitative estimates by software.

In contrast to the end-point measurements, RT-PCR allows for the monitoring of the amplification reaction as it actually occurs (in real time), and analysing data when the reaction is in its exponential phase (log phase). This phase is used because the amplification efficiency remains constant, and there is a direct correlation between the amount of PCR product generated and the initial amount of target template. It is therefore possible to quantify the DNA content of a sample by simple interpolation of its threshold cycle (Ct), defined as the number of cycles necessary to generate a signal statistically significant above the noise level, with a log standard curve plotting the initial amount of target molecules versus the Ct value. Several different fluorescent systems are available: intercalating dyes (e.g. SYBR Green) and hybridisation probes: hydrolysis probes

(Taqman), Fluorescent Resonance Energy Transfer (FRET), molecular beacon and scorpions probes.

Advantages of this methodology are the following: the applicability also in high processed foodstuffs due to amplification of very short DNA fragments, high specificity thank to the use of target specific probes, reduced time for analysis due to the elimination of gel electrophoresis step, possibility of automation. On the other hand, the major drawback is the high cost of instrumentation and reagents.

4. Method of Analysis

The application of GMO analytical methods for official control according to the present European regulation requires the use of validated procedures. Validation is the process of proving that an analytical method is acceptable for its intended purpose. In general, methods for must include studies on: applicability, practicability, specificity, robustness, accuracy, sensitivity, repeatability, detection limit (LOQ) and quantification limit (LOQ). Standardization within Europe is being undertaken by bodies such as European Committee for Standardization (CEN), which deals with the standardization of methods, and the International Organisation for Standardization (ISO), dealing with harmonisation of standards. Currently, a database of validated methods compiled by JRC is available (JRC database, 2003).

The operational procedure for the analysis of food potentially containing GMO consists of different steps: 1) sampling and homogenisation, 2) extraction of the analyte, 3) screening of GMOs, 4) identification of GMOs (number and if are authorized), 5) GMOs quantification (Figure 2).

5. Problems Related to Transgene Detection of Bt-maize in Food Products

5.1. Sampling

Obtaining a representative sample is critical for meaningful results. Thus, sampling procedure and sample preparation are crucial steps in the process of detection and quantification of GMOs. In general, sampling consists of: creation of the bulk sample, reduction (sub-sampling) of the bulk sample to produce the laboratory sample and, then, the test sample, from which it is produced the sample for the analysis. In a sampling strategy both the sampling plan and the sampling size are important. A sampling plan must take into account the distribution of sample in the lot (homogeneity/heterogeneity), the type of material (raw materials, primary ingredients, and final food products), and the number of samples to be taken (increments). On the other hand, the sample size must take in consideration the limit of detection of the selected testing method (essentially kernels or DNA copies) and the accepted risk (tolerance) to allow reliable detection at the desiderate acceptance.

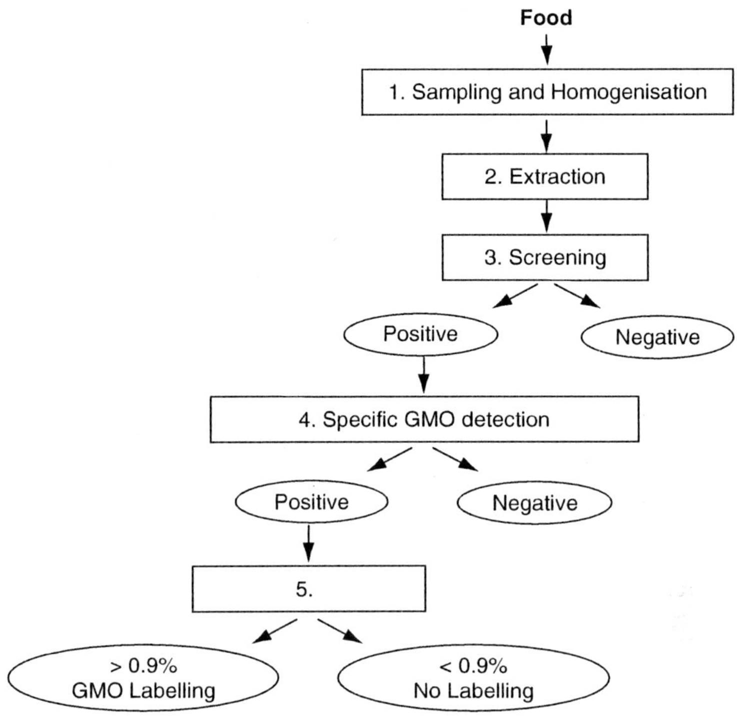

FIGURE 2. Scheme of the different steps for the analysis of potentially containing GMO.

Directive 98/53 on sampling and analysis of certain contaminants in foodstuffs (European Commission, 1998) was the first sampling plan applied to GMOs and is still used. Also in USA, as well as in EU, the demand for sampling guidelines has been acknowledged. GIPSA (Grain Inspection, Packers and Stockyards Administration) of USDA (US Department of Agriculture) developed plans used to specifically test GM seeds or grains. For example, for a single-step sampling procedure and qualitative testing the number of kernels that must be analysed can be determined with the following formula (binomial distribution):

$$n = \log [1-(G/100)]/\log[1-(P/100)]$$

where n is the sample size, G is the probability of rejecting a lot concentration and P is percent concentration in the lot. For example, in order to reject a lot for the presence of 1% GMO at 95% probability a laboratory sample of 299 kernels would be prepared (USDA/GIPSA, 2000).

Actually, although several sampling strategies are available and currently adopted, only some of them are specifically addressed to GMO and lack standardised sampling plans to be adopted in compliance with EU legislative requirements. In this context, ENGL, within KELDA (Kernel Lot Distribution

Assessment) (KELDA, 2003) project, is aimed to evaluate and implement sampling strategies. A new software tool is currently under development to estimate the sampling error associated to different sampling protocols as function of both number and size of samples according to a distribution-free statistical. In fact, a likely situation with respect to the adventitious presence of GM material in bulk commodities is a non-random distribution, for which standard statistical models cannot be applied.

5.2. DNA stability during food processing

Demand of traceability at all the stages of the placing on the market of GM products implemented by recent regulation will require, besides the adoption and documentation of appropriate systems for the differentiation of GMO and non-GMO (segregation and identity preservation systems), the necessity for food testing at different levels of the food chain.

In this respect, DNA stability during food processing is important, because is a prerequisite for successful GMO identification and quantification. Different factors affects DNA degradation during food processing and food storage, they are: physical forces, such as temperature, pressure, pH, shear forces, chemical agents, and enzymatic activities (nucleases). The action of one or more of these factors can led to DNA degradation (fragmentation) and damage (hydrolysis, depurination). On the other hand, it was observed that certain food matrices exert a protective effect against DNA degradation. This was for example demonstrated in complex food, like summer sausages, and in ingredients, such as arginine, polyamines, and biogenic amines (positively charged molecules) (van den Eede, 2004).

DNA stability in food was studied analysing the DNA content by agarose gel electrophoresis and monitoring DNA degradation taking advantage of PCR technology (minimum length for amplification is 70-100 bp). In this context several studies were conducted. For example, Straub et al. (1999) followed the detection of RR soybean in the wheat bread production reporting the successful detection of 35S promoter even in the final food product.

5.3. Extraction and purification of DNA

A critical prerequisite for PCR feasibility is the extracted DNA, both in terms of quality and quantity. Efficiency of PCR clearly depends on the availability of DNA of sufficient length and yield and on the presence of inhibitory substances (proteins, fats, polysaccharides, polyphenols, cocoa extracts, and caramelised sugar) which may negatively affect the result of the amplification. In this respect, proper suitable extraction methods must be employed in dependence of the different types of matrices analysed.

Several methods of DNA extraction have been developed and many commercial kits have been introduced into the market. Thus, the performance and validity of the different protocols needs to be addressed for the different matrices.

Currently, methods using nucleic acid binding resins and the hexadecyltrimethylammoniumbromide (CTAB)-based methods are preferred for GMO detection. A comparative study, conducted by Zimmerman et al. (1998), reported the validity of such methods, giving rise to low yield, but high quality DNA, respect to simpler and faster methods (e.g. chelex and rose-based methods) producing high yield, but low quality DNA.

In the reported example (Rizzi et al., 2003) five DNA extraction methods, namely the CTAB-method and four commercially available kits, were compared with respect to quality and yield of the DNA isolated from fix food products derived from maize. Here, extraction procedures were performed according to the manufacture's guidelines; a more accurate comparison should foresee, for example, the evaluation of DNA recovery and quality across different quantitative range and the optimisation of the protocol for each specific matrix.

The general characteristics and the principle of DNA purification on which the different extraction methods are based are summarized in Table.1. Different approaches were used to evaluate the DNA recovery: agarose gel electrophoresis, U.V. absorption at 260 nm and quantitative RT-PCR. Agarose gel electrophoresis gave indication of the degree of DNA fragmentation and permitted a rough estimate of DNA yield. Because of the limited amount of DNA that can be visualised in agarose electrophoresis after ethidium bromide staining, this technique gave no information in the case of highly processed foodstuffs (chocolate corn flakes and infant formula), containing low and degraded DNA. On the other hand, evaluation of DNA yield by U.V. absorption proved unreliable, as demonstrated by checking the A_{260}:A_{280} absorption ratio, due to the low DNA concentration or to the presence of impurities which affected the test. For example, samples of chocolate corn flakes yielded a very low A_{260}:A_{280} absorption ratio (<1.5) suggesting that a significant amount of contaminants was present. In order to gain further information on the quality and yield of "difficult" products, such as corn chip snacks, chocolate corn flakes and infant formula, a

TABLE 1. General characteristics and principle of DNA purification of the extraction methods (Rizzi et al, 2003)

Method	Principle	Characteristics
CTAB	Selective precipitation of nucleic acids	Widely used, cheap, official method of German Test method
Wizard method	Selective binding of nucleic acids to specific resin	Widely used, official method of Swiss Food Manual
Wizard Magnetic DNA Purification System	Selective capture of nucleic acids to a specific magnet	Tested protocols for raw and processed material
GenElute Plant Genomic kit	Selective binding of nucleic acids to specific resin	Protocols for different matrices
QIAamp DNA Stool Mini kit	Selective binding of nucleic acids to specific resin	Suitable for samples with high concentration of inhibitors

real time system amplifying very short DNA fragments of the zein gene was performed. The CTAB and WP methods produced similar amounts of DNA, while the other methods yielded much less DNA. With the chocolate corn flakes and infant formula very low yields of amplifiable DNA were obtained, suggesting that DNA was widely fragmented with low amounts of available targeting sequences.

Suitability of the extracted DNA for PCR amplification was tested, in addition to real time PCR, also by conventional PCR; in particular a PCR nested system targeting zein gene was used. All of the methods tested on flour samples gave the expected bands of 277 bp. The methods CTAB, WP and WM yielded amplified products in almost all reactions tested from canned maize, corn chip snacks and cheese corn puff snacks. With all the methods tested only few samples from chocolate corn flakes gave amplified fragments; in the case of infant formula some positive signals were obtained only with CTAB method. Control reactions carried out adding the PCR reaction of external maize DNA, confirmed that the failure in amplification was due to low quality/amount of DNA rather than to the presence of inhibitors.

Concluding, among the methods CTAB, WP and WM gave the best results with regard to quality and yield of DNA isolated from the various maize foodstuffs.

6. Case Study

An official laboratory for food quality control, on the basis of the recent European law analysed a sample of "ready to use polenta" and found that the amount of GMO present in this foodstuffs exceeded the limit of 0.9%.

The data have been contested by the manufacturing company and the judge has entrusted you to explain the found result.

From the laboratory report, it is known that:

- Five samples have been withdrawn at random by a supermarket by a lot of 1000 pouches of around 1Kg each.
- The five samples have been mixed and from this mixture samples of 50 g have been withdrawn to submit to analysis.
- The extraction and purification of the DNA has been conducted according to the Wizard method.
- GMOs have been analysed by the official method.

Could you give some scientifically sound reason, to the judge, that can explain the results of the laboratory and particularly if errors have been committed during the phase of collecting samples and of analysis.

The following papers can help you to answer the question:

Rizzi et al. 2003

Simon Kay and Claudia Paoletti -Sampling Strategy for GMO – Detection and/or quantification (JRC – Internet document)

TABLE 2. DNA yields determined by U.V. adsorption at 260 nm and by real time PCR (RT-PCR) after extraction of food products with the different extraction methods (Rizzi et al, 2003)

Food		CTAB	WP	WM	GE	QI
Flour	Yield by UV[a] (μg/100 mg)	19.6±2.7	9.3±0.7	1.8±0.2	0.8±0.1	4.3±0.7
	Amplifiability[c]	8/8	8/8	8/8	8/8	8/8
Canned maize	Yield by UV[a] (μg/100 mg)	8.0±0.3	3.5±0.6	0.5±0.07	0.6±0.2	0.6±0.1
	Amplifiability[c]	8/8	8/8	8/8	1/8	2/8
Cheese corn puff snacks	Yield by UV[a] (μg/100 mg)	0.2±0.01	1.3±0.2	0.4±0.02	0.8±0.2	0.3±0.08
	Amplifiability[c]	8/8	8/8	1/8	6/8	5/8
Corn chip snacks	Yield by UV[a] (μg/100 mg)	4.3±1.1	4.1±0.6	0.2±0.06	0.9±0.1	0.3±0.08
	Yield by RT-PCR[b] (μg/100 mg)	6.5±0.5	5.6±1.4	0.2±0.05	<0.01	0.2±0.1
	Amplifiability[c]	8/8	8/8	8/8	0/8	3/8
Chocolate corn flakes	Yield by UV[a] (μg/100 mg)	0.6±0.2	2.4±0.3	0.7±0.2	0.5±0.08	0.1±0.04
	Yield by RT-PCR[b] (μg/100 mg)	<0.01	n.d.	n.d.	<0.01	<0.01
	Amplifiability[c]	1/8	0/8	1/8	0/8	1/8
Infant formula	Yield by UV[a] (μg/100 mg)	0.4±0.06	1.0±0.1	0.1±0.07	0.7±0.1	0.1±0.03
	Yield by RT-PCR[b] (μg/100 mg)	0.067±0.01	<0.01	<0.01	<0.01	<0.01
	Amplifiability[c]	3/8	0/8	0/8	0/8	0/8

[a]Each value is the mean (± standard deviation) of eight samples
[b]Each value is the mean (± standard deviation) of four samples
[c]Number of positive samples out of eight samples each separately extracted
n.d.: not detected
N.D. not determined
The $A_{260/280}$ ratio for flour ranged 1.8-2, for canned maize and cheese corn puff snacks 1.6-1.9, for corn chip snacks and infant formula 1.6-1.8 and for chocolate corn flakes 1.3-1.5

7. References

Bindslev-Jensen, C., 1998, Allergy risk of genetically engineered foods, *Allergy*, **53**(45 supplement):58.

Convention on Biological Diversity, Cartagena Protocol on Biosafety to the Convention on Biological Diversity, Secretariat of the Convention on Biological Diversity, Montreal; http://www.biodiv.org/

CRL, Community Reference Laboratory, Joint Research Centre, Biotechnology and GMOs Unit, European Commission; http://gmo-crl.jrc.it/

Dale, P.J., Clarke, B., and Fontes, E.M.G., 2002, Potential for the environmental impact of transgenic crops, *Nature Biotech.* **20**:567.

Directive 90/220/EEC of the European Parliament and Council, Official Journal of the European Communities, L 117 of April 23, 1990.

EC Directive 2001/18/EC of the European Parliament and Council, Official Journal of the European Communities, L 106 of March 12, 2001.

EC Directive 98/53 of the European Parliament and Council, Official Journal of the European Communities, L 201 of July 16, 1998.

EC Regulation 1139/98 of the Council, Official Journal of the European Communities, L 159 of June 3, 1998.

EC Regulation 1829/2003 of the European Parliament and Council, Official Journal of the European Communities, L 268 of September 22, 2003.

EC Regulation 1830/2003 of the European Parliament and Council, Official Journal of the European Communities, L 268 of September 22, 2003.

EC Regulation 258/97 of the European Parliament and Council, Official Journal of the European Communities, L 043 of February 14, 1997.

EC Regulation 49/00 of the Council, Official Journal of the European Communities, L 006 of January 11, 2000.

EC Regulation 50/00 of the Council, Official Journal of the European Communities, L 006 of January 11, 2000.

EC-JRC, 2003, GMO Methods Database. Joint Research Centre, European Commission, Ispra, Italy; http://biotech.jrc.it/methodsdatabase.htm.

ENGL, European Network of GMO Laboratories, 2003, Institute for Health and Consumer Protection. Joint Research Centre, European Commission, Ispra, Italy; http://engl.jrc.it.

Holst-Jensen, A., Rønning, S.,B., Løvseth, A., and Berdal, K.G., 2003, PCR technology for screening and quantification of genetically modified organisms (GMOs). *Anal. Bioanal. Chem.* **375**:985.

James, C., 2003, Global review of commercialised transgenic crops. International Service for the Acquisition of Agri-Biotech Applications (ISAAA), Ithaca, NY.; http://isaaa.org/.

KELDA, Kernel Lot Distribution Assessment Project, 2003, Joint Research Centre, European Commission, Ispra, Italy; http://biotech.jrc.it/kelda/login.asp.

Lipp, M., Anklam, E., and Stave, J.M., 2000, Validation of an immunoassay for detection and quantitation of a genetically modified soybean in food and food fractions using reference materials: Interlaboratory Study. *J. AOAC Internat.* **83**:919.

Mayeno, A.N. and Gleich, G.J., 1994, Eosinophilia-myalgia syndrome and tryptophan production: a cautionary tale, *Tibtech*, **12**:346.

Rizzi, A., Panebianco, L., Giaccu, D., Sorlini C., and Daffonchio D., 2003, Stability and recovery of maize DNA during food processing, *Ital. J. Food Sci.* **15**:499.

Straub, J.A., Hertel, C., and Hammes W.P., 1999, Limits of a PCR-based detection method for genetically modified soy beans in wheat bread production. *Z. Lebensm. Unters. Forsch. A.* **208**:77.

USDA/GIPSA. Practical application of sampling for the detection of biotech grains, 2000; http://www.usda.gov/gipsa/biotech/sample1.htm.

Van den Eede, G., Aarts H.J., Buhk H.J., Corthier G., Flint H.J., Hammes W., Jacobsen B., Midtvedt T., van der Vossen J., von Wright A., Wackernagel W., and Wilcks A., 2004, The relevance of gene transfer to the safety of food and feed derived from genetically modified (GM) plants, *Food and Chemical Toxicol.* **42**:1127.

Zimmermann, A., Lüthy, J., and Pauli, U., 1998, Quantitative and qualitative evaluation of nine different extraction methods for nucleic acids on soy bean food samples. *Z. Lebensm. Unters. Forsch. A.* **207**:81.

SUBJECT INDEX

Page numbers followed by f and t indicate figures and tables, respectively.

Adhesion, 79, 137, 145, 255, 257
Aflatoxin, 27–30, 33–34, 36–39, 41, 44–45
Alcohol, 38–39, 75, 154, 156, 162, 169–173, 177, 184, 186, 265t
Allergenicity, 282
Amensalism, 178
Antagonism, 139, 178, 182
Antimicrobials, 157, 162f
Antioxidant, 5f, 13, 69, 156t, 166
Apple(s), 32, 34
  Elstar, 14
  juice, 44–47
Aspergillus carbonarius, 30–31, 33
Aspergillus flavus, 27, 29, 31, 37, 39, 45
Aspergillus ochraceus, 31, 33
Aspergillus parasiticus, 29, 45
Astroviruses, 202, 203t, 208–209

Bacillus thuringiensis, 114, 126–127, 281
Bacteria, 4–5, 6f, 8f–9f, 10–15, 17, 19–24, 46t, 50, 54, 60–61, 80, 82–83, 97, 106–107, 112–113, 120, 126, 128, 133–139, 142, 144–148, 155–161, 163, 165–166, 171–174, 177–187, 190–191, 192t, 193–195, 196t, 203, 209, 212–213, 215, 233–234, 235t, 240, 253, 257, 260–262, 265–270, 274, 282
Bacteriocins, 22–23, 156t, 157, 182, 187
Bifidobacterium, 136t, 137–138, 142, 146, 185, 200t
Bioactive ingredients, 134
Biocontrol agents (BCA), 30, 34
Biosafety Clearing House, 118, 124
Bioweapons, 47
Biscuits, 42t, 148
Breakfast cereals, 42t, 148
Breeding, 39, 113, 115–117, 130, 281

Cancer, 26–28, 34, 41, 140, 141f, 147, 157, 178, 196t, 200t
Carbohydrates, 136, 144–145, 156, 169t, 170, 181, 186
  non-digestible, 141–142, 146, 157
Carcinogenic mycotoxins, 28, 30–31, 140
Cardiovascular disease, 147
Cartagena Protocol on Biosafety, 112, 124, 284
Cationic surfactants, 261, 268–270
  amphoteric compounds, 269–270
  detergent and disinfectant activity, 270
  disinfectant activity, 269
  uses, 269
Cell fusion, 112
Cereal bars, 148
Cheese manufacturing, 91, 97, 99, 103, 108
Chlorine dioxide, 264–265
  chlorite, 264–265
  preparation, 264
Chocolate, 148, 196t, 291–292, 293t
Cholesterol, 140, 147, 156t
Cleaning and sanitizing procedures, 254
  frequency, 254
Cleaning disinfection application, 51f, 60
  water soft drinks bottling, 268, 273, 274t, 275
Cleaning, 253–279
Coconut, 44–45, 82

Colitis, 17, 100, 134, 200t, 237t
Colon, 134, 136–137, 142, 144–147, 197t
Colonisation, 134, 139, 183, 200t
Colo-rectal cancer, 140
Commensalism, 178, 180–183
Competitive exclusion, 139
Confectionary, 148
Constipation, 140, 144, 146, 200t, 236t
Contamination, 92, 99, 101–105
Convention on Biological Diversity, 112, 118, 125, 284

Dairy products, 12, 30, 32t, 99–100, 102, 161, 163–164, 174, 183–185, 199t, 237t
Deaths, 41, 179
Deliberate release of GMO into environment, 124
Deoxynivalenol (DON), 28, 31, 33–34, 37, 41
Detergent, 258–261
  acids, 258–259
  alkalis, 258, 260
  surfactants, 259
    amphoteric, 260
    anionic, 259
    non ionic, 259–260
Diabetics, 146
Diarrhoea, 114, 140, 141f, 179, 191t, 194–195, 196t–198t, 200t, 205–206, 208, 211, 215–216, 236t
Dietary fibers, 141, 146
Dirt removal, 254–256
Dirt, 59, 61, 97, 253–260, 270–272
Disease, 3–4, 21, 26–27, 30f, 32, 60, 81, 91, 100, 113–114, 117, 128, 130, 133–135, 140, 145, 147–148, 154–155, 169, 173, 179, 188–189, 190t, 191t, 193–195, 196t–198t, 203–208, 210, 212, 214–215, 233–235, 236t–237t, 239–240, 242–243, 247, 253, 257
Disinfectant activity, 263
Disinfection methods, 270–279
  CIP, 270–271, 273, 274f, 275, 276f–277f, 278t–279t
  cleaning disinfection protocols, 253–279
  COP, 270
  manual, 271, 273, 275, 278t
Disinfection, 14, 50, 51f, 60–61, 235t, 253–279
DNA, 113, 115–116, 119–120, 159, 160t, 166, 168–169, 194, 266, 281, 285–288, 293t
  extraction and purification, 290–292
  stability, 290

Enterotoxic activity, 137f
Ethical issues of biosafety, 124
EU legislation, 245, 282–285
European Corn Borer (*Ostrinia nubilalis;* ECB), 114, 126–128

Familiarity concept, 117–118
Fatty acids, 82, 114, 142, 148, 166, 173
Fermentation, 5, 12–13, 22–24, 38, 80, 82, 101t, 113, 115, 136, 138, 145, 147, 153–159, 161–163, 165, 167, 169–174, 177, 183–184, 186, 195, 199t
Fermented foods, 11, 138, 153, 155, 156t, 157–158, 161, 179, 187, 195
Fermented milks, 148, 153, 157
Feta, 95–99, 102, 104
Filamentous fungi, 26–27
Flow diagram, 45, 95–96, 98–99
Food ecosystems, 179–180, 182
Food legislation, 225
Food packaging
  active packaging, 69, 85
  aseptic packaging, 70, 267–268
  biobased materials, 75–80
  biobased packaging materials, 74
  chemical contamination, 84
  edible coatings, 69, 81
  microbial contamination, 16, 83–84, 209, 212
  modified atmosphere packaging, 69, 81, 83, 86
  packaging materials, 55, 69–70, 73–78, 80–81, 84–86, 95
  packaging safety considerations, 83–85

packaging systems, 70–72, 85
physical contamination, 27, 46, 50, 51, 52, 68, 77, 82
vacuum packaging, 14, 69, 72–73, 85
Food preservation
methods
thermobacteriology, 4, 5
kinetics, 5–6
microorganisms
dead-time of, 5–9
safety, 23–24
Food safety, 50–66, 83, 86, 94, 103–104, 112–131, 133–149, 162f, 187, 195, 202, 212, 225–227, 228f, 239–251, 283
assessment, 120, 130
training, 243–244
Foodborne and waterborne viral infections, 202
Foodborne illnesses, 99, 101t, 179, 194, 233–237
Foodborne pathogens, 23, 191, 193–195, 196t
Foodborne viruses, 202–217
astroviruses, 202, 203t, 208–209
Hepatitis A, 50, 194–195, 197t, 203, 205–207, 211–212, 234t, 237t
Hepatitis E, 203t, 206–207
noroviruses, 193–195, 202–205, 207–208, 211, 214, 217
rotaviruses, 202, 203t, 208, 212
sapoviruses, 202, 203t, 207–208
Fruit, 32, 43
Fumonisin, 31–34
Functional foods, 134–135, 145–149, 281
Functional properties of cultures, 162f
Functional starter, 161, 195, 198, 199t
Fungi, 12–13, 21, 26–34, 36–37, 39, 44, 45t, 83, 113, 117, 138–139, 178–179, 182–183, 186–187, 235, 269
Fusarium cerealis, 31
Fusarium culmorum, 31
Fusarium equiseti, 31
Fusarium graminearum, 31
Fusarium poae, 31
Fusarium sporotrichioides, 27, 31
Fusarium verticillioides, 27, 31–34
Gastroenteritis, 198t, 202–209, 211, 213, 215–217
Gastrointestinal infection, 135, 140, 146, 190t
Gastrointestinal physiology and function, 144, 146
Gene insertion stability, 119
Gene transfer, 118–119, 139, 282
Genetic engineering, 113–115, 124, 130, 159–161, 195, 225, 282
Genetic modification techniques, 112–113, 116, 118–119, 127, 159, 285
Genetically modified animals, 112–113, 118–121, 123–131
Genetically modified food, 117–120, 125–126, 128–129
Genetically modified microorganisms, 121
Genetically modified organism (GMO), *see* GMO (genetically modified organisms)
Genetically modified plants, 118
Genetically modified products, 113, 125, 131
Glufosinate (Basta®), 114
Glycaemia, 146
Glyphosate (Roundup®), 114, 128
GMO (genetically modified organisms), 112–118, 120–129
allergenicity, 119, 282
benefits, 130–131
detection methods, 285–288
ELISA, enzyme linked immunosorbent assay, 285–286
PCR-based method, 285–288, 290–292, 293t
GMO interaction with environment, 124
GMO potential impact on human health, 113
GMO's generations, 113, 117, 129, 131
GMP, genetically modified plants, 118, *see also* GMO
Grains, 30, 33, 42, 178, 289
Gut barrier function, 137f, 140, 146

HACCP implementation, 240–242, 245–246
strategy, 249–251
HACCP management 91, 94–95, 99, 108–110, 226–232, 239–249
HACCP plan, 45, 47, 95, 99, 108–110, 230, 235, 241, 242
HACCP, 27, 35, 43–47, 51–52, 86, 91, 94–95, 98, 99, 239–250
Halogens activity, 261–266
Hazard identification, 50, 122
Health benefits, 43, 138, 140–141, 146–147
Health claims, 125, 140, 145–146
Health, 26–27, 33, 40–41, 43, 50, 53, 74, 85–86, 91–92, 94, 100, 101t, 103, 108, 113, 115, 117, 121–122, 124–125, 131, 134–142, 145–148, 155–157, 161, 162f, 174, 185, 194–195, 198, 200t, 202–217, 225, 227, 230, 233, 240, 242–243, 253, 282–284
Hepatic encephalopathy, 144
Hepatitis A, 50, 194–195, 197t, 203, 205–207, 211–212, 234t, 237t
Hepatitis E, 203t, 206–207
High added value products, 155, 173–174
Human enteric viruses, 202, 203t, 209, 212
Hybridisation techniques, 112
Hydrogen peroxide, 266–267
industrial application, 267
mode of action, 266
preparation, 266
Hypochlorous acid, 261–264
by products, 262–263
chloroamines, 264t
mode of action, 261

Immune system, 26, 146, 179, 183, 194, 196t–197t, 266
Immunity, 137f, 183, 188, 190t, 206
Infant formulae, 41t, 148
Intestinal function, 138, 146, 156
Intestinal microflora, 133–149, 183
Inulin, 143t, 144, 146–148, 169t
Iodine, 265–266
commercial solutions, 265t
mode of action, 265
uses, 265, 266t
ISO 9000, 91

Lactic acid bacteria (LAB), 22–23, 97, 135, 138–139, 156t, 156–160, 160t, 163, 166, 172, 182–183, 186–187, 195
Lactobacillus, 14, 23, 97, 136t, 137–139, 142, 156, 174t, 184, 186, 199t, 200t
Lactose intolerance, 140, 141f, 200t
Legal framework, 145–146
Legislative standards for bivalve molluscan shellfish, 214t
Living modified organism (LMO), 112, 284–285

Macro-injection, 112
Microbial interactions, 180–195
Microbial relationships, 179–180
Microbiological analysis, 104, 107, 108
Microbiological criteria, 103, 106, 243
Microecology, 137, 182–189
Micro-encapsulation, 112
Micro-injection, 112
Microorganisms, 4–24, 39, 46t, 69–70, 80, 82–83, 84t, 89–200, 234, 237t, 253–254, 257, 261, 265–266, 273
pathogenic, 4, 83, 97, 102, 113, 134, 153, 178, 183, 188–195, 237t, 257, 261, 266
Milk, 13, 16, 18–19, 22, 30, 32–33, 38–39, 41, 52, 70, 92–93, 95–108, 138, 142, 145, 148, 153, 156t, 158t, 159, 163, 166, 167t, 168f, 182–185, 196t–197t, 237t, 259, 267–268, 270
Moulds, 4, 6, 10–12, 15, 19, 21, 98, 103, 153, 157–158, 160, 163, 172, 177, 181, 185, 267
Mutualism, 178, 180–181
Mycotoxin, 26–28, 30, 32, 34–36, 37, 38, 40, 42–44, 47, 94, 102, 178, 187
nivalenol (NIV), 28, 31, 34
ochratoxin A (OTA), 30–31

New biotechnology, 115–117
Non bacteriological gastroenteritis outbreak, 216t
Non-thermal food processing technology, 18
  combined methods, 23–24
  nutrition, 4
  conventional, 4–17
Noroviruses, 193–195, 202–205, 207–208, 211, 214, 217
Novel foods, 23, 120, 126, 283–285
Nutrients, 4, 100, 115, 116t, 119, 129, 137f, 141–147, 155, 157, 165, 177, 181–183, 187

Oestrogenic properties, 32
Oligofructose, 143t, 144, 147–148
Oligosaccharides, 141–142, 143t, 144–146, 148
Osteoporosis, 147, 156t

Parasitism, 178, 180, 181t, 183
Pasteurization, 92, 97, 100, 102–103, 106–108, 155
Pathogenic microorganisms, 4, 83, 97, 102, 113, 134, 153, 178, 183, 188–195, 237t, 257, 261, 266
Pathogenicity, 161, 188–189
Pathogens, 5, 15, 17, 23, 34–35, 86, 99, 100, 102–103, 113, 121, 134, 137, 146, 156–157, 188–190, 191, 193–196, 198, 202, 209, 212, 233, 235, 266
Patulin, 28, 31, 34–35, 39, 41, 45–48
PCR (polymerase chain reaction), 285–288, 290–292, 293t
  qualitative PCR, 287
  quantitative PCR, 287, 291
  real-time PCR, 212–213, 287, 291
Peanuts, 32, 36, 37
Penicillium expansum, 31, 34, 46–47, 182
Penicillium verrucosum, 31, 33
Peracetic acid, 267–268
  commercial solutions, 267–268
  sporicidal action, 268
  uses, 268
  virucidal action, 268
pH
  acid, 61, 62, 68, 69
  alkaline, 194, 255, 269
  low, 10–11, 13, 186
  neutral, 261, 268
Physicochemical properties, 145
Phytochemicals, 148
Plasmid, 112, 159–160
Polysaccharides, 141–142, 144, 173–174, 193, 290–291
Post-market surveillance, 126
Prebiotics, 133–149
Prerequisite programs, 91, 94, 102, 109
Pre-washing, 256–257, 270
Probiotic bacteria, 142, 146, 148, 158t, 185
Probiotic products plant, 275–279
Probiotic strain, 138–140, 166, 185, 200t
Probiotics, 133–149, 198
Process safety
  acceptance limits, 63–64, 131, 148, 288
  air, 54–56, 59–60
  barrier, 55–56, 72, 74–75, 78–79
  building element, 56
  certified equipment, 54, 61
  cleaning validation, 62–66
  door, 56–57, 230
  drainage, 56, 60, 66
  electrical cabling, 59
  elevators, 56
  framework, 56, 60, 66, 93
  hollow bodies, 59, 60, 62, 66
  hygienic design, 51–54, 60–61, 66
  insulation, 57, 59
  Legionella, 50, 53, 60
  light source, 58
  monitoring, 62–64, 66
  pipe, 59–60, 92, 107, 230, 278
  preventive measures, 51–52, 61, 91–92, 233
  product contact surface, 61, 64
  risk analysis, 63–64, 86, 120, 122, 124, 179
  sampling, 64–65
  sensors, 59, 62, 86
  utility installation, 54, 58, 66
  water, 60
  window, 55–57, 230
  zoning, 54–55, 66

Protoplast fusion, 112,160, 173
Public acceptance of new products, 115, 130

QC-PCR(quantitative-competitive PCR), 287, 291
Quality
  audit, 47, 86t, 95, 109, 229, 231
  contamination, 165, 171, 177, 185, 206, 208–213, 215, 216t, 230, 232t, 233–235, 251, 268, 270–271, 287
  foodborne, 3, 15, 23, 91, 99, 101t, 102, 133, 179, 191, 193–195, 196t, 202–217, 233–237, 242, 247, 251
  GHP-good hygiene practices, 214
  GMP-good manufacturing practices, 36–37, 45t–46t, 51–52, 101t, 103, 133, 228f, 229, 231, 246
  HACCP, 225–237
  hazard, 27, 46t, 50–51, 61, 86t, 91, 94, 99–104, 108, 110, 117–123, 129, 133, 209, 213–214, 226–227, 230, 234–235, 239–241, 246, 248f, 254, 272
  hygiene, 51–55, 58f–59f, 62, 93, 101t, 102–103, 107, 123, 155, 168, 177, 197, 207, 209–211, 213, 225–226, 228–233, 235, 240–243, 251, 254, 258, 260–261, 271–273
  illnesses, 99, 101t, 133, 177–179, 194, 202–203, 233–237
  infections, 135, 137, 140, 183, 190t, 193–195, 202–203, 205, 207–208, 211–212, 216, 225, 233–234, 235t–237t
  intoxication, 234
  ISO 22000, 227–229
  ISO 9001:2000, 227–229
  management, 91, 228–229, 232
  prerequisites, 92–93, 228–229, 231–232
  risk, 120, 122–123, 284
  safety, 18, 24, 73, 133, 180f, 182, 187, 195, 227–228
  SQA-supplier quality assurance, 229
  support network, 228, 229f
  TQM-total quality management, 91
  traceability, 232, 243, 283–284, 290
  training, 30, 55, 65, 91, 229f, 230, 240–246, 248–250

Recombinant DNA technology, 115, 159, 169
Rinsing, 60, 65, 256–258, 260, 269–271, 274–275, 278
Risk analysis, 64, 86, 120, 122, 124, 179
Risk communication, 120, 124
Risk evaluation, 284
Risk management, 120, 122–124, 284
Risks, 52, 54, 60, 64–65, 88–110, 117, 120–121, 123, 125, 130, 226–227, 231–232, 234–235, 240, 282, 284
Rotaviruses, 202, 203t, 208, 212
RTE-food, 210–211
RT-PCR(real-time PCR), 212–213, 287, 291
  screening methods, 159, 160t, 163, 173, 286, 288, 289f
  specific methods, 286–287
RT-PCR, 212–213, 216, 287

Safety assurance, 94, 99, 104, 110, 133, 226
Safety criteria, 228
Sampling, 35–36, 63t, 64–65, 106–109, 284, 288–290, 292
Sanitation procedures, 92, 93
Sanitization, 229f, 232t, 253, 260, 270–271, 273, 274t, 275–276, 279
sapoviruses, 202, 203t, 207–208
Selection, 64, 86, 113, 139, 159, 183, 162–163, 168, 174, 213, 281
Self-life, 14, 16, 147
Sensory properties, 147
Serving size, 140
Shellfish and fishery products, 213–214
Short chain fatty acids (SCFA), 136, 141–142, 146
Side effects, 139, 147
Soft drinks, 84t, 148, 269t, 270, 273
Solubility, 145, 147, 255–256, 260, 265, 269

contact time, 255, 263, 264t, 265, 268–270, 272
detergent concentration, 256
mechanical strength, 256
temperature, 256
Sourdough, 163, 186
Starter cultures, 96f, 97, 153–155, 157–159, 161–174, 182, 185, 195, 199t
for wine, 13, 23, 32t, 33, 38, 42t, 138, 153–154, 156t, 158t, 160, 162, 169t, 170, 177, 186–187
Sterilization, 5f, 84t, 167, 253, 267, 275, 279
Storage, 3, 13–14, 16–17, 19, 21–22, 24, 35–36, 45t–46t, 47, 51f, 60, 68, 70, 73, 75–76, 86, 91–95, 96f, 98, 101t, 102, 106–107, 109, 144, 147, 154, 161–162, 164–165, 178, 196t, 198, 213, 231, 264t, 290
Substantial equivalence, 117–119, 125, 128–129
Sugar, 10, 17, 24, 141, 145, 148, 168, 169t, 171, 186, 195, 199t, 255
Synbiotics, 148
Systemic infections, 139–140

T-2 toxin, 31–32, 34
Temperature in cheese manufacturing
constant, 5–6,
high, 24, 38,45, 92, 102, 162, 171, 255–256, 273
low, 15–17, 24, 46, 76, 100, 102, 162, 166
Thermal food processing, 20
Thermal treatment, 54, 92, 97, 100, 102, 106–108, 234
Toxins, 26–47, 94, 99, 101t, 102, 105t, 114, 116t, 126–127, 133, 178–179, 182, 187, 189–191, 192t, 193–194, 202, 234–235, 236t
Traceability, 232, 243, 283–284, 290
Transformation for disease resistance, 114
Transformation for herbicide tolerance, 114
Transformation for insect resistance, 114
Transformation with plant quality genes, 114
Trans-galactooligosaccharides, 143t, 144, 148
Transgene, 114, 282, 288–292
Transgenic herbicide-resistant crops, 114
Tricothecenes, 31

Vector, 112, 121, 123, 160, 286
Vegetable, 34, 70, 72, 74–75
Viral foodborne contamination, 209–211
Virulence factors, 188, 191
Vitamins, 136, 140, 148, 156t, 162f, 173, 181–184

Water conditioners, 260–261
Water, 5f, 9–11, 14, 17, 19–20, 22, 24, 36–39, 52–66, 74, 75t, 76–83, 91–96, 107, 136–137, 154, 166–167, 194, 197t, 203, 206–216, 230, 232t, 236t–237t, 254–271, 273, 274t–275t, 276f, 278, 279t
Weaning foods, 148
Wheat, 33–35, 41–44, 80
Wine, 13, 23, 32t, 33, 38, 42t, 138, 153–154, 156t, 158t, 160, 162, 169t, 170, 177, 186–187

Yeasts, 4, 6, 10–12, 15, 19, 21, 98, 103, 138–139, 153, 157–158, 160, 163, 171–173, 177, 179, 181–185, 195, 267, 269t
Yoghurt, 16, 20, 138, 148, 153, 156t, 184, 185, 198, 199t

Zearalenone (ZEA), 28, 31–33

AUTHOR INDEX

A
Aarts, H.J., 294
Abd El-Salam, M.H., 97, 110
Achour, L., 149
Adachi, N., 220
Adair, F., 269, 279
Adams, C.E., 242, 246, 250–251
Adams, M.R., 139, 149
Aggett, P.J., 149
Ahvenainen, R., 87
Aieta, E.M., 264, 279
Akihara, M., 217
Alichanidis, E., 110
Alsterlund, R., 220
Alter, M.J., 219
Alzamora, S.M., 25
Ananta, E., 166, 174
Ando, T., 218
Andrew Ebert, G., 87
Anifantakis, E.M., 103, 110
Anklam, E., 294
Appleton, H., 203, 205, 208, 214, 215, 221
Arai, S., 147–149
Arcot, J., 149
Arndt, G., 87
Arstall, S., 217
Ashwell, M., 149
Astwood, J., 131
Atmar, R.L., 205, 208, 217–219
Attali, M., 220
Attar, A., 149
Auriol, D., 174

B
Baati, L., 166, 174
Baird-Parker, T.C., 111
Baker, P.S., 48
Baker, R.J., 262, 279
Balayan, M.S., 218
Balazs, K., 250–251
Baldry, M.G.C., 268, 280
Balny, C., 25
Banks, J.N., 37, 44, 49
Bannon, G., 117, 131
Barardi, C.R.M., 217
Barataud, D., 218
Barbosa-Canovas, G.V., 18–19, 21–22, 25
Barbuti, G., 220
Baric, R.S., 221
Barker, G.C., 201
Bártová, J., 201
Barug, D., 49
Beauty, E.R., 150
Beergabel, M., 220
Beever, D.E., 120, 131
Begley, T.H., 85, 87
Bell, B.P., 205, 219
Bell, C., 5, 25
Belton, P.S., 226, 237
Benanti, F., 217
Benno, Y., 135, 149
Berdal, K.G., 294
Beresford, T.P., 163, 175, 185, 200
Berg, R.D., 134–137, 149
Bergman, N., 220
Bergmire-Sweat, D.A., 218
Berke, T., 218
Bery, D.J., 279
Beynena, A.C., 201
Bhalgat, M.K., 131
Bhatnagar, D., 37–38, 49
Bidawid, S., 205, 210, 217
Biggeri, A., 149
Bindslev-Jensen, C., 282, 293
Blanc, P.J., 174
Blanchfield, J.R., 243, 251
Block, S.S., 280

Blumenthal, M.M., 246, 252
Bones, A.M., 132
Booth, I.R., 10, 25
Borda, D., 225, 238
Bornet, F., 149–150
Borovská, D., 201
Bosschaert, M.A.R., 155, 175
Bouhnik, Y., 146, 149
Bouley, C., 151
Boutron-Ruault, M.C., 151
Boxman, I.L.A., 215–217
Brancati, G., 217
Branen, A.L, 22, 25
Brennan, N.L., 175, 200
Bresee, J.S., 220
Briggs, D.E., 175
Brock, T.D., 175
Brody, A.L., 87
Brown, B.E., 9, 25
Brown, I.L., 152
Brown, I., 144, 149
Brown, D., 220
Brunton, J., 221
Bryan, F.L., 210, 217
Buckenhuskes, H., 158, 175
Buhk, H.J., 294
Bureau, G., 87
Burkhardt, R.S., 221
Burton, B., 219

C
Cacopardo, B., 207, 217
Caderni, G., 149
Caldwell, G., 175
Campbell, D.S., 219
Capece, A., 175
Capuano, B., 219
Carlisle, D., 220
Carter, A.O., 221
Caruso, M., 175
Carvalho, A.S., 166, 175
Casadevall, A., 180–181, 188–190, 200
Caul, E.O., 211, 217, 220
Cebra, J.J., 137, 149
Cebula, T.A., 218
Cerdeira, A., 48
Chacon, D., 116, 221
Chadwick, P.R., 211, 217
Champagne, C.P., 175
Chan, T.Y., 206, 217
Chanes, J.W., 25
Channock, R.M., 219
Cheftel, J.C., 19, 25
Cheirsilp, B., 184, 200
Cherbut, C., 147, 149
Chiba, S., 208, 217, 220
Chironna, M., 220
Ciccozzi, M., 218
Clarke, B., 131, 293
Clarke, I.N., 218
Cliver, D.O., 206, 217
Clune, Y., 149
Coelho, C., 206, 217
Cogan, T.M., 175, 200
Coker, R., 35, 39, 44, 48
Coleman, P., 246–247, 251
Collins, K.J., 149
Collins, M.D., 138, 149
Connely, A.M., 220
Conner, A.J., 121–122, 131–132
Conway, P.L., 152
Corlett, D.A., 91, 110
Corthier, G., 294
Costin, G.M., 237
Coussement, P., 152
Cramer, E.H., 219
Croci, L.D., 206, 209, 218
Cronin, M., 219
Crosby, N.T., 85, 87
Crueger, A., 171, 174, 175
Crueger, W., 175
Cubitt, D.W., 219
Cubitt, W.D., 208, 217–218
Cukrowska, B., 201
Culioli, J., 19, 25
Cummings, J.H., 136–137, 145, 149–151
Cummings, J., 152
Cunningham-Rundles, S., 137, 149
Currier, M., 205, 218

D
D'Souza, H., 221
Daffonchio, D., 281, 294
Dale, P.J., 123, 131, 282, 293
Dally, H., 269, 280
Daniels, N.A., 215, 218
Darveau, A., 219
Dass, C.R., 159–160, 175

Davidson, P.M., 22, 25
Davis, J.G., 153, 175
De Leenheer, L., 152
de Medici, D., 218–219
De Vries, M.K., 250–251
de Vuyst, L., 186, 195, 198–201
de Wit, M.A.S., 208, 218
Deakin, D., 220
DeLeon, R., 220
Delzenne, N., 152
Demain, A.L., 161, 175
Derieux, J., 154, 175
Desai, B.B., 4, 25
Desenclos, J.-C.A, 206, 218
Di Pasquale, S., 218
Dietz, V., 220
Dillon, M., 227, 237
Diplock, A.T., 148–149
Dodds, K.L., 87
Doerfler, W., 120, 131–132
Dolara, P., 149
Dow, L., 220
Downes, T.W., 83, 87
Doyle, A., 205, 218
Doyle, J.J., 115, 123, 131
Doyle, M.P., 201
Drastich, P., 201
Dubois, E., 219
Duffey, J.L., 55, 67
Duizer, E., 219

E

Easter, M.C., 241, 246, 251
Eddy, C.R., 220
Ehiri, J.E., 240, 242, 251
Elliot, P.R., 218
Elsas, J.D., 132
Engel, D., 248, 251
Englyst, H., 152
Escaler, M., 132
Estes, M.K., 208, 217–221
Evans, A.J., 152

F

Fabre-Gea, 174
Fankhauser, R.L., 218, 220
Farber, J.M., 87, 217
Farkas, T., 219
Farr, D.R., 149
Feinstone, S.M., 206, 218
Feldman, R., 219
Felsot, A.S., 131
Femia, A.P., 147, 149
Fern, E.B., 149
Fernandes, A., 38, 48
Fey, R.E., 220
Field, C.J., 146, 151
Fields, B.N., 219
Filtenborg, O., 49
Fiore, A.E., 203, 218
Fiore, P., 175
Fitzsimons, N.A., 175, 200
Fleet, G.H., 179–180, 182, 187, 201
Flint, H.J., 294
Fliss, I., 219
Flourie, B., 149–150
Fonden, R., 175
Fontes, E.M.G., 131, 293
Forsythe, S.J., 280
Forward, K., 221
Fox, P.F., 110
Fralick, R.A., 221
Franck, A., 144, 148, 150–152
Frankhauser R.L., 203, 218, 221
French, M.S., 280
Frisvad, J.C., 49
Fryers, P.T., 220
Fu, T.Y., 219
Fukui, S., 217
Fuller, R., 138, 150
Funda, D.P., 201

G

Gaastra, W., 194, 201
Gadaga, T.H., 183, 201
Gailing, M.F., 150
Gainnini, A., 149
Gallay, A., 218
Galmiche, J.P., 150
Gardner, N.J., 163, 175
Garrido, J., 48
Gary, G.W., 219, 221
Gaskins, H.R., 150
Geisen, R., 201
Gelrer, J., 279
Gensheimer, K.F., 219
Gerba, C.P., 220
Germinario, C., 220

Geurts, T.J., 111
Ghazala, S., 25
Giaccu, D., 294
Gibbs, P., 175
Gibson, G.R., 134, 136–138, 142, 145–146, 148–151
Gilbert, B., 201
Gilmour, S.G., 150
Ginovart, M., 184, 201
Giraffa, G., 179, 201
Glaa, R.L., 221
Glare, T.R., 131
Glass, R.I., 218–220
Gleich, G.J., 282, 294
Goff, J.W., 87
Goldman, C., 221
Goldstein, S.T., 219
Gomes, A.M.P., xvi–xvii, 186, 201
Goodman, R., 131
Gorris, L.G.M., 24–25
Goswami, B.B., 205, 218
Gotanda, Y., 221
Gould, G.W., 12–14, 16, 21–23, 25,111
Gray, J.J., 208, 218
Green, J., 158, 207, 220
Green, K.Y., 218
Greenberg, H., 219
Greene, S.R., 203, 219
Griffin, M.R., 205, 219
Griffin, P.M., 220
Griffith, C., 227, 237
Griffiths, S., 240, 251
Grosso, L., 219
Guibert, A., 150

H

Hackney, C.R., 219
Hahn-Hagerdal, B., 171, 175
Halliday, M.L., 206, 219
Hammes, W.P., 145, 150, 294
Hammes, W., 294
Hamosh, M., 151
Hanon, F.X., 220
Hansen, E.B., 155, 159, 175
Haswell, P., 220
Haugarreau, L., 219–220
Hauser, G., 67
Hayashi, R., 25
Hayes, P.R., 280
Hedlund, K.-O., 220
Hefle, S., 131
Heggum, C., 230, 237
Heinert, A.P., 217
Heldman, D.R., 4, 25
Hemard, M.T., 219
Henderson, B., 181, 201
Hendricks, K.A., 218
Henson, S., 241, 251
Hernández, F.R., 206, 219
Hertel, C., 145, 150, 294
Herwaldt, B.L., 205, 219
Ho, P., 175
Hoban, T.J., 119, 132
Hoebregs, H., 152
Hoekstra, E.S., 49
Hofman, J., 201
Holah, J.T., 257–258, 280
Holah, J., 52, 67, 280
Hollifield, H.C., 85, 87
Holst-Jensen, A., 286, 294
Holt, G., 251
Holzapfel, W.H., 201
Holzapfel, W., 175
Hopkins, M., 152
Horino, K., 217
Hose, H., 175
Hotchkiss, J.H., 3, 15, 17, 25, 83, 85, 87
Hough, J.S., 154, 175
Howley, P.M., 219
Hu, M.D., 219
Hu, S.L., 219
Huang, P.W., 219
Huang, Y.S., 219
Hudcovic, T., 201
Hudson, B.J.F., 24–25
Huggett, A., 48
Huis in't Veld, J., 155, 175
Humphery, C.D., 219–220
Hunter, J.O., 134, 150
Hutin, Y.J.F., 206, 219
Hutson, A.M., 204–205, 219
Hutten, H., 67

I

Illman, R.J., 149
Ingram, T., 220
Inoue, Y., 221

Isolauri, E., 151
Istafanos, P., 218

J
Jackson, K.G., 147, 152
Jacobsen, B., 294
James, C., 181–282, 294
Javanainem, P., 175
Jay, J.M., 133, 150
Jaykus, L.A., 205, 209–210, 219, 221
Jean, J., 205, 219
Jellema, A., 111
Jensen, B.B., 150
Jian, R., 150
Jiang, X., 207, 212, 219, 221
Jiménez, C., 219
Johns, N., 248, 251
Johnson, M.G., 137, 152
Jonker, M.A., 40, 49
Jouve, F., 230, 237
Jouve, J.L., 241, 251
Joynson, D., 220

K
Kaferstein, F., 252
Kailasapathy, K., 246–247, 251
Kang L.-Y., 219
Kapikan, A.Z., 208, 219
Kaplan, J.E., 211, 219, 221
Keeley, J., 48
Kemp, C.F., 120, 131
Kersters, K., 175
Khandke, S.S., 241, 245–246, 252
Kim, W.J., 137, 150
Kingery, T.B.III,
Kinoshita, Y., 221
Kirby, R., 241, 252
Kirjavainen, P.V., 151
Kitajima, N., 221
Klaver, F.A.M., 201
Kleter, G.A., 121, 123, 132
Klontz, K.C., 218
Klungness, J.H., 84, 87
Knipe, D.M., 219
Knorr, D., 18, 25, 174
Knowles, M.J., 252
Knowles, M., 244, 252
Koff, R.F., 209, 219
Kogasaka, R., 217
Kogawa, K., 220
Kohli, E., 218
Kojima, K., 148, 150
Kok, E.J., 132
Kok, N., 152
Kokešová, A., 201
Kolida, S., 144, 150
Koningb, C.J.M., 201
Koopmans, M.P.G., 211, 215, 218–220
Korpela, R., 152
Kozáková, H., 201
Kramer, J., 201
Kuiper, H.A., 118, 121–123, 125–126, 132
Kulka, M., 218
Kunz, C., 145–146, 150
Kyrakides, A., 5, 25

L
Lachlan, H., 243, 252
Langlois, G.W., 221
Larousse, J., 9, 25
Larson, R.A., 220
Lartigue, S., 150
Lawrence, C.A., 280
Lax, A.J., 201
Le Guyader, F.S., 205, 217–220
Leaper, S., 267, 280
Lecannu, G., 149
Lee, B., 156, 160, 169, 174, 175
Lee, Y.K., 145–146, 150
Lees, D., 209, 219
Legein, J., 151
Lehr, J., 48
Leistner, L., 24–25
Lelieved, H.L.M., 280
Leroy, F., 195, 198–199, 201
Lettman, C., 132
Lew, J.F., 219
Lewis, D.C., 219
Lieb, S., 205, 220
Lilly, D.M., 138, 150
Lima, N., 35, 48
Lin, C.H., 87
Lin, D.H.L., 137, 149
Linko, S., 163, 175
Linko, Y-T., 175
Lipp, M., 286, 294
Litopoulou-Tzanetaki, E., 92, 111

Lo, S.V., 205, 220
Lodinová-žádníková, R., 201
Lookabaugh, C., 219
Lopalco, P.L., 220
López, V., 201
Lopez-Garcia, R., 49
Lopman, B.A., 203–204, 220
Løvseth, A., 294
Low, D.E., 221
Luceri, C., 149
Ludwig, W., 175
Lund, B.M., 111
Lund, D.B., 4, 25
Lüthy, J., 294
Lyytikäinen, O., 220

M
Macfarlane, G.T., 136–137, 145, 149–150
Macfarlane, G., 134, 152
Mackie, R.I., 134, 150
Mackinnon, I.H., 269, 280
Macrae, R., 235–237
Madden, J.A.J., 134, 150
Magan, N., 49
Mager, K., 67
Maitin, V., 147, 151
Malakar, P.K., 187, 201
Malcata, F.X., 175, 201
Malnick, S., 220
Mantis, A.G., 100, 103, 111
Margolin, A.B., 212, 220
Margolis, H.S., 219
Marks, P.J., 210, 220
Marteau, P., 134, 139, 149–150
Masters, K., 67
Matson, D.O., 218–220
Matsumoto, K., 151
Mattila-Sandholm, T., 151, 200–201
Matto, J., 151, 201
Maunula L., 220
Mayeno, A.N., 282, 294
Mayes, T., 241, 245–246, 252
Mazzocco, M.A., 247, 252
McBain, A.J., 134, 150
McCaig, L.F., 220
McCartney, A.L., 150
McCloskey, D.I., 217–219
McEwen, J., 251
McNab, R., 201
Mead, P.S., 202, 208, 218, 220
Mechnik, L., 207, 220
Meesters, G.M.H., 67
Mele, A., 218
Melnick, J.L., 219
Metaxopoulos, J., 176
Metcalf, T.G., 217, 219, 221
Metchnikoff, E., 150
Metz, P.L.J., 132
Mexico, D.F., 127, 132
Michel, C., 149
Midtvedt, T., 294
Miki, M., 221
Mikkelsen, L.L., 150
Miller, G.D., 175
Miller, J., 220
Mintzberg, H., 246, 252
Miossec, L., 219
Mishiro, S., 221
Mitsuoka, T., 149
Mittelholzer, C., 220
Miyagawa, S., 252
Miyagishima, K., 252
Mizuo, H., 221
Moe C.L., 219
Mogensen, G., 172, 175
Moineau, S., 198, 201
Molis, C., 146, 150
Monath, T.P., 219
Monge, R., 155, 219
Monroe S.S., 218–221
Moore, W., 217
Moraru, C., 226, 238
Moreau, M.C., 151
Morris, G.P., 242, 251
Morrison, Y., 241, 252
Mortimore, S.E., 91, 111, 229, 237, 239, 251–252
Mosenkis, B., 220
Mostert, M.A., 280
Motarjemi, Y., 252
Mountzouris, K.C., 133, 135, 142, 144, 147, 150–151
Moura, F., 48
Moy, G., 252
Mulderc, L., 201
Mullendore, L.J.L., 209, 220
Multon, J.L., 87
Murphy, J.J., 221

N
Nainan, O.V., 219
Nakao, T., 217, 220
Nakata, S., 208, 220
Nap, J-P., 120, 131–132
Narvhus, J.A., 183, 201
Nataska, S., 218
Naughton, P.J., 150
Neill, F.H., 217
Neill, J.D., 218
Newton, D., 152
Neysens, P., 186, 200
Ngo, D., 218
Nielsen, K.M., 120, 132
Nishizawa, T., 221
Noel J.S., 218, 220
Noomen, A., 111
Norrung, B., 202, 220
Northen, J., 251
Noteborn, H.P.J.M., 132
Numata, K., 220
Nunnari, A., 217

O
O'Brien, J., 175
O'Connor-Shaw, B., 24–25
Obermeier, T., 175
Ohara, S., 221
Okamoto, H., 221
Olano-Martin, E., 144–145, 151
Olsen, M., 49
Olsson, L., 171, 175
Oosterom, J., 67
Ottaway, P.B., 148, 150–151
Ouwehand, A.C., 139, 151, 175

P
Palmer, S.R., 220
Palou, E., 25
Pan, Q.C., 219
Panebianco, L., 294
Panisello, P.J., 245, 252
Pappa, M., 176
Paraggio, M., 175
Parashar, U.D., 205, 220
Parkin, W.E., 219
Parlierani, M., 149
Paterson, R.R.M., 26–28, 30–31, 34–37, 40, 46–48
Patterson, W., 205, 210, 220
Patton, S., 151
Pauli, U., 294
Pead, P.J., 218
Persley, G.J., 115, 123, 131–132
Peters, R., 239, 252
Peterson, J.A., 145, 151
Philips, A., 221
Pierson, M.D., 219
Pierzynski, B., 219
Pinsky, P.F., 221
Pirofski, L-A., 180–181, 188, 200
Pitt, J.I., 26, 48
Pittet, A., 38, 48
Playne, M., 149
Pletincx, M., 140, 151
Pochart, P., 149
Pommepuy, M., 219–220
Pon, E.W., 219
Pönkä, A., 205, 220
Pool, V., 219
Pot, B., 161, 175
Pothakamury, U.R., 25
Potter, N.N., 25
Prato, R., 205, 220
Preiser, W., 217
Puong, K.Y., 145–146, 150

Q
Quantick, P.C., 252
Quarne, F., 150
Quarto, M., 220
Quigley, M., 152
Quinn, M., 219

R
Rambaud, J.C., 149
Rastall, R.A., 147, 150–151
Ray, B., 3, 15, 17, 25
Reacher, M.H., 220
Reddy, S., 218
Reece, I., 243–244, 249, 252
Rees, A.E., 217
Reháková, Z,
Richards, G.P., 221
Richardson, D.P., 148, 151
Richardson, K.J., 220
Ridley, W.P., 131
Rizzi, A., 281, 291–294

Roberfroid, M.B., 135–137, 141–142, 146, 148–151
Roberfroid, M., 151–152
Robinson, R.K., 237
Rogers, A., 244, 252
Roizman, B., 219
Romalde, J.L., 217
Romano, P., 162, 175
Romboutsd, F.M., 201
Rønning, S.B., 294
Rooney, M.L., 14, 25, 87
Rossmann, P., 201
Rotaru, G., 225–226, 237–238
Rowe, S.H., 218
Rowland, I.R., 140, 151
Rowlands, R.E., 87
Rozum, J., 13, 25
Rudloff, S., 145, 150
Rumpf, W., 67
Russel, N.J., 12, 25
Russo, R., 217

S
Saarela, M., 139–140, 151, 201
Sadler, M.J., 237
Saeed, A.A., 218
Sair, A.I., 221
Sair, H.S., 219
Sako, T., 144, 151
Sakuma, Y., 217
Salfati, J., 149
Salminen, S., 135–136, 138, 142, 151, 175
Salvadori, M., 149
Salyers, A.A., 134, 151
Samelis, J., 163, 176
Samson, R.A., 49
Sanderson, I.R., 134, 151
Saski, N., 221
Sattar, S.A., 217
Sava, N., 238
Savage, D.C., 134, 151
Savard, T., 175
Sawada, K., 149
Sawyer, L.A., 211, 221
Scalfaro, C., 218
Schaafsma, G., 151, 175
Scheinbach, S., 140, 151
Schillinger, U., 187, 201
Schleifer, K.-H., 175
Schleining, G., v, xiii, xv, xviii, 50, 67
Schley, P.D., 146, 151
Scholz-Ahrens, K.E., 147, 151
Schrezenmeir, J., 151
Schronberger, L.B., 221
Schubbert, R., 120, 131–132
Schwab, K.J., 218
Seiber, J.N., 131
Seksik, P., 150
Septic, S., 279
Seymour, I.J., 221
Sghir, A., 150
Shahidi, F., 25
Shapiro, C.N., 219
Shapiro, C., 220
Shapton, D.A., 7, 10–11, 25
Shapton, N.F., 7, 10–11, 25
Shieh, Y.C., 220–221
Shimizu, H., 200
Shioya, S., 200
Shoji, H., 200
Shortt, C., 151, 175
Silbert, M., 201
Silia, H., 175
Silva, Joana, 175
Simões, C.M.O., 217
Singleton, P., 180–181, 201
Sinha, K.K., 37–38, 49
Šinkora, J., 201
Skudamore, K.A., 37, 44, 49
Slater, D., 280
Slomka, M.J., 221
Slutsker, L., 220
Smalla, K., 132
Smith, C.R., 269, 280
Smith, J.E., 34, 49
Smith, J.L., 221
Smith, J., 175
Smits, G., 152
Snyder, O.P.Jr., 247, 249, 252
Sobsey, M.D., 219–220, 264, 280
Sokol, D., 201
Sokolowski, N., 220
Solomons, G.L., 34, 49
Sorlini, C., 281, 294
Southgate, D.A.T., 134, 151
Sperber, W.H., 251

Sperner, B., 206, 221
Stave, J.M., 294
Stepánková, S., 201
Stevens, R., 175
Stier, R.F., 246, 252
Stillwell, R.H., 138, 150
Stolle, A., 221
Straub, J.A., 290, 294
Straus, S.E., 219
Studdert, J.M., 218
Surak, J.G., 95, 99, 104, 108–109, 111
Surowiec, J.J., 219
Svensson, L., 220
Swanson, B.G., 25

T
Takahashi, K., 221
Takahashi, M., 221
Tamura, Z., 136–137, 152
Tanaka, R., 151
Tapia de Daza, M.S., 13, 24–25
Tauxe, R.V., 220
Taylor, E., 246, 252
Taylor, L., 219
Taylor, S., 131
te Loeke, N.A.J.M., 217
Tei, S., 221
Teixeira, P., 175
Terashima, H., 220
Teuber, M., 100, 111
Teuri, U., 146, 152
Thiel, H-J., 218
Thiolet, J.M., 218
Thomas, P.D., 220
Tilburg, J.J.H.C., 217
Timmermana, H.M., 201
Tlaskalová-Hogenová, H., 201
Todar, K., 201
Topping, D.L., 149
Tornare, D., 48
Toti, L., 218
Touhy, K., 150
Tourte, Y., 113, 132
Tucková, L., 201

U
Ukae, S., 220
Unger, H., 247, 252

V
Vahedi, K., 149
Vaillant, V., 218
Valls, J., 201
Van Aarle, P., 219
Van Boekel, M.A.J.S., 111
van den Eede, G., 282, 290, 294
van den Heuvel, E.G.H.M., 151–152
van der Stegen, G.H.D., 48
van der Vossen, J., 294
van Duijnhoven, Y., 220
van Duynhoven, Y.T.H.P., 218
van Egmond, H., 39, 40, 49
Van Loo, J., 144, 146, 152
van Osenbruggen, T., 49
van Vliet, T., 152
van't Riet, K., 201
Vandamme, E.J., 175
Vandenplas, Y., 151
Venâncio, A., 26, 48
Viani, R., 48
Viljoen, B.C., 184–185, 201
Vinjé, J., 219
Vipond, I.B., 220
Visconti, A., 49
Vlassi, M., 176
Volkert, M., 174
von Bonsdorff, C-H., 220
von Wright, A., 294
Vuyst, L.De, 175

W
Wackernagel, W., 294
Wainwright, M., 176
Walker, G., 169–170, 173, 176
Walker, M., 217
Walker, S., 243–244, 249, 252
Wallace, C., 91, 111, 211, 228, 238
Wallace, J., 239–240, 251
Wallace, L.A., 262, 280
Walmsley, S., 221
Walstra, P., 97, 111
Wang, X., 144, 150
Warhurst, M., 149
Wassenaar, T.M., 194, 201
Weber, C.J., 87
Weth, J., 219
White, B., 280

Whitt, D.D., 134, 151
Wigley, R.C., 164, 167, 176
Wilcks, A., 294
Wilder, M.H., 218
Williams, C.M., 147, 152, 228, 238
Williams, I.T., 219
Williams, K.S., 220
Wilson, M., 201
Wojnarski, R., 219
Wolf, I.D., 83, 87
Woodley, C.M., 217
Wreghitt, T.G., 218
Wright, D., 220

Y
Yamabe, H., 221
Yamatani, T., 221
Yazaki, Y., 221
Yildirim, Z., 137, 152
Young, J., 148, 152
Young, T.W., 175

Z
Zerfiridis, G.K., 92, 97–98, 110–111
Zhong, W.M., 219
Zhou T.-K., 212, 219
Zhou, Y.J., 221
Zimmermann, A., 294
Zwietering, M.H., 201

Printed in the United States of America